Lecture Notes in Computer Science 11756

More information about this series at http://www.springer.com/series/7407

Carlos Paternina-Arboleda ·
Stefan Voß (Eds.)

Computational Logistics

10th International Conference, ICCL 2019
Barranquilla, Colombia, September 30 – October 2, 2019
Proceedings

 Springer

Editors
Carlos Paternina-Arboleda (iD)
Universidad del Norte
Barranquilla, Colombia

Stefan Voß (iD)
Universität Hamburg
Hamburg, Germany

ISSN 0302-9743   ISSN 1611-3349   (electronic)
Lecture Notes in Computer Science
ISBN 978-3-030-31139-1   ISBN 978-3-030-31140-7   (eBook)
https://doi.org/10.1007/978-3-030-31140-7

LNCS Sublibrary: SL1 – Theoretical Computer Science and General Issues

This Springer imprint is published by the registered company Springer Nature Switzerland AG
The registered company address is: Gewerbestrasse 11, 6330 Cham, Switzerland

Preface

The international transport and management of goods and other support areas for logistics are experiencing a new wave of digitalization and digital transformation. The use of digital technologies to support the management of operations is rising at an unprecedented pace. New emerging transportation equipment technologies, trade and business documents using blockchain technologies, tracking and tracing of cargo conditions (not just the position), and congestion management in real time, as well as many problems that are more practical and in constant evolution are now addressed with the use of digital technologies. This creates significant changes in business models and processes and the emergence of new products and services.

Computational logistics gives support to all these processes and makes them work their best. We are still, at this time, radically (or gradually) improving solutions to optimization problems related to the broad spectrum of the supply chain, but we are also tackling novel problems of growing interest, such as electric vehicle fleet management, big data or automation (and autonomy) on ports and transport assets. We deal with the design of transport equipment, mode shift selection, and decision support systems to plan the operations. We are showing great insights on how infrastructure helps in the development of regional competitiveness and, with the use of technology, we enhance the capabilities of heuristics and exact methods to improve our quest for logistics problem solving.

Moreover, the use of novel techniques to analyze existing and growing data sets to gather wisdom and make decisions is gaining great popularity and relevance nowadays. In a more social context, we see now how the shared economy works to our best interest as well as where we will go in the future with the use of smart technologies. In all cases, we can see the connection between ICT tools and solving complex problems in logistics and supply chain management.

The International Conference on Computational Logistics (ICCL) is an academic forum where recent advances and innovative case studies within the field are presented and discussed. This year marked the 10th International Conference on Computational Logistics (ICCL 2019), held in Barranquilla, Colombia, during September 30 – October 2, 2019. We received 60 submissions, of which 27 peer-reviewed papers out of almost 50 full paper contributions were selected to be published in these proceedings. The papers display various directions of importance in computational logistics, classified into five topic areas, reflecting the interest of researchers and practitioners in the field. The articles in this volume are grouped as follows:

1. *Freight Transportation and Urban Logistics:*
   Starting with a comprehensive description on emerging issues of city logistics, this section analyzes the impact that decision support systems have on improving cargo compensation as well as the environmental impact. It then presents a case study which addresses the reduction of the current inbound/outbound freight imbalance that enters a marketplace through seaports and airports. Then, it shows how

collaboration in urban transportation networks positively affects the environment. We end this section with a paper that uses metaheuristics to solve a problem of territorial planning.

2. *Maritime and Port Logistics:*

Concerning a major mode in transportation, we see several articles presented in the area of maritime and port logistics. Over the years, ICCL has been strong in addressing issues of maritime and port logistics and it was not different for this year's conference. The papers in this field address various problems of interest in areas, such as vessel capacity planning, port internal dynamics, information systems for port integration, port governance, as well as the impact that the expansion of the Panama Canal has on potential shifts in trade routes.

3. *Vehicle Routing Problems:*

Albeit a classical area of research, vehicle routing problems still need quite a bit of attention, especially when they include more practical constraints such as handling split demands and incompatibility between products. Novel rich applications, such as those in which the problem involves electric vehicles, have received a lot of attention recently and we see those considered here. Papers presented deal with different solution approaches such as heuristics, metaheuristics, and exact methods.

4. *Network Design and Distribution Problems:*

We range from problems that determine the location of facilities, the distribution of products, and the aside multi-echelon inventory management problems to those which find the best allocation of resources for maintaining service operations, both active and business-running. Some of the papers deal with the integrated modeling of supply chain operations, such as production, inventory, and distribution. Others, with security issues underlying the distribution of goods. We see here papers that deal with methods, frameworks, and algorithms that give improved solutions to supply chain problems, as well as others that focus on how computational tools help in providing good solutions to real-life problems.

5. *Selected Topics in Decision Support Systems and ICT Tools:*

In this section, we included papers that treat problems by making use of software platforms to improve their operations planning. They range from papers that deal with helping in the humanitarian response to disaster relief, the design and use of computational platforms for e-shared logistics services, as well as those designed for optimized transportation operations planning. All these problems are still in need of better algorithmic developments and online connectivity for real-time performance. The papers that we have selected for this volume handle these matters with quite some depth.

ICCL 2019 was the tenth edition of this conference series, following the earlier ones held in Shanghai, China (2010, 2012), Hamburg, Germany (2011), Copenhagen, Denmark (2013), Valparaiso, Chile (2014), Delft, The Netherlands (2015), Lisbon, Portugal (2016), Southampton, UK (2017), and Salerno, Italy (2018).

The editors thank all the authors for their contributions and the reviewers for their invaluable support and feedback. We trust that the present volume supports the continued advances within computational logistics and inspires all participants and readers to its fullest extent.

Besides editors, authors, reviewers and committee members there is more. In that sense, we thank Daniela Cassandro, Julia Bachale, and all the staff members from the Organizing Committee, both in Barranquilla and Hamburg, for their continuous support. Finally, we thank Springer for supporting ICCL again and for publishing the proceedings in their *Lecture Notes on Computer Science* (LNCS) series.

September 2019

Carlos Paternina-Arboleda
Stefan Voß

Organization

Program Committee Chairs

Carlos D. Paternina-Arboleda	Universidad del Norte, Colombia
Stefan Voß	University of Hamburg, Germany

Program Committee

Michele Acciaro	Kühne Logistics University, Germany
René A. Amaya	Universidad del Norte, Colombia
Panagiotis Angeloudis	Imperial College, UK
Michael Bell	Sydney University, Australia
Jürgen Böse	Ostfalia, Germany
Víctor Cantillo	Universidad del Norte, Colombia
Buyang Cao	Tongji University, China
Raffaele Cerulli	University of Salerno, Italy
Joachim R. Daduna	HWR Berlin, Germany
René de Koster	Erasmus University Rotterdam, The Netherlands
Yezid Donoso M.	Universidad de los Andes, Colombia
Carsten Dorn	Hochschule Bremerhaven, Germany
Kjetil Fagerholt	NTNU, Norway
Monica Gentili	University of Kentucky, USA
Rosa González	Universidad Los Andes, Chile
Zoila Guerra de Castillo	Universidad Tecnológica de Panamá, Panama
Angappa Gunasakeran	California State University, USA
Hans-Otto Günther	TU Berlin, Germany
Hans-Dietrich Haasis	University of Bremen, Germany
Ana Ximena Halabi	Universidad de la Sabana, Colombia
Richard Hartl	University Vienna, Austria
Geir Hasle	Sintef, Norway
Leonard Heilig	University of Hamburg, Germany
Sin Ho	Hong Kong University, SAR China
José Holguín-Veras	Rensselaer Polytechnic Institute, USA
Carlos Jahn	Hamburg University of Technology, Germany
Miguel Jaller	UC Davis, USA
Andreas Jattke	University of Applied Sciences Ingolstadt, Germany
Miguel A. Jiménez	Universidad de la Costa, Colombia
Raka Jovanovic	HBKU, Qatar
Herbert Kopfer	University of Bremen, Germany
Ioannis Lagoudis	MIT, Malaysia
Eduardo Lalla-Ruiz	University of Twente, The Netherlands

Jasmine Lam	NTU, Singapore
Gilbert Laporte	Cirrelt, Canada
Venus Lun	City University, UK
Dirk Mattfeld	TU Braunschweig, Germany
Andrés Medaglia	Universidad de los Andes, Colombia
Christopher Mejía	MIT, USA
Jairo R. Montoya-Torres	Universidad de la Sabana, Colombia
Rudy Negenborn	TU Delft, The Netherlands
Marcos José Negreiros Gomes	State University of Ceara, Brazil
Theo Notteboom	Shanghai Maritime University, China
Dario Pacino	DTU, Denmark
Julia Pahl	Syddansk University, Denmark
Anna Paias	Universidade de Lisboa, Portugal
Jana Ries	University of Portsmouth, UK
Mario Ruthmair	University of Vienna, Austria
Ivan Saavedra	Competitive Insights, LLC, USA
Dirk Sackmann	Hochschule Merseburg, Germany
Jose J. Salazar	Universidad de La Laguna, Spain
Alcides Santander	Universidad del Norte, Colombia
Xiaoning Shi	WMU, Sweden
L. Douglas Smith	University of Missouri, USA
Elyn Solano	Universidad de la Sabana, Colombia
Dong-Wook Song	WMU, Sweden
Maria Grazia Speranza	University of Brescia, Italy
Shunji Tanaka	Kyoto University, Japan
Kevin Tierney	University of Bielefeld, Germany
Ding Yi	Shanghai Maritime University, China

Additional Reviewers

Julián Arellana Ochoa	Universidad del Norte, Colombia
Juan Pablo Escorcia Caballero	Universidad del Norte, Colombia
Gina Galindo	Universidad del Norte, Colombia
Luceny Guzman Acuña	Universidad del Norte, Colombia
César A. Henao	Universidad del Norte, Colombia
Qing Liu	University of Hamburg, Germany
Javier Maturana-Ross	Pontificia Universidad Católica de Chile, Chile
Gonzalo Mejía	Universidad de la Sabana, Colombia
Abtin Nourmohammadzadeh	University of Hamburg, Germany
Seckin Ozkul	University of South Florida, USA
Michael Palk	University of Hamburg, Germany
Diana Ramirez-Rios	Rensselaer Polytechnic Institute, USA
Daniel Romero	Universidad del Norte, Colombia

Oscar Rubiano	Universidad del Valle, Colombia
Frank Schwartz	University of Hamburg, Germany
Milton Soto Ferrari	Indiana State University, USA
Jorge Velez	Universidad del Norte, Colombia
Mario C. Velez-Gallego	Universidad EAFIT, Colombia
Ruben Yie-Pinedo	Universidad del Norte, Colombia
Jingjing Yu	University of Hamburg, Germany

Contents

Selected Topics in Decision Support Systems and ICT Tools

Freight Transportation and Urban Logistics

Developments in City Logistics - The Path Between Expectations and Reality

Joachim R. Daduna[✉]

Berlin School of Economics and Law,
Badensche Str. 52, 10825 Berlin, Germany
daduna@hwr-berlin.de

Abstract. Discussions in the context of city logistics relate to two areas, the supply of inner-city retail trade stores and service facilities as well as the delivery of parcels in the context of mail order and online retail business. These two problems must be considered separately due to the different requirements, even if at some points in the processes links can occur. However, the common problem is the increase in traffic congestion caused by delivery trips and the associated environmental impacts. The objective must be to find sustainable solutions to reduce these negative effects. Basis for this are, among others, a demand-oriented adaptation of the structures of the logistics facilities, improvements in routing, the use of new vehicle concepts and technologies as well as the expansion of real-time-based information management. We consider current developments and ideas regarding the two problems and related options and limitations. It turns out that not all solutions are effective and not necessarily sustainable. It is also unclear whether fundamental changes in economic and social structures will have far-reaching effects on future developments.

Keywords: City logistics · Delivery concepts · New technologies

1 Developments in City Logistics

In recent decades, there have been many and sometimes highly controversial discussions about possible solutions to solve transport problems in inner-city areas. A key point here was, and still is, to improve the retail delivery processes there. The focus was (and will continue to be) on the reduction of road congestions, which were often caused by delivery vehicles during loading and unloading stops. At the same time, however, a reduction in mileage was also sought, with the objective of saving energy and reducing traffic-related environmental pollution.

With increasing efforts to restrict the further expansion of large-scale shopping malls in suburban areas and a concurrent revitalization of the inner-city areas, the pressure grew stronger, to install efficient delivery logistics in order to reduce the resulting negative effects. A large number of projects have been started worldwide, but these have often been given up due to lack of success (see, e.g. [15]). The causes for this are complex, amongst other things diverging interests of the participants, differently motivated resistances against cooperative solutions as well as organizational and technical problems.

© Springer Nature Switzerland AG 2019
C. Paternina-Arboleda and S. Voß (Eds.): ICCL 2019, LNCS 11756, pp. 3–21, 2019.
https://doi.org/10.1007/978-3-030-31140-7_1

The main focus of these developments was the organization of freight transport. However, this purely logistics-oriented approach cannot lead to the objective of improving the attractiveness of inner-city areas. For this purpose, the public transport services must be adapted accordingly, and the motorized private transport must be passably integrated. In particular, this also refers to the accessibility for customers from outside in the case of cities with a high importance for the supply of the population in the surrounding regions.

For a number of years, the tasks in city logistics have expanded considerably. It is no longer just about supplying commercial customers (retail trade and service facilities), but also about delivering (even short-term) orders in the booming online retail trade. This results in requirements that have led to an extreme increase in demand for parcel services. Because at this point it is no longer about the delivery of (mainly long term) fixed locations with usually larger demand volumes, but it is the delivery of constantly changing locations (of private and business customers) with very limited shipment sizes. There are also very different expectations among customers with respect to delivery forms, delivery times and delivery conditions.

The following comments deal with these two fields of city logistics in detail. In Sect. 2 the supply of retail trade stores and service facilities is in the foreground. The focus here is on delivery concepts, the problems caused by regulatory intervention, the question of vehicle selection, as well as the discussions on integrating freight transport in public mass transit and applying alternative transport solutions. Section 3 examines the field of inner-city parcel delivery. Presented here are the essential delivery concepts and the used vehicle systems. Moreover, the impact of new vehicle technologies is discussed as well as possible approaches relating to the sharing economy.

2 Deliveries in the Retail Trade Sector

In the case of retail trade deliveries, a differentiated approach is necessary because the integration of a number of sales facilities into a concept for city logistics is not possible or only possible to a limited extent. These are large-scale facilities as well as stores of bigger retail trade chains with their own logistics structures. These usually have a delivery volume that is handled within the scope of *full truck loads* (FTL), so that in such cases there is no relevant potential for cooperation. For this reason, city logistics concepts have the focus on smaller-scale sales and service facilities whose delivery volume is *less than truck loads* (LTL) (see, e.g. [21]). The following discussions in this section are therefore based on this limitation.

In the foreground of the considerations on city logistics were (and still stand today) the efforts to achieve a reduction of the (local) traffic volume by *bundling* the *delivery operations*. The objective is to reduce the traffic in the inner cities on the one hand and to improve the delivery processes on the other (see, e.g. [27, 29, 87]). One of the main reasons for many of the initiated (and often publicly funded) projects was the negative impact of delivery trips in the inner-city areas. These are less a problem of traffic volume, but they were (and still are) mainly perceived as a disturbance factor. Significant disruptions of traffic flows appear, e.g., based on inefficient unloading and

loading processes on lanes. The negative effects quickly can increase, if there exist a higher demand and an inadequate infrastructure especially in shopping streets.

In the previous considerations, usually only the delivery transport is in the focus, an approach that makes sense in principle, both from an economic and an environmental point of view, since at least partial relief effects are achieved. However, this cannot be considered sufficient, because apart from the logistical needs, a significant point is ignored in most cases: the accessibility of retail trade facilities by the customer. Efforts to increase logistical efficiency make little sense if due to insufficient customer potential not enough sales are generated. It follows that, in addition to an efficient organization of delivery processes, considerations of city logistics must include passenger transport as an essential factor, also in order to be able to open sufficient demand potential in the long term.

This claim is in particular in the focus, if (for example in major regional centers and mid-sized centers) at a revitalization of inner-city centers shopping activities should be connected with attractive leisure offers (theater, cinema, restaurant visits, etc.) in particular for customers from the surrounding region. The underlying considerations of a differentiated importance in the basic structures go back to the *theory of central places* formulated by [28], in which the scope of services in the locations is hierarchically structured. An additional effect resulting from the bundling of different activities is from a macroeconomic perspective the *reduction of traffic volume* and for rural residents the *reduction of mobility costs*.

In order to install a successful city logistics concept in economic and ecological terms, a broad cooperation is required. This applies to inner-city retail trade companies, their suppliers, the logistics service providers involved and, in particular, the local municipalities concerned (see, e.g. [53, 97]). The latter play an essential role in the development of the urban planning and transport framework conditions, for example through infrastructure measures in shopping streets (installation of loading zones, etc.) (see, e.g. [85]). Added to this is the creation of suitable legal framework conditions, amongst others in the urban land use planning by including transport logistics requirements. Ultimately, however, responsible for the realization of cooperation are the companies concerned, because only in this way a long-term success can be achieved.

2.1 Delivery Concepts

The delivery concepts are based in their basic structures on multi-tier (mostly hierarchically structured) networks. The dimensions of these networks are determined by a number of parameters, such as the extent of the urban area, the transport infrastructure and settlement structures, the topographical conditions as well as the economic structures. The main objective here is to reduce the negative effects of transport as well as the environmental impacts, but also to reduce the occurring delivery costs (from the retail trade company's perspective).

The basis for an (economically and ecologically) efficient design of the delivery processes is the *bundling of orders*, independent of formal organizational structures. The essential point here must be to improve low productivity of the distribution vehicles used, among other things, which result from tight goods receiving time

windows and delivery time restrictions, traffic restrictions and obstructions as well as in part from low bundling interests of the shippers, logistics service providers and clients.

The basis for multi-tier delivery processes is a modified transport chain based on a hierarchical system of logistics facilities (see, e.g. [34], pp. 10–12). The top level of these (usually) is formed by *freight villages*, followed by *regional* and *local logistics facilities*. In urban logistics concepts, the regional level is essentially formed by different types of *Urban Consolidation Centers* (UCC) (see, e.g. [14, 15, 66, 71]) or *City Distribution Centers* (CDC) (see, e.g. [12]). At the last level, different forms occur, such as *City Distribution Terminals* (CDT) (see, e.g. [32]), *Urban Distribution Centers* (UDC) (see, e.g. [10]), *Shipping Terminals* (see, e.g. [34], pp. 10–12) and Satellites (see, e.g. [12]) whose functional tasks are, however, largely similar. These logistics facilities at the final stage present the starting point for the *consolidated delivery routes*. Most of these are stationary facilities, but in some cases also *temporary cross-docking points* can be used (see, e.g. [16]).

The main advantages of such a multi-stage concept lies in the flexible adaptation of *vehicle sizes* on the different stages, on the one hand in the supply to the respective logistics facilities and on the other hand in the adaptation to the road networks in the service areas. The basic structures of a one-tier and two-tier terminal concept are shown in Figs. 1 and 2.

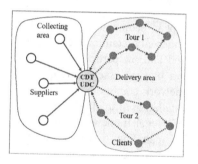

Fig. 1. One-tier structure applied in the case of smaller cities.

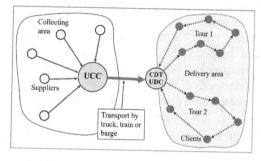

Fig. 2. Two-tier structure applied in the case of large cities and metropolitan areas.

For smaller cities, a one-tier structure may be useful, that means, the logistics facilities within (or close to) the inner-city area are directly supplied. In the case of cities with historic urban centers, it makes sense, as far as suitable areas are available, to realize a location in their close surrounding area, also with regard to the possible (capacitive) restrictions on delivery vehicles (see e.g. [32]). For large cities and metropolitan areas, a two-tier structure is needed to attain additional bundling effects and to reduce the number of trips to inner-city areas for serving the logistics facilities. As far as an appropriate infrastructure is available, *rail transport* (see. e.g. [69]) and *inland waterway transport* (see, e.g. [56, 96]) can also be used, with a view to reduce traffic congestion and negative environmental impacts.

In addition to the organizational problems in structuring the delivery processes, appropriate procedures for efficient (*dynamic*) vehicle routing and a real-time information management play an essential role at the last stage of the delivery structure (see, e.g. [21, 35, 41, 102]). However, the planning on the customer side sometimes results in highly restrictive constraints due to *positive* (permissible) and *negative* (non-permissible) *delivery time windows*. This problem is also often intensified by administrative measures (see, e.g. [29, 81]).

2.2 Regulatory Interventions

Interventions in the logistical processes are primarily the introduction of temporary *access restrictions* in certain areas, as well as *size limitations* (in terms of the dimensioning and the permissible gross vehicle weight) of the vehicles to be used for deliveries. These measures necessitate an increase in the number of required tours within given (positive) time windows and thus also the vehicles needed. In addition, the effects of various measures can accumulate, so that the achievable solutions sometimes have considerable efficiency losses. This results in higher costs for the supply of the retail trade stores as well as an increase in environmental impacts (see, e.g. [81, 98]), that means, the actual effects are in contrast to some of the main objectives that should be associated with the concepts in city logistics. A key problem (see, e.g. [1, 2]) is considered a too general formulation of the (political) objectives, which are not questioned and whose sustainable achievement cannot be adequately checked by the absence of sufficiently concrete guidelines.

A useful form of administratively given time windows can be the overnight delivery (see, e.g. [87]). This is not only suitable from an economic and traffic-related perspective, but also with a view to a significant reduction of the negative environmental impacts compared to deliveries during the daytime (see, e.g. [43, 59]). In addition, this approach usually results not in an increase in vehicle requirements resulting from unfavorable delivery processes. However, the acceptance of directly and indirectly affected persons is not guaranteed (see, e.g. [55]). One problematic aspect are the working time regulations of the employees in the retail and service facilities, which may prove to be a restriction unless the entire delivery process is externalized, including any rack services that may be necessary. In addition, residents resist (mostly already in principle) against possible noise pollution by nighttime delivery operations.

2.3 Selection of Delivery Vehicles

The selection of the vehicles to be used regarding the available transport capacity is determined by two factors, regardless of the logistical requirements. These are, on the one hand, the above-mentioned administrative restrictions regarding vehicle dimensions and the permissible gross vehicle weight. On the other hand, there are restrictions due to the urban street network structures that exist especially in cities with historic centers (see, e.g. [3, 5]). In addition, there are function-related criteria that are required for the transport of certain goods due to technical and legal conditions.

Regardless of possible restrictions, there is a trend in city logistics for the use of smaller vehicle sizes, also with a view to better mobility and greater flexibility. Research by [86] shows that such a solution is both economically efficient and environmentally meaningful.

Due to the recurrent discussions about the negative environmental influences of delivery transport, the question of vehicle drive is increasingly coming to the fore. Despite significant technical developments, the diesel drive in delivery vehicles is often considered to be an undesirable technique due to the emissions. In addition to *Liquefied Natural Gas* (LNG), *biodiesel* is also discussed as an alternative because the local emissions are lower compared to the use of diesel fuel. However, the biodiesel in particular is highly controversial due to the lack of sustainability in terms of production, especially with regard to the competition with the food sector.

At present, therefore, *electric traction* is the preferred solution, although this technology is not uncontroversial (see, e.g. [38, 51, 73, 76, 103]). At a first view, it seems to be an appropriate step in order to reduce emission loads, but this effect only occurs locally in the affected urban areas. Moreover, it does not take into account the negative environmental effects, amongst other things, in connection with the procurement of raw materials for the production of the required batteries. Therefore, an implementation of electro mobility will ultimately depend on long-term sustainability and economic benefits. In addition, it must be waited whether in the next years with the *fuel cell* a better alternative will be available (see, e.g. [64]), especially in the commercial vehicle sector for vans and trucks.

Another aspect is the use of *autonomous vehicles* (see, e.g. [24, 45, 61, 89]). Their advantages are considerable in city logistics, especially in connection with environmentally friendly drives. Moreover, not only the substantial savings in personnel costs must be seen, but also the independence of legal working time regulations in their use. The operational capability from a technical point of view is given but the legal framework has to be adopted.

2.4   Integration of Public Transport Operations into Delivery Processes

An again and again discussed concept for solving transportation problems in city logistics is the (partial) integration of freight transport into public mass transit (see, e.g. [8, 88]). In the foreground, however, is not the supply of private customers, but the transport between logistics facilities within the overall system. Here, two fundamentally different forms must be distinguished. On the one hand, there are solutions in which passenger and freight are carried by the same vehicles and on the other hand there are infrastructure sharing concepts, especially for track-based networks, which are intended primarily for public transport services in urban areas.

The combined vehicle solutions make sense from an economic as well as ecological point of view in *rural areas*. These have been practiced in various European regions for many years, in particular in connection with regional (post)bus transport services (see, e.g. [58, 65]). This concept is also applied worldwide in the context of *paratransit services* with small buses, mainly in developing and emerging countries (see, e.g. [22, 23, 75]). However, for some years there are proposals for applying this concept in urban areas for road transport (see, e.g. [93]). One solution is a two-stage concept for

bus transport, which is presented by [63]. This is based on roll containers, which are collected in *Consolidation and Distribution Centers* (CDC) and then loaded there in city buses. At defined stops these are transshipped into smaller vehicles (*city freighters*) and delivered to customers. With this flexible approach, a simple adaptation to the constantly changing demand structures should be made possible. A similar concept is proposed by [77]. [13] suggest the use of autonomous vehicles for on-demand door-to-door services, which are equipped for goods transport with compartments. However, no realized project is known here in connection with urban public road-based transport services.

Applications related to sharing track-based infrastructure in *urban areas* are constantly propagated, mainly for city rail, subway and tram systems (see, e.g. [42], pp. 49–107; [33, 48, 84, 110]). However, information regarding a concrete implementation is also in these cases not available.

[7] and [60] describe different examples of tram integration. The *CityCargo Tram* in Amsterdam (The Netherlands) was a project that was realized but discontinued after a short time as it conflicted with the requirements of public transport and the lack of funding. Further examples are concepts that relate to other logistical tasks. These are the *CargoTram* in Zurich (Switzerland) for collecting bulky waste, the *CarGo Tram* in Dresden (Germany), which supplies the "*Gläserne Manufaktur*" of Volkswagen AG from a spatially separated logistics center, and the *GüterBim* in Vienna (Austria), which was designed for inner-urban heavy goods transport, but it was operated only a short time. In earlier years, however, there were many transport solutions based on freight trams in different cities (see the examples from Germany in [18], pp. 175–201), but these were mainly used for in-company transport or for external freight transport for which no sufficiently efficient alternatives were available.

The problem of combined solutions in urban areas is that, especially in the larger cities, public transport has to play an essential role, both in the transport of persons and for ecological reasons. The integration of freight transport will inevitably lead to delays at stops or stations due to loading and unloading processes. This leads to a negative impact on the quality of public transport, because possible delays due to longer waiting times can cause the loss of connections at transfer nodes, especially in multi-tier mashed networks. This inevitably results in a loss of attractiveness and thus possibly also leads to negative impacts in the choice of means of transport and thus on the modal split. For the public transport companies, there are additional costs by an increase in the running times, since the required vehicles to operate a given timetable can grow significantly. Apart from this, more drivers are needed, as the examples outlined by [88] show. In addition, as a rule, especially in rail-based transport systems, necessary infrastructure measures are required to enable the delivery or removal of freight containers to or from stops and station platforms.

A use during off-peak hours can also not be a solution, since such an increase of offered services in this time is associated with additional costs and sometimes also has a negative impact on operating processes (among other things regarding the organization of the cleaning and workshop stays). All in all, it has to be said that the integration of freight transport into (urban) public transport is currently not or only partially compatible with its priority objectives, as the failure of the already mentioned Amsterdam project *CityCargo Tram* showed (see also [7]).

2.5 Alternative Developments

In view of the traffic loads and emissions in inner-city areas, alternative transport systems are discussed again and again (see, e.g. [26, 100, 104]). One approach here is the use of *underground transport systems* that have their own driveway in a tube network, such as the *CargoCab* transport system (see, e.g. [47]). This is an automated track-guided system for the transport of palletized goods for the supply of commercial customers. However, there are in most cases significant problems in the construction of the necessary tube network, since the underground is occupied in many cities by other forms of use (among other things public transport infrastructure, water and waste water networks, cable networks, etc.). In addition, there are problems in the monitoring and control, since for loading and unloading vehicles must enter and exit the tube system, whereby the achievable transport speed is significantly limited.

Such systems could be of interest to connect logistics facilities directly with each other, as far as there is the possibility to build an appropriate tube network. However, there are up to now no realized systems, not even the *Ondergronds Logistiek Systeem Schiphol* (OLS-ASH) system in The Netherlands (see, e.g. [7, 104]), which was planned as a system of direct connections between three locations.

Another approach is the inclusion of inland waterways in the city logistics transport processes (see, e.g. [56, 94, 96]). Even if this seems unusual at first glance, it should not be forgotten that for many centuries inland waterway transport, even at the local level, was of considerable importance for urban transport, especially due to the lack of alternatives. In the meantime, this transport mode is adjusted with a view to be applied in delivery transport. Smaller transshipment points are installed at the waterways from which the supply of (commercial) customers can take place.

3 Deliveries in Mail Order and Online Retail Trade

The delivery of goods to customers in the *mail order* and *online retail trade* differs in the basic structures of those in the supply of shopping and service facilities (see Sect. 2). Here, in most cases, the shipments to be delivered are strictly limited in terms of weight and volume to a (standard) parcel size. Orders beyond this scope (and possibly also associated with additional delivery services) are handled separately via freight forwarders. Deliveries in the field of e-grocery are largely excluded, since specific parts are out of range due to specific (legal and technical) transport regulations and weight and volume often exceed the parcel service limits.

The structures required for implementation exist, as they correspond to the established processes of the (classical) *parcel services* as well as to the delivery processes in (traditional) *mail order selling* (see, e.g. [74]). The *logistics facilities* (e.g. CDT and UDC) that exist for the supply of commercial clients can also be integrated into the delivery structures to serve customers on the *last mile*.

However, due to the huge increase in online retail trade in recent years, with a significant increase in traffic as well as environmental pollution, adjustments have become necessary. In addition, the concepts discussed primarily focus on solutions for *urban areas. Rural areas*, on the other hand, are largely excluded, although due to the

often inadequate supply of shopping possibilities there is a special need. However, in consequence of specific demand structures, for rural areas it is necessary to find other ways for an economically justifiable and for customer's acceptable form of delivery.

3.1 Delivery Concepts

An important point in the delivery concepts is the design of the processes as well as the handover to the customer (see, e.g. [101, 105]), especially under the aspect of cost allocation between supplier and customer. The customer's preferences are the direct delivery to themselves or a representative at a determined location (at home, at the workplace, etc.). However, delivery costs that are unacceptable from the customer's point of view can influence the selection and therefore may lead to the acceptance of other but less expensive delivery concepts. The following remarks will deal with the currently existing variants in detail.

Attended Home Delivery: This variant (which is preferred in most cases) has the advantage that, from the customer's point of view, it involves the smallest amount of effort, but this is normally associated with corresponding costs (see, e.g. [50]). The disadvantage, however, is a temporal binding of customers to a self-determined location. The design options here are very diverse (see, e.g. [49, 68]), the determination of a *time period* or a *fixed date* for delivery (for example, the following day, on the same day, at a certain time of day, within a given time window, possibly with a differentiated pricing) as well as the number of delivery attempts (see, e.g. [49]). The problem to which extent in this case the resulting delivery costs incurred by the supplier are passed on to the customer, will not be discussed here.

Interesting, however, is the differentiated pricing, in which a customer can choose between different service levels or that he can receive financial benefits in the case of deviations from the desired delivery dates/delivery windows. In this respect, this is an important instrument for the supplier, as this offers the possibility of making the delivery processes more cost-effective according to a greater degree of freedom in planning. In routing, the efficiency of the processes receives a stronger weighting, while the customer's objectives are less weighted. The question here is which expectations the customer has and how high the willingness of the supplier is to make financial concessions. That means, in this case it is a game-theoretical decision problem.

Unattended Home Delivery: The basic idea here is the decoupling of the delivery process from the personal presence of the customer. A frequently discussed solution is the provision of lockable boxes at the customer's location in which the shipments can be stored. From the supplier's point of view, this concept is advantageous because it does not require a personal receiving of deliveries and, relating to vehicle routing, the possibility for cost reduction can be given greater importance.

The main problem here is to provide *suitable* and *legally permissible places*, especially in (bigger) apartment buildings, and the question of access for the delivery personnel. Regarding the latter point, there exist possible solutions through intelligent key systems, with the help of which the suppliers can obtain the (each one-time) access to houses via appropriate codes. However, the problem here is the acceptance on the customer side, which may not be given or only conditionally. But, assuming the

situation in suburban (and also rural) areas, the situation is different. Due to the settlement structures, the spatial requirements are normally given here in order to provide suitable boxes which can be used via corresponding codes.

A specific variant of unattended home delivery is the integration of customer's private car in the delivery process (see, e.g. [83, 88]). The supplier receives in this case the current position of the vehicle and a code to open the trunk to store the shipment there. However, this involves a number of risks for customers and it must also be asked whether they will accept such an approach.

Integration of Pick-up-Points: The advantages of an integration of pick-up points between the supplier and the customer are for him in the (far-reaching) elimination of the temporal binding, but provided that the facilities are easily accessible. However, suppliers also benefit from this approach, as it results in an improved planning of transport processes as well as a reduction in delivery costs due to the occurrence of greater bundling effects. Another advantage of such structures also arises in an efficient handling of returns. The disadvantage for the customer is usually the additional share of travel cost and time for the collection and also the associated infrastructure costs on the supplier side.

- *Parcel shops*: These are, for example, *shopping* and *service facilities* with suitable opening hours, where deliveries can be made available for collection and also for returns (see, e.g. [99]). An efficient solution also may be the linking of a parcel shop in connection with the workplaces of customers (see, e.g. [109]).
- *Locker systems*: This concept is based on publicly accessible facilities with *lockers* of different sizes, which can be opened via codes (*parcel stations*) (see, e.g. [30, 52, 80, 95]). Packages can be picked up here as soon as the code has been transmitted, as well as return shipments.
- *Automated Storage and Retrieval Systems*: These are partly larger automated storage systems where access is regulated by codes. A real example of this was the *Tower24* (see, e.g. [39]), but this project ultimately failed due to a lack of cost-effectiveness.
- *Drive-in-Concepts*: In this case, customer's orders are picked in a warehouse or in a shopping facility that is easily accessible to the customer. For the customer, the main advantage is the time savings due to the not necessary order picking and by the sellers also in their own decision regarding the collection date.

Which of the different options or others (see, e.g. the self-collection services described by [107]) can be used depends on the respective requirements and framework conditions. In addition, the question of acceptance of the various solutions by the customers must also be taken into account.

3.2   Actual Vehicle Systems

The processing of the deliveries is based on the basic concept of the classic *parcel service providers*, in which diesel-powered vans were used in the distribution (and collection) of parcels. With a substantial growth of e-commerce and the resulting increase in last-mile transport demand, claims were made to adapt to the situation in the frequently congested delivery areas and reduce transport-related emissions. The solutions discussed in recent

years (and those already implemented) prefer the use of (smaller) *vans* with *electric drive* and the use of *cargo bikes* (see, e.g. [54]).

Application of Vans: The transition from classic vans to electrically driven vehicles is largely unproblematic, as no significant structural changes in the processes are required. However, a limiting factor may be the (currently) small operating range due to the lack of performance of the batteries currently in use. In terms of the local region, there is an emission reduction, but there are considerable doubts relating to the entire product life cycle (see Sect. 2.3).

Application of Cargo Bikes: The basic idea behind the use of cargo bikes is based on two points: *emission-free operation* and *greater flexibility* in the movements in delivery areas (see, e.g. [57]). However, some disadvantages have to be seen; the low capacity of the cargo bikes increases the staff requirements and the structures of the logistics facilities must be adopted. Here it is necessary to install an additional level below the CDT or UDC level, to keep the required mileage of the cargo bikes low. These may be stationary facilities or else mobile systems (see, e.g. [62]) whose respective position is aligned with the requirements for reloading the cargo bikes.

Comparative studies (see, e.g. [4, 91, 108]) on parcel delivery with the help of cargo bikes with other delivery concepts on the last mile show a different picture, also due to the respective framework conditions. However, it is clear that the logistics facilities to transship parcels on cargo bikes should be in the close proximity of the destination area. In the case of stationary facilities, the parcels may also be shipped by rail or by inland waterway, if an appropriate infrastructure is available. However, a solution based mobile system seems to make more sense, since a permanent adaptation to the current demand structures is possible. This means that it is at this point a specific form of a dynamic location routing problem where the movements of the vehicles must be synchronized with those of the cargo bikes.

3.3 Automated and Autonomous Vehicle Systems

With the rapid development of *drones*, the use of these *unmanned aerial vehicles* (UAVs) has been propagated as an alternative solution for bypassing urban traffic problems (see, e.g. [9, 40, 67]). Moreover, *autonomously* or *automatically operating delivery robots* with an electric drive were also developed for ground-based operations and tested in various projects.

Drone Deliveries: The use of drones in deliveries on the *last mile* has been discussed for several years and is often seen as a major step towards more environmentally friendly delivery processes (see, e.g. [72, 88]). Even if the technical conditions are met, there is considerable debate as to whether a wide market penetration can be achieved. The focus is on *performance parameters* (for example capacity and range), *cost considerations*, *legal questions* and *safety regulations*.

First, the capacity is limited to a few kilograms; an increase will require other drone sizes, with then partly changed aviation regulatory framework. Whether in the cases in which drones are used for parcel transport, a cost-covering price is used, is unlikely. However, it can be assumed that in this case marketing aspects are in the foreground. But this does not mean that in certain situations such a transport makes sense. Applications at DHL are the supply of temporarily difficult to reach islands of the

German North Sea coastal area (*Paketcopter* 2) and in inaccessible mountain landscapes in the Alps (*Paketcopter* 3). An interesting variant could be the concept described by [20], in which the delivery of the parcels for the customer takes place from a moving truck with a drone.

Nevertheless, the main problems are in the field of legal regulations and public acceptance (see, e.g. [9, 67]). For example, there are extensive no-fly zones and overflight bans, which considerably limit the drone movements. The scope of the expected increase in drone applications also raises the question of ensuring suitable airspace control. There must be a consistent and area-wide surveillance, including the ground-level airspace, to ensure the necessary safety. In addition, there are possible threats to terrorist actions, so there is also a considerable need for regulation under this aspect.

In spite of the question of the regulatory framework, it should be noted that the use of drones will be of considerable importance in many areas in future (see, e.g. [82, 70]). This applies, e.g., to possible applications in disaster situations, in particular in the case of destroyed or impassable transport infrastructure. The delivery of shipments by parcel service providers will rather not be a dominating field of application (see, e.g. [70, 82]).

Ground-based Vehicles: For some years, attempts have been made to include *automated* or *autonomous robots* in the form of small vehicles in the delivery processes on the last mile (see, e.g. [44, 46, 78]). In these concepts (see, e.g. [17]), the robots are transported by vans to a designated location in the planned delivery area where they are dropped off. The way to the customer's site is carried out automatedly or autonomously on the sidewalks, whereby the handover point can be flexibly coordinated.

Previous tests have shown the technical feasibility. However, the main problems here also lie in the field of the legal framework and the lack of acceptance by the population due to security aspects caused by concurrent use of the sidewalks. Swiss Mail, for example, has discontinued its test project because, for security reasons, the robots may only be moved in the public area when accompanied by an on-sight person.

3.4 Inclusion of Sharing Concepts

The basic concept here is to link activities of passenger mobility with logistical tasks (see, e.g. [19]). This approach, referred to as *crowdsourced delivery* (see, e.g. [6, 11, 31, 79, 106]), is a platform-based real-time communication system that helps to organize and control pick-up and delivery processes. The objective here is to link private car trips with the transport of parcels to reduce inner-city traffic and also traffic-induced emissions. Another approach is described by [90] and [37], in which public transport users are integrated instead of car drivers.

That such a concept can be realized in a sharing economy is demonstrated by the existing offers for (public) passenger transport (see, e.g. [92]). However, such a mobility solution is not uncontroversial both in passenger transport and in logistics (see, e.g. [25]). On the one hand, it is the often inadequate income structure of drivers and, on the other hand, the competitive distortion to the disadvantages of, for example, (commercial) taxi companies. Only if these problems are solved, a permanent and sustainable acceptance and thus the existence of this delivery form can be achieved.

In addition, various studies have also shown that the propagated reduction of traffic in cities is not reached by sharing offers in passenger transport, but on the contrary, the traffic is in many cases increased significantly (see, e.g. [36]). Moreover, the increase in such services also comes at the expense of public transport and therefore ultimately has a counterproductive effect on transport policy. If these experiences are transmitted to the transport of parcels, a similar development may possibly be expected, as in this case a partial de-bundling will occur. From the point of view of the individual customer, this may be advantageous regarding its possibly faster delivery, but with a view to the total traffic volume, it is to be rated negatively.

4 Outlook and Future Developments

Our paper has shown that essential problems in city logistics have not yet been solved to a satisfying extent so that further developments are necessary. Key points here are a further automation and a greater reduction of negative environmental impacts. Not only automation to the vehicle systems is necessary, but also the entire delivery processes, whereby the possibilities of using humanoid and semi-humanoid robots will be of considerable importance. This is not only a question of personnel costs, but it is also about much larger scope in the process design. The basis for this is the elimination of a restrictive legal framework, in particular regarding to working time regulations. In addition, it is necessary to attain an improvement and expansion of the data supply, also with regards to the quality of the planning results. An increased collection of real-time data and an efficient (process-related) information management are a mandatory basis for this. However, it must also be made clear that a strict objective orientation is necessary. Not every concept is ultimately effective and useful, even if this is repeatedly brought forward.

References

1. Akgün, E.Z., Monios, J.: Institutional influences on the development of urban freight transport policies by local authorities. In: Shiftan, Y., Kamargianni, M. (eds.) Advances in Transport Policy and Planning, Volume 1, Preparing for the New Era of Transport Policies - Learning from Experience, pp. 169–195. Elsevier, Amsterdam (2018)
2. Akgün, E.Z., Monios, J., Rye, T., Fonzon, A.: Influences on urban freight transport policy choice by local authorities. Transp. Policy **75**, 88–98 (2019)
3. Alves, R., da Silva Lima, R.S., Silva, K., Gomes, W., González-Calderón, C.A.: Challenges in urban logistics - a research study in São João Del Rei, a historical Brazilian city. Transp. Res. Rec. **57**(4), 1–13 (2017)
4. Anderluh, A., Hemmelmayr, V.C., Nolz, P.C.: Synchronizing vans and cargo bikes in a city distribution network. CEJOR **25**(2), 345–376 (2017)
5. Antún, J.P., Reis, V., Macário, R.: Strategies to improve urban freight logistics in historical centers - the cases of Lisbon and Mexico City. In: Taniguchi, E., Thompson, R.G. (eds.) City Logistics 3, pp. 349–366. Wiley, Hoboken (2018)
6. Arslan, A.M., Agatz, N., Kroon, L., Zuidwijka, R.: Crowdsourced delivery - a dynamic pickup and delivery problem with ad hoc drivers. Transp. Sci. **53**(1), 222–235 (2019)

7. Arvidsson, N., Browne, M.: A review of the success and failure of tram systems to carry urban freight - the implications for a low emission intermodal solution using electric vehicles on trams. Eur. Transp. **54**(5), 1–18 (2013)
8. Arvidsson, N., Givoni, M., Woxenius, J.: Exploring last mile synergies in passenger and freight transport. Built Environ. **42**(4), 523–538 (2016)
9. Aurambout, J.-P., Gkoumas, K., Ciuffo, B.: Last mile delivery by drones - an estimation of viable market potential and access to citizens across European cities. Eur. Transp. Res. Rev. **11**(30), 1–21 (2019)
10. Awasthi, A., Chauhanb, S.S., Goyal, S.K.: A multi-criteria decision making approach for location planning for urban distribution centers under uncertainty. Math. Comput. Model. **53**(1–2), 98–109 (2011)
11. Ballare, S., Lin, J.: Preliminary investigation of a crowdsourced package delivery system - a case study. In: Taniguchi, E., Thompson, R.G. (eds.) City Logistics 3, pp. 109–128. Wiley, Hoboken (2018)
12. Barceló, J., Grzybowska, H., Orozco, J.A.: City logistics. In: Martí, R., Pardalos, P.M., Resende, M.G.C. (eds.) Handbook of Heuristics, pp. 887–930. Springer, Cham (2017)
13. Beirigo, B.A., Schulte, F., Negenborn, R.R.: Integrating people and freight transportation using shared autonomous vehicles with compartments. IFAC PapersOnLine 51-9, 392–397 (2018)
14. Björklund, M., Johansson, H.: Urban consolidation centre - a literature review, categorisation, and a future research agenda. Int. J. Phys. Distrib. Logist. Manag. **48**(8), 745–764 (2018)
15. Borghesi, A.: City logistics - is deregulation the answer? In: Bilgin, M.H., Danis, H., Demir, E., Can, U. (eds.) Financial Environment and Business Development, pp. 385–400. Springer, Cham (2017). https://doi.org/10.1007/978-3-319-39919-5_28
16. Boysen, N., Fliedner, M., Scholl, A.: Scheduling inbound and outbound trucks at cross docking terminals. OR Spectr. **32**(1), 135–161 (2010)
17. Boysen, N., Schwerdfeger, S., Weidinger, F.: Scheduling last-mile deliveries with truck-based autonomous robots. Eur. J. Oper. Res. **271**(3), 1085–1099 (2018)
18. Bauer, G., Wiegard, H.: Straßenbahn Archiv 7 - Arbeits- und Güterstraßenbahnfahrzeuge. Transpress, Berlin (1989)
19. Buldeo Rai, H., Verlinde, S., Merckx, J., Macharis, C.: Crowd logistics - an opportunity for more sustainable urban freight transport? Eur. Transp. Res. Rev. **9**(3/39), 1–32 (2017)
20. Carlsson, J.G., Song, S.: Coordinated logistics with a truck and a drone. Manage. Sci. **64**(9), 4052–4069 (2017)
21. Cattaruzza, D., Absi, N., Feillet, D., González-Feliu, J.: Vehicle routing problems for city logistics. EURO J. Transp. Logist. **6**(1), 51–79 (2017)
22. Cervero, R.: Paratransit in Southeast Asia. Rev. Urban Reg. Dev. Stud. **3**(1), 3–25 (1991)
23. Cervero, R., Golub, A.: Informal transport - a global perspective. Transp. Policy **14**(6), 445–457 (2007)
24. Chan, C.-Y.: Advancements, prospects, and impacts of automated driving systems. Int. J. Transp. Sci. Technol. **6**, 208–216 (2017)
25. Chee, F.: An Uber ethical dilemma - examining the social issues at stake. J. Inf. Commun. Ethics Soc. **16**(3), 261–274 (2018)
26. Chen, Z., Dong, J., Ren, R.: Urban underground logistics system in China - opportunities or challenges? Undergr. Space **2**(3), 195–208 (2017)
27. Crainic, T.G., Ricciardi, N., Storchi, G.: Models for evaluating and planning city logistics systems. Transp. Sci. **43**(4), 432–454 (2009)
28. Christaller, W.: Die zentralen Orte in Süddeutschland. Gustav Fischer, Jena (Neudruck 1968 Wissenschaftliche Buchgesellschaft, Darmstadt) (1933)

29. Dablanc, L.: City distribution, a key element of urban economy - guidelines for practitioners. In: Macharis, C., Melo, S. (eds.) City Distribution and Urban Freight Transport - Multiple Perspectives, pp. 13–36. Edward Elgar, Cheltenham and Northampton (2011)

30. Deutsch, Y., Golany, B.: A parcel locker network as a solution to the logistics last mile problem. Int. J. Prod. Res. **56**(1–2), 251–261 (2018)

31. Devari, A., Nikolaev, A.G., He, Q.: Crowdsourcing the last mile delivery of online orders by exploiting the social networks of retail store customers. Transp. Res. Part E **105**, 105–122 (2017)

32. Di Bugno, M., Guerra, S., Ambrosino, G., Boero, M., Liberato, A.: A centre for eco-friendly city freight distribution - urban logistics innovation in a mid-size historical city in Italy (2007). https://www.researchgate.net/publication/298078750_A_centre_for_eco-friendly_city_freight_distribution_Urban_logistics_innovation_in_a_mid-size_historical_city_in_Italy

33. Dong, J., Hu, W., Yan, S., Ren, R., Zhao, X.: Network planning method for capacitated metro-based underground logistics system. Adv. Civ. Eng. **2018**, 1–14 (2018)

34. Ehmke, J.F.: Integration of Information and Optimization Models for Routing in City Logistics. Springer, New York (2012). https://doi.org/10.1007/978-1-4614-3628-7

35. Ehmke, J.F., Campbell, A.M., Thomas, B.W.: Optimizing for total costs in vehicle routing in urban areas. Transp. Res. Part E **116**, 242–256 (2018)

36. Erhardt, G.D., Sneha Roy, S., Cooper, D., Sana, B., Chen, M., Castiglione, J.: Do transportation network companies decrease or increase congestion? Sci. Adv. **5**(5), 1–11 (2019)

37. Gatta, V., Marcucci, E., Nigro, M., Patella, S.M., Serafini, S.: Public transport-based crowdshipping for sustainable city logistics - assessing economic and environmental impacts. Sustainability **11**(1), 145 (2019)

38. Giordano, A., Fischbeck, P., Matthews, H.S.: Environmental and economic comparison of diesel and battery electric delivery vans to inform city logistics fleet replacement strategies. Transp. Res. Part D **64**, 216–229 (2018)

39. Goldman, T., Gorham, R.: Sustainable urban transport - four innovative directions. Technol. Soc. **28**, 261–273 (2006)

40. Goodchild, A., Toy, J.: Delivery by drone - an evaluation of unmanned aerial vehicle technology in reducing CO_2 emissions in the delivery service industry. Transp. Res. Part D **61**, 58–67 (2018)

41. Groß, P.-O., Ehmke, J.F., Mattfeld, D.C.: Cost-efficient and reliable city logistics vehicle routing with satellite locations under travel time uncertainty. Transp. Res. Procedia **37**, 83–90 (2019)

42. Heckler, W.: Der Einsatz von Stadtschnellbahnen für die Abwicklung des innerstädtischen Güterverkehrs. Veröffentlichungen des Verkehrswissenschaftlichen Instituts der Rheinisch-Westfälischen Technischen Hochschule Aachen (Heft 28) (1978)

43. Holguín-Veras, J., et al.: Direct impacts of off-hour deliveries on urban freight emissions. Transp. Res. Part D **61**, 84–103 (2018)

44. Hoffmann, T., Prause, G.: On the regulatory framework for last-mile delivery robots. Machines **6**(3), 33 (2018)

45. Hulse, L.M., Xie, H., Galea, E.R.: Perceptions of autonomous vehicles - relationships with road users, risk, gender and age. Saf. Sci. **102**, 1–13 (2018)

46. Jennings, D., Figliozzi, M.A.: A study of sidewalk autonomous delivery robots and their potential impacts on freight efficiency and travel. Transp. Res. Rec. **2673**(6), 317–326 (2019)

47. Kersting, M., Klemmer, P., Stein, D.: CargoCap - Wirtschaftliche Transportalternative im Ballungsraum. Internationales Verkehrswesen **56**(11), 493–498 (2004)
48. Kikuta, J., Ito, T., Tomiyama, I., Yamamoto, S., Yamada, T.: New subway-integrated city logistics system. Procedia Soc. Behav. Sci. **39**, 476–489 (2012)
49. Köhler, C., Ehmke, J.F., Campbell, A.M.: Flexible time window management for attended home deliveries. Omega (2019, in Press)
50. Köhler, C., Haferkamp, J.: Evaluation of delivery cost approximation for attended home deliveries. Transp. Res. Procedia **17**, 67–72 (2019)
51. Lebeau, P., Macharis, C., van Mierlo, J., Lebeau, K.: Electrifying light commercial vehicles for city logistics? A total cost of ownership analysis. Eur. J. Transp. Infrastruct. Res. **15**(4), 551–569 (2015)
52. Lemke, J., Iwana, S., Korczak, J.: Usability of the parcel lockers from the customer perspective - the research in Polish Cities. Transp. Res. Procedia **16**, 272–287 (2016)
53. Lindholm, M., Behrends, S.: Challenges in urban freight transport planning - a review in the Baltic Sea Region. J. Transp. Geogr. **22**, 129–136 (2012)
54. Machado de Oliveira, C., De Mello Bandeira, R.A., Vasconcelos Goes, G., Schmitz Gonçalves, D.N., De Almeida D'Agosto, M.: Sustainable vehicles-based alternatives in last mile distribution of urban freight transport - a systematic literature review. Sustainability **9**(8), 1–15 (2017)
55. Macharis, C., van Hoeck, E., Verlinde, S., Debauche, W., Wilox, F.: Multi-actor multi-criteria analysis - a case study on night-time delivery for urban distribution. In: Macharis, C., Melo, S. (eds.) City Distribution and Urban Freight Transport - Multiple Perspectives, pp. 101–119. Edward Elgar, Cheltenham and Northampton (2011)
56. Maes, J., Sys, C., Vanelslander, T.: City logistics by water - good practices and scope for expansion. In: Ocampo-Martinez, C., Negenborn, R.R. (eds.) Transport of Water Versus Transport Over Water, pp. 413–437. Springer, Cham (2015). https://doi.org/10.1007/978-3-319-16133-4_21
57. Maes, J., Vanelslander, T.: The use of bicycle messengers in the logistics chain, concepts further revised. Procedia Soc. Behav. Sci. **39**, 409–423 (2012)
58. Mager, T.J.: Mobilitätslösungen für den ländlichen Raum. Standort **41**(3), 217–223 (2017)
59. Marcucci, E., Gatta, V.: Investigating the potential for off-hour deliveries in the city of Rome - retailers' perceptions and stated reactions. Transp. Res. Part A **102**, 142–156 (2017)
60. Marinov, M., et al.: Urban freight movement by rail. J. Transp. Lit. **7**(3), 87–116 (2013)
61. Martínez-Díaz, M., Soriguera, F.: Autonomous vehicles - theoretical and practical challenges. Transp. Res. Procedia **33**, 275–282 (2018)
62. Marujo, L.G., et al.: Assessing the sustainability of mobile depots - the case of urban freight distribution in Rio de Janeiro. Transp. Res. Part D **63**, 256–267 (2018)
63. Masson, R., Trentini, A., Lehuédé, F., Malhéné, N., Péton, O., Tlahig, H.: Optimization of a city logistics transportation system with mixed passengers and goods. EURO J. Transp. Logist. **6**(1), 81–109 (2017)
64. Moriarty, P., Honnery, D.: Prospects for hydrogen as a transport fuel. Int. J. Hydrogen Energy **44**(31), 16029–16037 (2019)
65. Muschwitz, C., Monheim, H., Reimann, J., Sylvester, A., Pitzen, C.: kombiBUS-Modell Uckermark - Kombinierter Personen- und Güterverkehr zur Stabilisierung ländlicher ÖPNV-Systeme. Internationales Verkehrswesen **67**(2), 37–38 (2015)
66. Nataraj, S., Ferone, D., Quintero-Araujoc, C., Juan, A.A., Festa, P.: Consolidation centers in city logistics - a cooperative approach based on the location routing problem. Int. J. Ind. Eng. Comput. **10**, 393–404 (2019)
67. Nentwich, M., Horváth, D.M.: The vision of delivery drones. Zeitschrift für Technikfolgenabschätzung in Theorie und Praxis **27**(2), 46–52 (2018)

68. Nguyen, D.H., de Leeuw, S., Dullaert, W., Foubert, B.P.J.: What is the right delivery option for you? Consumer preferences for delivery attributes in online retailing. J. Bus. Logist. (2019, in Press)

69. Nuzzolo, A., Crisalli, U., Comi, A.: Metropolitan freight distribution by railways. In: Taniguchi, E., Thompson, R.G. (eds.) Recent Advances in City Logistics, pp. 351–367. Elsevier, Amsterdam (2008)

70. Otto, A., Agatz, N., Campbell, J., Golden, B., Pesch, E.: Optimization approaches for civil applications of unmanned aerial vehicles (UAVs) or aerial drones - a survey. Networks **72**(4), 411–458 (2018)

71. Paddeu, D.: Sustainable solutions for urban freight transport and logistics: an analysis of urban consolidation centers. In: Zeimpekis, V., Aktas, E., Bourlakis, M., Minis, I. (eds.) Sustainable Freight Transport. ORSIS, vol. 63, pp. 121–137. Springer, Cham (2018). https://doi.org/10.1007/978-3-319-62917-9_8

72. Park, J., Kim, S., Suh, K.: A comparative analysis of the environmental benefits of drone-based delivery services in urban and rural areas. Sustainability **10**(3), 1–15 (2018)

73. Pelletier, S., Jabali, O., Laporte, G.: Goods distribution with electric vehicles - review and research perspectives. Transp. Sci. **50**(1), 3–22 (2016)

74. Perboli, G., Rosano, M.: Parcel delivery in urban areas - opportunities and threats for the mix of traditional and green business models. Transp. Res. Part C **99**, 19–36 (2019)

75. Phun, V.K., Yai, T.: State of the art of paratransit literatures in Asian developing countries. Asian Transp. Stud. **4**(1), 57–77 (2016)

76. Piatkowski, P., Puszkiewicz, W.: Electric vehicles - problems or solutions. J. Mech. Energy Eng. **2**(1), 59–66 (2018)

77. Pimentel, C., Alvelos, F.: Integrated urban freight logistics combining passenger and freight flows - mathematical model proposal. Transp. Res. Procedia **30**, 80–89 (2018)

78. Poeting, M., Schaudt, S., Clausen, U.: Simulation of an optimized last-mile parcel delivery network involving delivery robots. In: Clausen, U., Langkau, S., Kreuz, F. (eds.) ICPLT 2019. LNL, pp. 1–19. Springer, Cham (2019). https://doi.org/10.1007/978-3-030-13535-5_1

79. Punel, A., Stathopoulos, A.: Modeling the acceptability of crowdsourced goods deliveries - role of context and experience effects. Transp. Res. Part E **105**, 18–39 (2017)

80. Quak, H., Balm, S., Posthumus, B.: Evaluation of city logistics solutions with business model analysis. Procedia Soc. Behav. Sci. **125**, 111–124 (2014)

81. Quak, H.J., de Koster, M.B.M.: Delivering goods in urban areas - how to deal with urban policy restrictions and the environment. Transp. Sci. **43**(2), 211–227 (2009)

82. Rao, B., Goutham, A., Maione, R.: The societal impact of commercial drones. Technol. Soc. **45**, 83–90 (2016)

83. Reyes, D., Savelsbergh, M., Toriello, A.: Vehicle routing with roaming delivery locations. Transp. Res. Part C **80**, 71–91 (2017)

84. Rien, W., Roggenkamp, M.: Can trams carry cargo? New logistics for urban areas. World Transp. Policy Pract. **1**(1), 32–36 (1995)

85. Romano Alho, A., de Abreu e Silva, J., Pinho de Sousa, J., Blanco, E.: Improving mobility by optimizing the number, location and usage of loading/unloading bays for urban freight vehicles. Transp. Res. Part D **61**, 3–18 (2018)

86. Ruesch, M., Schmid, T., Bohne, S., Haefeli, U., Walker, W.: Freight transport with vans - developments and measures. Transp. Res. Procedia **12**, 79–92 (2016)

87. Russo, F., Comi, A.: Measures for sustainable freight transportation at urban scale - expected goals and tested results in Europe. J. Urban Plan. Dev. **137**(2), 142–152 (2011)

88. Savelsbergh, M., van Woensel, T.: City logistics - challenges and opportunities. Transp. Sci. **50**(2), 99–110 (2016)

89. Schwarting, W., Alonso-Mora, J., Rus, D.: Planning and decision-making for autonomous vehicles. Annu. Rev. Control Robot. Auton. Syst. **1**, 187–210 (2018)
90. Serafini, S., Nigro, M., Gatta, V., Marcucci, E.: Sustainable crowdshipping using public transport - a case study evaluation in Rome. Transp. Res. Procedia **30**, 101–110 (2018)
91. Sheth, M., Butrina, P., Goodchild, A., McCormack, E.: Measuring delivery route cost trade-offs between electric-assist cargo bicycles and delivery trucks in dense urban areas. Eur. Transp. Res. Rev. **11**(11), 1–12 (2019)
92. Standing, C., Standing, S., Biermann, S.: The implications of the sharing economy for transport. Transp. Rev. **39**(2), 226–242 (2019)
93. Trentini, A., Malhéné, N.: Flow management of passengers and goods coexisting in the urban environment - conceptual and operational points of view. Procedia Soc. Behav. Sci. **39**, 807–817 (2012)
94. Trojanowski, J., Iwan, S.: Analysis of Szczecin waterways in terms of their use to handle freight transport in urban areas. Procedia Soc. Behav. Sci. **151**, 333–341 (2014)
95. Vakulenko, Y., Hellström, D., Hjort., K.: What's in the parcel locker? Exploring customer value in e-commerce last mile delivery. J. Bus. Res. **88**, 421–427 (2018)
96. van Duin, J.H.R., Kortmann, L.J., van de Kamp, M.: Toward sustainable urban distribution using city canals - the case of Amsterdam. In: Taniguchi, E., Thompson, R.G. (eds.) City Logistics 1, pp. 65–84. Wiley, Hoboken (2018)
97. van Lier, T., Meers, D., Buldeo Rai, H., Macharis, C.: Evaluating innovative solutions for sustainable city logistics - an enhanced understanding of stakeholder perceptions. In: Macharis, C., Baudry, G. (eds.) Decision-Making for Sustainable Transport and Mobility, pp. 149–163. Edward Elgar Publishing, Cheltenham and Northampton (2018)
98. van Rooijen, T., Groothedde, B., Gerdessen, J.C.: Quantifying the effects of community level regulation on city logistics. In: Taniguchi, E., Thompson, R.G. (eds.) Recent Advances in City Logistics, pp. 351–399. Elsevier, Amsterdam (2008)
99. Verlinde, S., Rojas, C., Buldeo Rai, H., Kin, B., Macharis, C.: E-consumers and their perception of automated parcel stations. In: Taniguchi, E., Thompson, R.G. (eds.) City Logistics 3, pp. 147–160. Wiley, Hoboken (2018)
100. Visser, J.G.S.N.: The development of underground freight transport - an overview. Tunn. Undergr. Space Technol. **80**, 123–127 (2018)
101. Visser, J.G.S.N., Nemoto, T., Browne, M.: Home delivery and the impacts on urban freight transport - a Review. Procedia Soc. Behav. Sci. **125**, 15–27 (2014)
102. Wang, K., Shao, Y., Zhou, W.: Metaheuristic for a two-echelon capacitated vehicle routing problem with environmental considerations in city logistics service. Transp. Res. Part D **57**, 262–276 (2017)
103. Watróbski, J., Małecki, K., Kijewska, K., Iwan, S., Karczmarczyk, A., Thompson, R.G.: Multi-criteria analysis of electric vans for city logistics. Sustainability **9**(8), 1453 (2017)
104. Wiegmans, B.W., Visser, J., Konings, R., Pielage, B.-J.A.: Review of underground logistic systems in the Netherlands - an ex-post evaluation of barriers, enablers and spin-offs. Eur. Transp. **45**, 34–49 (2010)
105. Winkenbach, M., Janjevic, M.: Classification of last-mile delivery models for e-commerce distribution - a global Perspective. In: Taniguchi, E., Thompson, R.G. (eds.) City Logistics 1, pp. 209–229. Wiley, Hoboken (2018)
106. Yildiz, B., Savelsbergh, M.: Service and capacity planning in crowd-sourced delivery. Transp. Res. Part C **100**, 177–199 (2019)
107. Yuen, K.F., Wang, X., Ting, L., Ng, W., Wong, Y.D.: An investigation of customers' intention to use self-collection services for last-mile delivery. Transp. Policy **66**, 1–8 (2018)

108. Zhang, L., Matteis, T., Thaller, C., Liedtke, G.: Simulation-based assessment of cargo bicycle and pick-up point in urban parcel delivery. Procedia Comput. Sci. **130**, 18–25 (2018)

109. Ziegler, M.: Corporate Startups - Unternehmensinterner Nährboden für digitale Geschäftsmodelle - Dargestellt am Beispiel der Last Mile Lösung pakadoo. In: Proff, H., Fojcik, T.M. (eds.) Mobilität und digitale Transformation, pp. 393–403. Springer Gabler, Wiesbaden (2018). https://doi.org/10.1007/978-3-658-20779-3_24

110. Zych, M.: Identification of potential implementation of the Cargo tram in Warsaw - a first overview. Procedia Soc. Behav. Sci. **151**, 360–369 (2014)

Using Advanced Information Systems to Improve Freight Efficiency: Results from a Pilot Program in Colombia

Adriana Moros-Daza[1,2]([⊠]), Daniela Cassandro-De La Hoz[2],
Miguel Jaller-Martelo[3], and Carlos D. Paternina-Arboleda[2]

[1] Institute of Information Systems (IWI), University of Hamburg,
Hamburg, Germany
`adriana.moros.daza@uni-hamburg.de`
[2] Department of Industrial Engineering, Universidad del Norte,
Barranquilla, Colombia
`{dcassandro,cpaterni}@uninorte.edu.co`
[3] Department of Civil and Environmental Engineering,
University of California, Davis, USA
`mjaller@ucdavis.edu`

Abstract. The efficiency of the urban and inter-urban freight transport sector is crucial for Colombia's economic competitiveness, and the government continually engages in projects to improve freight transport activities. Recently, as part of a program towards the reduction of greenhouse gases (GHGs) and the improvement of efficiency from the transportation sector, the Ministry of Transport pilot tested advanced information services using a freight brokerage system. This work discusses the results of this pilot program. Specifically, the work discusses the data collection methodology, provides details about the program, and describes the findings. The study had three stages. First, a socio-economic analysis of at least 500 vehicles/drivers, based on information from both vehicles and drivers. Second, the implementation of an ESD tool (Engine System Diagnostic Tool) device on the vehicle computer to measure some control variables of importance. Last, the data analysis of potential reduction on empty trips. The results show that the program can reduce the number of empty and non-revenue generating trips of the over-the-road freight transport (45% of trips in Colombia are empty) and improve the system. Specifically, the program reduced a third of these trips and generated new revenue trips to the participating carrier companies. However, the system reduced empty trips by 20%, and increased the number of trips per month, therefore, yielding more return on investment to drivers/vehicles and to the carrier.

Keywords: Freight transport · Freight brokerage · Empty trips · Compensated trips · Engine System Diagnostic Tool

C. Paternina-Arboleda and S. Voß (Eds.): ICCL 2019, LNCS 11756, pp. 22–38, 2019.
https://doi.org/10.1007/978-3-030-31140-7_2

1   Introduction

14% of the global inventory of greenhouse gases (GHGs) comes from the transport sector [12,13]. In Colombia, this proportion is 12% of the emissions, and the sector has the highest energy demand in the country [16]. More importantly, freight transport, with a small share of the vehicle movements, is responsible for 43% of the total emissions of greenhouse gases in the sector [21].

During the last few years, the country has initiated efforts to reduce GHGs [20]. As part of these efforts, the country have tested the improvement potential of advanced freight information systems. In December 2016, the Ministry of Transport started to measure the effect of a freight brokerage systems on GHGs emissions reduction. Freight brokerage systems are information management platforms that facilitate trade and transaction between the economic agents involved in freight transport operations. These systems link cargo shippers with carrier (transport) companies. The expectation is that the system can improve capacity utilization of the vehicles by fostering a matching between demand and capacity at the system level, and provide benefits for the economic agents. For example, by reducing the number of empty or non-revenue generating trips by "balancing or matching" the flows between origins and destinations. The framework could also offer a secure, reliable, modern, and formal marketplace.

The objective of this study is to discuss the results of the pilot test and evidence the potential benefits of such advanced information systems to improve freight system efficiency. The paper discusses the pilot program, the data collection methodology before and during the pilot, and the main findings. The pilot program included three stages of system characterization, data collection, and implementation. However, several obstacles, mainly due to driver unions, limited the scope of the governmentally funded program to the first stage. Consequently, the program required private funding for the data collection and implementation. For the first stage, the research team collected data from more than 500 vehicle/drivers, about the vehicles used, and the drivers/company socio-economic characteristics. For the second stage of measuring vehicle and system performance, the team instrumented the vehicles with an Engine System Diagnostic Tool (ESD) to measure control variables. The last stage, or implementation, measured changes in the system performance. The pilot included a large shipper, 2 carrier companies, and about 100 vehicles owners/drivers. The team found that a major inefficiency in the system resulted from poor management, and lack of adequate communication between shippers and carriers. The results show that the program can significantly improve freight efficiency. For instance, it reduced a third of the empty trips. In Colombia, empty trips are between 30 and 45% of movements (depending on the geographic locations and origin-destinations pairs).

The paper first provides background about the transport sector in Colombia, discusses different initiatives for environmental emissions mitigation in the freight sector, and concentrates on freight brokerage. The second part focuses on describing the data collection methodology, followed by a section discussing the empirical and numerical analysis framework. Section four discusses and analyses the empirical results. The paper ends with a conclusions section.

2   Background

The transport sector contributes 4% to the national gross domestic product (GDP) and has the highest energy consumption in the country, demanding 35% of the total oil derivatives (373,000 TJ in 2009). In terms of GHG emissions, the sector contributes 12% to the national inventory (20 million tons, in 2009), with over-the-road transport responsible for 90% [1,26], and freight transport contributes 43%. This is not an isolated problem to Colombia; most of the countries around the world are experiencing similar patterns, where the freight transport sector accounts for the majority of the transport emissions of GHGs, and criteria pollutants. This is mainly because of the use of diesel vehicles and an aged fleet.

Academics, researchers, practitioners, and official agencies around the world have developed, pilot tested, or evaluated a number of strategies and initiatives to improve system's efficiency and reduce the environmental impacts [1–7,9–11,14,17–19,22–27]. Some of these initiatives include:

- Exclusive and segregated roads;
- Delivery time windows (pick-up and deliveries);
- System monitoring and use of dynamic signage to optimize infrastructure and system;
- Advanced traveler information systems (ATIS);
- Low emission zones;
- Cross-docking facilities;
- Delivery, servicing and logistics plans;
- Fuel standards, and vehicle technologies;
- Integrated logistics, management, and transport services (e.g., 3PLs, 4PLs, fleet management);
- Eco-driving, and eco-routing systems; and,
- Considering of transport externalities in policy and practice.

2.1   Freight Brokerage Systems

One of the strategies that allows reducing empty and non-revenue trips is the use of freight brokerage systems. These have become common during the last two decades because they facilitate the procurement and contracting processes resulting in efficient transactions. Many companies, where the system is available, have adopted the model to automate their operations. In general, freight brokerage systems can improve information exchange and transfer along the supply chain, improve the matching for compensation loads, lower logistics costs, provide access to a larger pool of transport services and demand loads, improve driver working conditions, optimizing energy consumption, and reduce environmental emissions.

Freight brokerage systems are sophisticated systems that collect large amounts of data such as: location and status, delivery arrival, volume of contents, container empty, speed, and potential late delivery, among others. The information aids in the planning, operation, and management of freight transportation systems, to

manage the transportation in real time and develop an appropriate service. Therefore, this tool can be used in two main areas: Commercial Operations, to improve the monitoring of the freight flows, and Fleet Management Systems, dealing with the fleet management operations and transport planning [15].

The freight brokers usually seek fewer empty trips (back-hauls) in search of cargo due to the planning of operations of the vehicles based on the supply, and the capacity of the vehicle. The empty return trips are an effect of the regional economy. Not all regions in a country offer closed cargo circuits at the same level (trade deficit). The implementation of freight brokers would allow load generators, transport companies and transporters, to hire real-time goods between different cities in the country. One of the objectives of freight brokers is to provide an efficient market that facilitates the healthy competition at lower prices among the different suppliers of road freight transport in the country.

Therefore, the net economic effect of the use of freight brokers is to increase return cargo, and thus optimize the use of the vehicle. From the environmental perspective, reducing the number of empty return trips implies that a high percentage of trips made today would not be made, due to better load factors and fleet management between the shippers and transport companies.

Freight brokers were born in the 80's with the need to optimize the resources used in the transport of goods. The first freight broker was founded in France in 1985 with the name of Teleroute using communications tools for the control of their operations [8]. There are currently many alternatives in Europe, the United States and more recently in Latin America that offer freight broker services such as [8]:

- Teleroute (Europe) was the first freight broker system founded in France in 1985, using the Minitel system for its communications. The Dutch multinational Teleroute has updated its freight broker on the Internet with an intelligent agent, which also searches a radio near the selected cities, a route planner and the immediate transmission of cargo information to the driver via sms. This service has 45,000 members that access 75,000 offers introduced each day and that can have their origin or destination in 40 countries. The Affiliate pays between 20 and 30 cents (Euro) for each consultation in which the contact details appear.
- Cargoplace.com (Colombia) Cargoplace.com is the first experience in Colombia of a freight broker, and is a product of business Place S.A.S., a company founded in 2007 to provide Internet-enabled business solutions, with focus on logistics. It is the first on-line tool in the country that enables the interaction between companies that need to move their products and/or raw materials and those that have the vehicles to do so. It is considered as an independent information operator whose central purpose is to consolidate information of the freight transport market in Colombia to make its negotiation and management more efficient and to increase the rates of use of the fleet. Since its inception, CargoPlace.com has focused on providing its service nationwide with the presence of commercial agents in the main cities of the country. Cargo place to date has 63 transport companies and 14 cargo generators affiliated in Colombia.

- Suruta.com (Venezuela) is a WEB-based cargo and transport market place that provides users with transporters and load generators a means of inter-communicating with each other and finding solutions to their cargo and transportation needs.
- Wtransnet (Europe): a Spanish service created in 1997 by Wotrant (owned by Catalana D'Iniciatives) has 4,300 associates that pay 85 euros per month. During the last year it published 2.5 million offers; the daily transaction is now 12,000 and each month incorporates 150 new partners, which can operate in several languages.
- Moving Freight Broker (Europe): is the first dedicated portal for a collaboration between moving companies. The idea of this site was to take advantage of the availability of empty vans and moving trucks.
- Bag Trans.eu (Europe) is a dedicated Internet platform to exchange information about the available loads and trucks. Every day more than 200 000 transport companies use this platform. It facilitates the access to offers, simple communication, and full security.

Mobile transmission technologies, Global Positioning System, availability of positioning information via GPS, route guidance, navigation systems, on-board sensors, and smart cards, are some essential instruments that help the tools to work efficiently. The results of this type of system, developed over the last 20 years in the European freight transport industry, are starting to become visible, including:

- Integrated information and services, based on the advanced technologies.
- Freight matching, transport auctions, transport exchanges for road transport.
- Interconnected computer systems, generation of emergency re-routing in case of incidents.
- Creative fleet management and monitoring capabilities.

3   Methods and Procedures

The first phase of the project was the integration of drivers to a freight brokerage system. In doing so, the team partnered with 4 transport companies. Moreover, the team we worked with the logistics and infrastructure management of ANDI to involve large cargo generating companies.

The team created a platform for independent drivers (owners) or under a confirmed programming scheme (drivers from one of the companies), with the participation of more than 500 drivers. Figure 1 shows the platform scheme.

The team studied users behavior (in the case of the driver). The objective was to evaluate the success of the system and marketplace based on: the transaction mode by the agent, system's stability, such as en-route tracking and additional connectivity services. The system also offer services related to the driver, such as agreements with service stations on the route and maintenance workshops which tend to the formalization of the sector by offering all costs concerned with the operation.

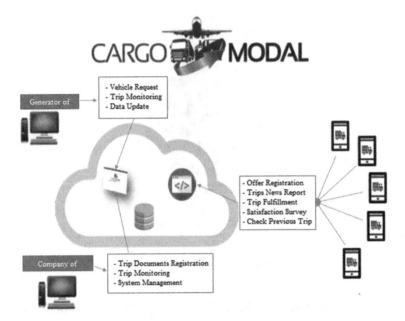

Fig. 1. Cargo Modal scheme

On the data analysis phases of the project, the data collected was used before and after the implementation of the cargo system to determine the behavior of the management indicators.

4   Experimental/Numerical Setting

During 2016, the data show that some companies made a total of 10,686 net trips, however the efficiency of these is not close to 100%. Figure 2 shows that of the total trips, 55% were loaded, equivalent to approximately 4926 trips. The remaining 45% were empty trips along the different corridors in the country. These empty trips are equivalent to 4758 trips, which means that almost half of the trips return empty to their place of origin (or to other relocation places), making transportation efficiency in Colombia very poor.

Also, Fig. 3 shows that the percentage of compensated trips was only 6%, which indicates that 94% of the trips returned without cargo to their place of origin.

Considering the 9 most important corridors of the country, it was possible to identify which were the routes with more empty trips in the country. The corridors were labeled as: *Costa*, *Antioquia*, *Bogota*, *Valle del Cauca*, *Tolima Grande*, *Llanos*, *Santanderes*, *Narino*, *Boyaca* and *Eje Cafetero*. According to Fig. 4 the corridor of the Coast (*Costa*) (trips leaving this node) is the corridor with the highest number of empty trips with approximately 3914 empty trips, representing 82% of the total empty trips.

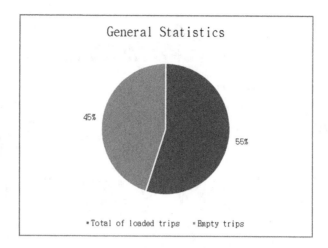

Fig. 2. General statistics

Following the **Bogota** corridor (trips departing from this node) with 442 empty trips, representing 9% of the total empty trips. A statistic quite distant from the previous one, but nevertheless significant. Another node with a high number of empty trips is **Antioquia**, with 226 empty trips, this node represents only 5% of the total empty trips. It can also be seen in Fig. 4 that the node with the lowest number of empty trips is the **Eje Cafetero** corridor, with only 5 empty trips, which represent only 0.1% of the total empty trips.

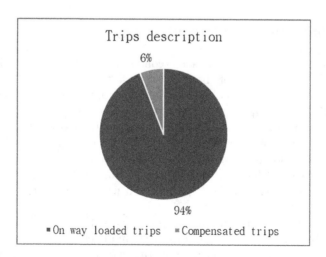

Fig. 3. Trips description

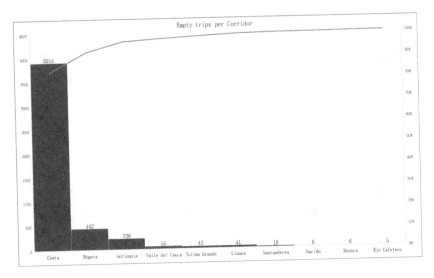

Fig. 4. Empty trips per corridor

Following is a detailed analysis of the three corridors with more empty trips in the country. This analysis includes exact routes with the largest number of empty trips, costs, and area specifications.

4.1   Costa Corridor

The *Costa* corridor has 13 routes that present 80% of its empty trips (Fig. 5). This Figure shows that the routes with the highest number of empty trips are: *La Jagua - Cartagena* with 1744 empty trips, *Cartagena - Barranquilla* with 646 empty trips, *Galapa - Barranquilla* with 197 empty trips and *Santa Marta - Barranquilla* with 156 empty trips.

Figure 6 Route *La Jagua - Cartagena* shows on the map the exact route between these two nodes. At the same time it can be observed that there are 4 tollbooths between the destination and the route is mostly made up of category 4 (flat) slope.

4.2   Bogota Corridor

This corridor contains 8 main routes with empty trips. Figure 7 shows that the routes with more empty trips are: *Bogota - Barranquilla* with 342 trips, *Bogota - Cartagena* with 52 empty trips and *Bogota - Buenaventura* with 32 empty trips.

Taking into account that the route with the highest number of empty trips of this node is *Bogota - Barranquilla*, general specifications of this route will be detailed below.

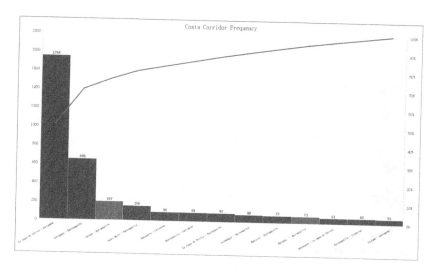

Fig. 5. Costa corridor frequency

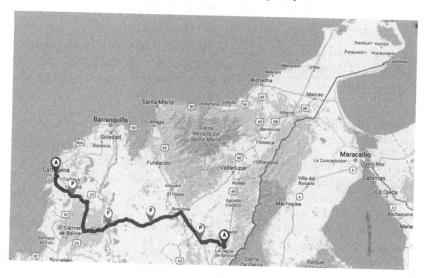

Fig. 6. Route with more trips (Costa Corridor - La Jagua - Cartagena)

Figure 8 *Bogotá - Barranquilla* route shows that there are 13 tollbooths on this route and the category of its slopes are approximately one of the most outstanding type 4 and 5 (medium planes and descents).

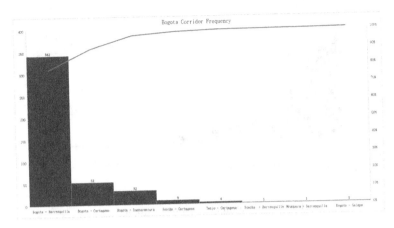

Fig. 7. Bogota corridor frequency

Fig. 8. Route with more trips (Bogotá Corridor - *Bogotá - Barranquilla*)

4.3 Antioquia Corridor

The **Antioquia** corridor consists of 15 routes with empty trips, the routes with the highest number of empty trips are: *Apartado - Barranquilla* with approximately 64 empty trips, *Medellin - Barranquilla* with 56 empty trips and *Apartado - Cartagena* with 52 empty trips (Fig. 9).

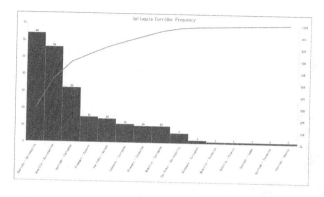

Fig. 9. Antioquia corridor frequency

The route with the highest number of empty trips in the **Antioquia** corridor-node, *Apartado - Barranquilla*, will be presented in detail in the following figure.

Fig. 10. Route with more trips (Antioquia Corridor - Apartado - Barranquilla)

Figure 10 *Apartado - Barranquilla* shows that this route is made up of 6 tollbooths and most of flat slopes and slight descents. This is a route of approximately 574 km with an average of 10 h of travel.

To conclude this section it is important to highlight that almost half of the trips of some of the allied companies of the Colombian coast are returned to their node of origin empty. Also, only 6% of the trips have a compensated load. This gives us a vision of what is happening in Colombia if we take this as a sample

at the scale of the country's reality. The corridors with the highest number of empty trips of the allied companies are **Costa**, **Bogotá** and **Antioquia**. At the same time it can be pointed out that the node with the highest number of empty trips is BARRANQUILLA, whether it is considered as origin or destination.

5   Analysis

We were able to run a sample with 37 trucks and 58 drivers (out of the 534-universe collected in phase 1) for running phase 2 and 3 of the experiments, to implement optimization solutions and analytics to reduce back-hauls (therefore greenhouse gas emissions) and to increase the utilization of freight. This represents around 6.4% of the total original universe collected for the first phase of our study. No further samples were collected due to increasing pressure from the truck drivers unions to not continue in the study.

The process runs as follows (Fig. 11):

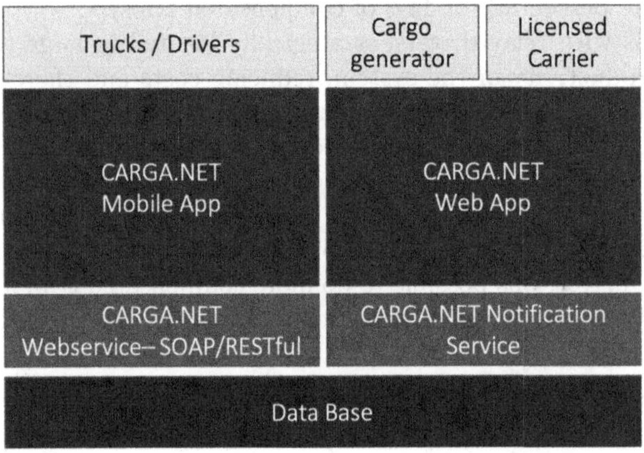

Fig. 11. Process scheme

Figure 11 shows that the system provide a demand management module that identifies potential cargo to be accommodated for LTL truck fulfillment in the direct distribution sense of a current load. Also, there is a demand planning module that fits location (geo-coded) with truck route compensation which guarantees all truck drivers regulations and optimizes for cost allocation and last. Furthermore, the system has a freight-brokering app were assigned trucks/drivers are given offers to be accepted or rejected.

The study guaranteed there is a licensed carrier company that handles all transportation needs and sets all conditions under which they want to provide the service. The system will provide tools to quickly select the most suitable vehicle to meet the customer's needs. If our anchor company is a load generator, the system provides tools to select the transport company whose service offer meets the need.

The architecture used to build-up the system for this pilot consists of 5 levels or layers

- Business layer: integrates the different applications that the system provide, including the privacy of the information.
- Application layer: all the necessary applications so that users can access and view the data collected by the IoT devices (including cargo conditions).
- Process layer: layer in which the data captured by the IoT devices is processed.
- Transport layer: transfers the data from the perception layer to the upper layers for processing.
- Perception layer: all the necessary applications so that users can access and view the data collected by the IoT devices.
- Only three (Business, application and transport layers) of all five layers were used since no information was collected at this phase on driving behavior or cargo conditions.

Once finished, the study showed a significant reduction in empty back-hauls (an approximate percentage of 44% of compensated trips).

We use trips with relay drivers, strategically located in large trip routes to enhance the productivity of the system, reducing costs and therefore tariffs to customers, which in turn resulted in more contracts for the pilot and therefore a better enhancement of the compensated trips to an approximate 73% of total trips made by the 27 trucks. The sample ran with only one company and our results show a large improvement in the operational measurements of the company. Figure 12 shows that with the use of the freight brokering platform, for the case of just one driver, was possible to do more compensated trip, increasing the general statistics of compensated trips from 6% to 48%. Meanwhile, for the case

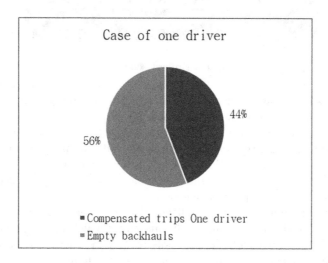

Fig. 12. Percentage of compensated trips after initial use of the freight brokering platform: case of only one driver/truck

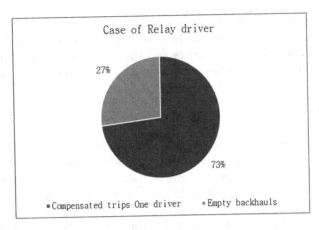

Fig. 13. Percentage of compensated trips after initial use of the freight brokering platform: case of only relay driver

of the use of relay driver, Fig. 13 exposes an increase of the compensated trips from 6% to 73%. Is important to highlight that compensated trips were possible in the routes: *Bogotá - Cartagena* and *Cali - Barranquilla*. Also, in terms of fuel consumption and equivalent emissions it was possible to obtain an approximate reduction of 38% and 34% subsequently.

To conclude this section it is important to highlight that the study gave the Ministry of Transport insights on how to regulate the use of information services from freight-brokering platforms so that the economic relationship between all stakeholders (i.e. drivers, cargo owners, carriers) should be able to interact.

Figure 14 shows how this interaction should work. Firstly, the proposal is based on the regulation of sectoral economic relations through the creation of contracts between authorized carriers and drivers/vehicles and cargo generators.

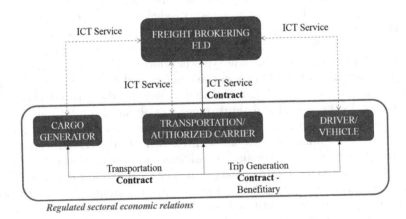

Fig. 14. Freight-brokering regulation proposal

Each contract should ensure the generation of trips in order to protect the main stakeholders of the platform. Secondly, it is proposed the creation of fixed contracts between the authorized carriers and the freight brokering ELD with the purpose of eliminate third parties and offer continuity through the whole system.

6   Conclusions

The development of the pilot on the use of freight brokers' information services to measure its impact on the emission of greenhouse gases (GHGs) was a joint project with the Colombian government to increase the competitiveness of the country and in turn have a positive impact on the environmental challenges of the country.

The study was assumed to be defined in three stages. First, a socio-economic analysis of at least 500 vehicles/drivers, where we collected information on both vehicles and drivers. Second, the implementation of an EDS apparatus (Engine System Diagnostic Tool) on the vehicle computer to measure some control variables of importance. Last, the data analysis of potential reduction on empty trips due to inefficiencies in vehicle cargo scheduling derived from poor demand communication between cargo owners, carriers and independent transporters (vehicle owner/drivers). Within the contract with the ministry of transportation, only stage one was undertaken, mainly due to nonconformities of driver unions in the country. Second and third stages were developed with private funding, and several very interesting ideas came out from this practice. The alliance between a large cargo generator, 2 carriers and 100 out of the 534 vehicles/drivers in the first stage yield interesting results on two main aspects.

In the first phase, more than 534 drivers were technically and socio-economically characterized to the same number of vehicles with gross vehicle weight greater than or equal to 10.5 tons. Most of the analyzed fleet corresponded to 3-axle tractors (more than 70%), leaving a little more than 28% rigid trucks. Likewise, a series of logistics and operation indicators have been evaluated for those vehicles, in relation to loaded trips, travel times, loading, unloading, and in general the operation times, number of empty trips, distances traveled, used fuel, depending on the type of vehicle and for the analysis of empty trips were developed by route and for 7 different types of vehicle. The statistics showed that empty trips are 45% of total trips in the country and that compensated trips are only 6% of total trips, a situation that is causing the greatest challenges of the sector. In the second and third stages was obtained a total reduction of empty trips (more than 28%) and second, an augmented number of trips per month therefore yielding more return on investment to drivers/vehicles and to the carrier (transportation company). In addition, we could transfer some of the gains to the cargo owner therefore long-term contracts are now in place.

The experiment on phases 2 and 3 consisted on a sample size of 37 trucks and 58 driver to implement optimization solutions and analytics to reduce backhauls (therefore greenhouse gas emissions) and to increase the utilization of freight. Once we ran the experiment on phases 2 and 3, the results show a large

improvement on the reduction of empty trips. The sample showed that with simple transportation good decisions, we go as high as 73% on compensated trips therefore yielding only 27% on empty trips. Also, it was possible to obtain a fuel consumption reduction of 38% and subsequently a reduction of emission of approximately 34%.

It can be concluded that the pilot proved that the use of freight brokering IT platforms mitigates greenhouse emissions. Numbers (cost, fuel consumption, etc.) demonstrate this postulate. However, it is important to highlight that we did not measures factors like oil or tired consumption (all these related to either direct or indirect emissions), so we propose to take them into account for future research.

References

1. Álvarez Espinosa, A.C., et al.: Compromiso de reducción de emisiones de gases de efecto invernadero: Consecuencias económicas. Departamento Nacional de Planeación DNP. Documento 440 (2015)
2. Aronsson, H., Huge Brodin, M.: The environmental impact of changing logistics structures. Int. J. Logist. Manag. **17**(3), 394–415 (2006)
3. Barth, M., Younglove, T., Scora, G.: Development of a heavy-duty diesel modal emissions and fuel consumption model (2005)
4. Bauer, J., Bektaş, T., Crainic, T.G.: Minimizing greenhouse gas emissions in intermodal freight transport: an application to rail service design. J. Oper. Res. Soc. **61**(3), 530–542 (2010)
5. Bektaş, T., Laporte, G.: The pollution-routing problem. Transp. Res. Part B Methodol. **45**(8), 1232–1250 (2011)
6. Bigazzi, A., Bertini, R.: Adding green performance metrics to a transportation data archive. Transp. Res. Rec. J. Transp. Res. Board **2121**, 30–40 (2009)
7. Browne, M., Allen, J., Nemoto, T., Patier, D., Visser, J.: Reducing social and environmental impacts of urban freight transport: a review of some major cities. Procedia Soc. Behav. Sci. **39**, 19–33 (2012)
8. Contento, J.P., Martínez, L.A., Cardozo Sáenz, D., et al.: Caracterización bolsa de carga para optimización del transporte de carga por carretera (2012)
9. Demir, E., Bektaş, T., Laporte, G.: The bi-objective pollution-routing problem. Eur. J. Oper. Res. **232**(3), 464–478 (2014)
10. Demir, E., Bektaş, T., Laporte, G.: A review of recent research on green road freight transportation. Eur. J. Oper. Res. **237**(3), 775–793 (2014)
11. Díaz, C.A., Galetovic, A., Sanhueza, R.: La regulación del transporte de carga en santiago: características, evaluación y propuestas. Cuadernos de economía **40**(119), 5–46 (2003)
12. Edenhofer, O., et al.: Summary for policymakers climate change 2014, mitigation of climate change. IPCC 2014, Climate Change 2014: Contribution of Working Group III to the Fifth Assessment Report of the Intergovernmental Panel on Climate Change (2014)
13. Edenhofer, O., et al.: IPCC, 2014: summary for policymakers. Climate change (2014)
14. Ericsson, E., Larsson, H., Brundell-Freij, K.: Optimizing route choice for lowest fuel consumption-potential effects of a new driver support tool. Transp. Res. Part C Emerg. Technol. **14**(6), 369–383 (2006)

15. Giannopoulos, G.A.: Towards a European ITS for freight transport and logistics: results of current EU funded research and prospects for the future. Eur. Transp. Res. Rev. **1**(4), 147–161 (2009)
16. Ideam, P., MADS, D.: Cancillería (2015). Nuevos escenarios de cambio climático para Colombia 2100 (2011)
17. Marquez F., G., Marsiglia Fuentes, R.M., Martinez B., W.J., Lopez, A., Tornet, J., Paternina-Arboleda, C.D.: The insertion of a multimodal transportation system to improve the logistics competitiveness of agricultural commodities in a Colombian region. Foreign commerce and value added contemporary. Eng. Sci. **11**(36), 1757–1769 (2018)
18. Marzuez, L., Cantillo, V., Paternina-Arboleda, C.D.: Accessibility variables in freight generation models for agricultural products: an elastic nationwide model for evaluating the impact of infrastructure projects. Technical report (2017)
19. Minett, C.F., Salomons, A.M., Daamen, W., Van Arem, B., Kuijpers, S.: Eco-routing: comparing the fuel consumption of different routes between an origin and destination using field test speed profiles and synthetic speed profiles. In: 2011 IEEE Forum on Integrated and Sustainable Transportation System (FISTS), pp. 32–39. IEEE (2011)
20. Ministerio de Ambiente y Desarrollo sostenible (MADS): Informe de Gestión 2018 (2019). http://www.minambiente.gov.co/index.php/component/content/article/105-informes-de-gestion#informes-de-gesti%C3%B3n-del-mads
21. Pachón, M.d.P.G., Navas, Ó.D.A.: Retos y Compromisos Juridicos de Colombia Frente al Cambio Climático. U. Externado de Colombia (2017)
22. Pan, S., Ballot, E., Fontane, F.: The reduction of greenhouse gas emissions from freight transport by merging supply chains. In: International Conference on Industrial Engineering and Systems Management (IESM 2009), p. 6 (2009)
23. Pan, S., Ballot, E., Fontane, F.: The reduction of greenhouse gas emissions from freight transport by pooling supply chains. Int. J. Prod. Econ. **143**(1), 86–94 (2013)
24. Ramos, T.R.P., Gomes, M.I., Barbosa-Póvoa, A.P.: Minimizing CO_2 emissions in a recyclable waste collection system with multiple depots. In: EUROMA/POMS Joint Conference, pp. 1–5 (2012)
25. Van Wee, B., Hagenzieker, M., Wijnen, W.: Which indicators to include in the ex ante evaluations of the safety effects of policy options? Taps in evaluations and a discussion based on an ethical perspective. Transp. Policy **31**, 19–26 (2014)
26. Vieda, V., María, M., Espinosa, M., Virgüez Rodríguez, É.A., Behrentz Valencia, E.: Incertidumbre en modelos de emisiones de gases de efecto invernadero: caso de estudio en el sector transporte colombiano. In: II CMAS South American Conference, p. 252 (2015)
27. Yang, C., McCollum, D., McCarthy, R., Leighty, W.: Meeting an 80% reduction in greenhouse gas emissions from transportation by 2050: a case study in California. Transp. Res. Part D Transp. Environ. **14**(3), 147–156 (2009)

Alternative Scenarios in Analyzing Florida's Freight Imbalance

Seckin Ozkul[1](✉) ⓘ, Donna Davis[1] ⓘ, Abdul Pinjari[2] ⓘ,
Iana Shaheen[3], and Prachi Gupta[3]

[1] Marketing Department, Muma College of Business,
University of South Florida, Tampa, USA
sozkul@usf.edu
[2] Civil Engineering, IISc, Bengaluru, India
[3] Muma College of Business, University of South Florida, Tampa, USA

Abstract. Florida struggles with a serious imbalance between inbound and outbound freight. With the third largest population in the U.S., the state offers a large and attractive consumer market. Statistics from the Federal Highway Administration (FHWA) reveal that the total freight tonnage entering the state is nearly double the tonnage leaving the state. This imbalance suggests that many ships, containers, rail cars, and trucks leave the state empty or only partially loaded, which raises the cost for shippers and reduces the effectiveness of freight movements. One solution to this problem is for Florida seaports to secure a higher percentage of shipments bound for Florida markets.

Therefore, the major goal of this project was to evaluate opportunities to reduce the current inbound-outbound freight imbalance by increasing the percentage of imports consumed in Florida that enter the marketplace through Florida seaports and airports. To achieve this goal, the research team performed a multitude of different tasks in order to come up with an action plan to alleviate Florida's freight imbalance. In return, the research team was able to determine specific scenarios where direct shipments to Florida has better results for landed times and landed costs for the shippers.

Keywords: Freight imbalance · Landed cost · Landed time

1 Introduction

In order to reduce the freight imbalance in Florida, we have observed two specific project objectives as part of the methodology. These are as follows:

1. Quantifying the latest volumes and categorizing the types of freight consumed in Florida that currently enter the U.S. through port gateways outside of Florida.
2. Evaluating logistics efficiency and effectiveness gains for shippers (e.g., landed costs, delivery times, etc.) by changing shipments from competing ports to Florida ports.

© Springer Nature Switzerland AG 2019
C. Paternina-Arboleda and S. Voß (Eds.): ICCL 2019, LNCS 11756, pp. 39–52, 2019.
https://doi.org/10.1007/978-3-030-31140-7_3

2   Literature Review

The literature review portion of the Evaluation of Florida's Inbound and Outbound Trucking Freight Imbalance project is divided into four major components: (1) Florida's Freight Challenges and Backhauling Studies, (2) Backhauling and Freight Imbalance Research by Selected States, (3) Backhauling and Freight Imbalance Research by Selected Countries, and (4) Review of Backhauling Problem and Solutions from the Academic Literature.

The first task that was performed was to conduct a literature review to identify previous research studies that deal with empty backhauling and/or freight inbound-outbound imbalances. A review of the literature revealed that Florida is not alone in facing this problem. Indeed, multiple states and countries are concerned about empty backhauling. Within the U.S., Arizona, California, and Nevada are among the top states with an outbound/inbound freight imbalance, with empty running in those states of 41%, 40%, and 39%, respectively (Hanh 2003; Arizona State Freight Plan 2014; Michael Gallis and Assoc. 2016). To date, most research has concentrated on optimizing vehicle routing problems with backhauls, whereas strategies to reduce and minimize existing outbound/inbound freight imbalances have received little attention (Ropke and Pisinger 2006; Wassan et al. 2013). One promising strategy for the empty backhaul problem in Florida is to increase the volume of imports bound for Florida markets through its seaports. Florida has 15 deep-water ports, including Port Miami, which has the capacity to handle Super Post-Panamax vessels.

2.1   Florida's Freight Challenges and Backhauling Studies

As an important study highlighting Florida's freight challenges and imbalance of trade flows, the *Florida Trade and Logistics Study (FTLS)* was initiated by the Florida Chamber Foundation in partnership with the Florida Department of Transportation (FDOT) (Florida Trade and Logistics Study 2010). As the first study of its kind that analyzes trade flows and relates logistics activity in the state, the main goal was to attract the attention of public and private investors to the trading opportunities from the widening of the Panama Canal. Empty backhauls are among the challenges faced by Florida. Since a significant share of international imports destined for Florida markets enter the country through seaports and airports in other states, the study proposed the need to maximize Florida's ability to serve its businesses through Florida Gateways.

Table 1 presents an analysis of the transportation and logistics costs involved in moving an imported container from Hong Kong to distribution centers located in northeast, central, or southeast Florida via three paths for entering the U.S. The results suggest that "all water" direct services from Hong Kong to Florida seaports could compete on a cost basis with the Pacific Coast routings. Also, a direct "all water" service to Florida would be cost-competitive with an "all water" service to the Port of Savannah, followed by a truck or rail trip from Savannah to Florida.

Table 1. Estimated cost of moving a container from Hong Kong to serve Florida import market using alternative ports of entry

Distribution center location	Port of entry to U.S.			Cost savings for Florida seaports compared to	
	Florida	Savannah	Los Angeles	Savannah	Los Angeles
Northeast Florida	$3,090	$3,345	$3,170	8%	3%
I-4 Corridor	$2,994	$3,521	$3,156	15%	5%
Southeast Florida	$2,974	$3,588	$3,579	17%	17%

Three years later, in a continuation of the 2010 study, the *Florida Trade and Logistics Study 2.0* was launched. The study identifies the trends in the expansion of the global marketplace and highlights current problems and future opportunities for our fast-growing state. Among other problems (Islam and Ozkul, 2019), the study looked in more detail at the existing imbalance between inbound and outbound freight flows in Florida. Figure 1 shows that the total tonnage of domestic and international goods moving into Florida is almost twice that of the tonnage leaving the state. This results in an increase in transportation costs, as carriers bringing consumer goods to Florida often return empty or partially-loaded.

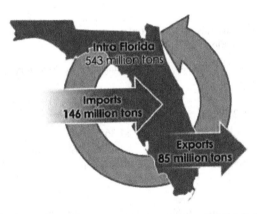

Fig. 1. Florida Freight Flows (recreated from FTLS 2.0, 2013)

In 2014, the Florida Seaport Transportation and Economic Development Council conducted a study titled *Analysis of Global Opportunities and Challenges for Florida Seaports (AGOC)*. The analysis assessed problems and challenges for Florida seaports and identified opportunities for the state to expand its position in global trade. Among other things, the study quantified a significant opportunity for Florida to take advantage of empty backhaul in northbound trucking (Kocatepe et al. 2019). According to Fig. 2, every year, Florida has approximately 500,000 northbound trucks leaving empty. The imbalance of equipment flows provides an opportunity for businesses to be able to negotiate more favorable backhaul truck rates.

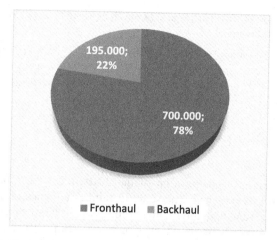

Fig. 2. Florida Northbound Flows (recreated from AGOC for Florida Seaports, 2014)

The study also looked at landed cost comparisons and identified the specific factors that cause BCOs and shipping lines to use non-Florida ports. Table 2 summarizes the logistics cost analysis to serve the Orlando market on the Southeast Asia trade lane.

Table 2. Logistics cost analysis to serve Orlando market on Southeast Asia Trade Lane (recreated from AGOC for Florida Seaports, 2014)

Hong Kong to Orlando	Voyage	Terminal	Inland	Total	Days	Inventory carrying cost	Total cost of move
Los Angeles	$1,990	$406	$3,660	$6,057	28.5	$1,423	$7,479
South Florida	$3,932	$235	$705	$4,872	30.0	$1,499	$6,371
Central Florida	$ 3,974	$216	$175	$4,365	30.3	$1,517	$6,881
Northeast Florida	$ 3,993	$216	$446	$4,655	30.6	$1,530	$6,185
Savannah	$4,005	$216	$660	$4,881	30.8	$1,539	$6,419

The *Florida Freight and Mobility Trade Plan* (FMTP) is another major study recently undertaken by FDOT. Objective 3 of the FMTP Policy Element focuses on minimizing costs along the supply chain. The study highlighted that the issue of empty backhaul is prevalent enough that it was ranked 5th among the top 10 outbound freight commodities for rail in 2014. Empty backhaul tends to increase supply chain costs, because trucking and rail carriers need to pass along the cost of empty backhaul to shippers and customers. Of the five strategies established to achieve this objective, Strategy 3.5 focuses on "support manufacturing and assembly that reduces empty backhauling." The sub-strategies associated with Strategy 3.5 support the development of Intermodal Logistics Centers (ILCs) as well as coordination with freight forwarders. The response to this issue consists of the following steps:

1. Invest in infrastructure that supports industries to create more outbound freight and thus reduce empty backhaul movements.
2. Develop ILCs that will support the growth of manufacturing industries in the state.
3. Develop public-private partnerships to support freight-generating economic development, including site selection and development, cross-modal connection, land use protection, and marketing.

In 2015, 151 million tons of freight were imported into Florida from other states, whereas only 63 million tons left the state within the U.S. Forecasts for 2045 show a similar pattern, with almost 198 million tons of freight being imported into the state and exports around 104 million tons.

2.2   Backhauling and Freight Imbalance Research for Selected States

Several U.S. states such as Arizona, California and Nevada are also facing the issue of Inbound-Outbound Freight imbalance. In Arizona, the empty running in 2013 was estimated to be 40%. Recently, the state acknowledged the problem of empty backhauling; however, the *Arizona State Freight Plan* (2014) does not provide any strategies to reduce empty truck running. In California, in 2003, ocean carriers shipped an average of 40–50% of import containers from Asia to the U.S. West Coast. These containers were eventually shipped back to Asia empty, and this repositioning of empty containers adds significant operational costs to ocean carriers. A study by the University of Southern California (Hanh 2003) suggested the following strategies to re-engineer the logistics of empty cargo containers in the SCAG region:

- Reuse of empty containers ("triangulation" or "street-turn" interchanged approaches)
- Off-dock empty return depots (ERD)
- Depot direct off-hire of empty leased containers
- Use of Internet-based support systems such as InterBox, eModal, SynchroNet, and Virtual Container Yard.

After almost 15 years since the report was published, Los Angeles still faces the problem of empty cargo containers. In 2015, at the Ports of LA, empties accounted for only 2.8% of imported containers but accounted for 57.3% of all exported containers (2.2 million TEUs). To remediate this issue, the Port is implementing load matching strategies that can provide several benefits (FESDG 2016).

In Nevada, empty running in 2013 was more than 42%. Following proposals were suggested by the Nevada Department of Transportation:

- Replace the dependence on inbound freight with locally manufactured goods.
- Invest in goods producing jobs (e.g., Tesla plant for electric car manufacturing) to ensure an aggressive growth in outbound freight in the state.
- Encourage Nevada's shippers and trucking firms to participate in empty equipment (truck, railcar, etc.) reduction strategies.

Moreover, the state is planning to build a stronger freight infrastructure to strengthen its multimodal and multidirectional services to improve its links to western

ports and eastern markets that will provide more opportunities for inbound shipments (Mardani et al. 2019a).

2.3 Backhauling and Freight Imbalance Research for Selected Countries

This section provides a review of backhauling and freight imbalance by selected countries/regions. Various studies have suggested following reasons to reduce the freight imbalance in countries such as UK, Norway, Thailand, Central America, China etc.:

- Outsourcing of road haulage operations
- Geographical imbalance in traffic flow
- Average length of haul
- Increased cost of road transport (Mardani et al. 2019b)
- Other strategies could be Trip structure, Reverse logistics, Use of Load matching services and adoption of new management initiatives.

In addition, the authors identified constraints that make it impossible to remove completely empty running:

- Priority given to outbound delivery service
- Unreliability of collection and delivery operations
- Inadequate knowledge of available loads
- Lack of coordination between purchasing and logistics departments
- Incompatibility of vehicles and products
- Resource constraints.

2.4 Review of Backhauling Problems and Solutions from Academic Literature

In its general form, the objective of Vehicle Routing Problems (VRP) is to design a set of minimum cost routes that serves some geographically-dispersed places and fulfills specific constraints of the problem. Since its first formulation in 1959, many publications have expanded its scope. VRP with Backhauls (VRPB) is one of the types of VRP and includes both a set of customers to whom products are to be delivered and a set of vendors whose goods need to be transported back to the distribution center. In VRPB, the total traveled distance is minimized, and linehaul and backhaul customers are allowed to be served on the same routes. According to the recent study (Ropke and Pisinger 2006), the VRPB has the following limitations:

- If a route contains both, backhaul customers must be served after the linehaul customers.
- A route is not allowed to consist entirely of backhaul customers.
- The capacity of the vehicle should be obeyed.
- The number of vehicles to use is given in advance.
- All customers are serviced from a single depot.
- All vehicles have the same capacity. Quantifying Volumes and Categorizing Types of Freight.

3 Methodology

3.1 Quantifying Volumes and Categorizing Types of Freight

The U.S. Census Bureau estimates that the population of Florida was more than 20 million people in 2015, making it the third most populous state in the U.S. Florida has 22 Metropolitan Statistical Areas (MSAs) defined by the U.S. Office of Management and Budget (OMB) and 67 counties (U.S. Census Bureau 2014).

Florida is home to the nation's second largest foreign trade zone network and would have the 21st largest economy in the world if it were a separate nation. It has the second largest number of exporters in the U.S. after California. More than 58,000 of Floridian companies export goods and services, accounting for 20% of all U.S. exporters (House Committee on Transportation and Infrastructure 2013). Florida's manufactured exports grew by 148% from 2000 to 2012, and U.S. manufactured exports grew by only 90% for the same period.

Taking growth into account, the transportation system in Florida includes a variety of options, including transit, auto (interstate highways, U.S. and state roads), rail (Amtrak, commuter, and cargo services), air (airports), and seaborne (seaports). According to FDOT, the state's highway system contains 1,473 miles of interstate highways and 9,934 miles of non-interstate highways, such as state highways and U.S. highways (FDOT 2016b). In addition, the state has 780 public airports and 15 seaports, and its rail system contains 2,753 railway miles (FDOT 2016a).

According to FDOT, transportation is a critical factor affecting the state's economic growth and development (FDOT 2015a, 2015b). The cost and quality of transportation directly impact the decisions of businesses, residents, and visitors. Florida must have a well-planned and adequately funded transportation system that addresses accessibility and mobility needs for all users, including freight transportation.

This project task analyzed freight movement volumes into and out of Florida so that an evaluation could be conducted to alleviate Florida's freight inbound and outbound freight imbalance. This paper includes quantification of the volumes and categorization of the types of freight using three databases—the Freight Analysis Framework (FAF), TRANSEARCH, and the Port Import/Export Reporting Service (PIERS). In the following subsections, a detailed analysis of Florida's freight flows is provided based on the information from these three databases. These commodity flow datasets are used to help answer questions regarding freight movements, including the amount of freight produced or consumed, the origin-destination patterns, and the modes used.

It should be noted that the percentages used in the discussion in this document are rounded up to whole numbers using the actual raw data from the tables reported in the report.

3.1.1 Freight Analysis Framework (FAF)

In 2012, approximately 728 million tons of freight, valued at approximately $910,000 million ($910 billion) were moved to, from, and within Florida. The majority of the freight flow in total weight (64.34%) originated within the state itself. In addition, 15.39% (by weight) of the total freight came to Florida from other states. This can be supported by the fact that Florida is the third-largest consumption market in the U.S.

However, a small percentage of 6.44% (by weight) of freight originated from Florida to other states. Although Florida is located close to international markets, international exports and imports accounted for only 8.07% of total freight tons. Moreover, out of this 8.07%, 5.42% (39.49 million tons) were received through Florida. This import percentage that is directly received to Florida is at the lower end and points to significant opportunities to attract more imports directly to Florida ports in the future.

3.1.2 Transearch

For this project, data were obtained through FDOT for the "actual data" year 2011. Similar to the FAF, in addition to the base year, the TRANSEARCH database also provides projections for the years 2015 to 2040. According to Table 3, in 2011, approximately 455 million tons of freight, valued at $1,405,512 million ($1.41 trillion) were moved to, from, and within Florida. Much of the freight flow (49% in weight) originated within the state. In addition, almost 35% of freight (in weight) came to Florida from other states. Florida imported almost 6 million tons of freight, and only 2.4 million tons of goods are exported by the state.

Table 3. Total tonnage and value by direction to/from Florida (recreated from TRANSEARCH 2011)

Direction	Origin	Destination	Total weight, MTons (%)	Total value $M (%)
Domestic	Florida	Florida	222.51 (48.95)	212,165.50 (15.10)
	Florida	Rest of USA	60.35 (13.28)	122,966.20 (8.75)
	Rest of USA	Florida	159.38 (35.06)	298,501.80 (21.24)
Import	Foreign	Florida	5.95 (1.31)	760,049.80 (54.08)
Export	Florida	Foreign	2.39 (0.53)	5,087.66 (0.36)
Through	Outside Florida	Outside Florida	3.98 (0.87)	6,740.91 (0.48)
Total	–	–	454.55 (100)	1,405,511.87 (100)

In terms of value, Rest of USA to Florida represents the highest value among the domestic shipments. In 2011, $300 million of commodities came to Florida, and only $210 million of goods left the state. Import accounts for the highest share of value. More than 50% of the total freight value generated in 2011 was imported to Florida from other countries. However, less than 0.5% was exported from Florida to other countries. In 2025, the total tonnage and value of goods are expected to increase for domestic shipments. In terms of value, domestic, export, and import freight flows are expected to grow at similar trends and increase by more than 50% before 2025. In terms of total freight weight, domestic freight flows are expected to reach 600,000k tons by 2025.

3.1.3 Port Import/Export Reporting Service (PIERS)

The PIERS U.S. Waterborne Import database uses the Harmonized Commodity Description and Coding System, generally referred to as "Harmonized System" (HS). HS is a multipurpose international product nomenclature developed by the World

Customs Organization (WCO). According to World Customs Organization, HS comprises about 5,300 commodity groups, each identified by a 6-digit code, arranged in a legal and logical structure and supported by well-defined rules to achieve a uniform classification. Additionally, 5,300 commodity groups are arranged in 99 chapters that are grouped in 21 sections.

The United Nations International Trade Statistics Knowledgebase indicates that the six digits can be broken down into three parts. The first two digits (HS-2) identify the chapter into which the goods are classified (e.g., 09 = Coffee, tea, maté and spices). The next two digits (HS-4) identify groupings within that chapter (e.g., 09.02 = Tea, whether or not flavored). The next two digits (HS-6) are even more specific (e.g., 09.02.10 Green tea [not fermented]). Up to the HS-6 digit level, all countries classify products in the same way (a few exceptions exist where some countries apply old versions of the HS). Table 4 presents the first two-digit (HS-2) commodity classification divided by major groups.

According to ISO (2014), the commodities used in PIERS are reported using three classification groups—containerized, non-containerized, and forklift pocket (FLP) non-containerized. Containerized commodities come to the U.S. using intermodal containers (also called shipping containers) made of weathering steel. Non-containerized commodities are transported in a ship's hold and represent any loose or unpackaged goods, such as bulk products FLP non-containerized are goods that are transported on the forklift panels. Most seaports in the U.S. transport containerized commodities. Non-containerized and FLP non-containerized commodities are transferred only through several specific ports in the U.S.

The PIERS database uses the short ton as a unit of weight instead of the regular ton used by FAF and TRANSEARCH. Since FAF and TRANSEARCH used tonne as a unit of weight, all values in PIERS were recalculated to match the numbers in other datasets.

3.2 Evaluation of Logistics Efficiency and Effectiveness Gains for Shippers by Changing Shipments from Competing Ports to Florida Ports

According to the Florida Chamber Foundation, trucking is the dominant form of goods movement in the state, accounting for more than 73% of all tonnage. Waterborne goods movement accounts for about 15% of all freight flows, followed by rail at 12%. Air accounts for less than 1% by volume, but it is an important mode for high value/perishable goods.

According to FDOT, Florida seaports facilitate the flow of more than 103 million tons of waterborne commerce and 15.2 million cruise passengers, supporting more than 700,000 jobs throughout the state. Similarly Florida has 15 public seaports supporting cargo, cruise, and other industry sectors. Of these, 10 container seaports service Mexico and the Caribbean, Central and South America, Africa, Europe, the Middle East, Australia, and Asia. Around 3.5 million TEUs (20-foot equivalent units– containers) crossed the docks of Florida's ports using the state's seaport infrastructure, highway, and rail networks.

Additionally, Florida is home to three waterways on the fuel-taxed system: The Gulf Intracoastal Waterway (GIWW) system, including two sub-sections, the Northern Gulf Intracoastal Waterway (NGIWW) and the Western Gulf Intracoastal Waterway (WGIWW); the Atlantic Intracoastal Waterway (AIWW) system; and the Apalachicola, Chattahoochee, and Flint Rivers (ACF) system.

For the purposes of this paper the top 10 commodities (by weight) transported to the Port of Savannah and the Port of NY/NJ from India (Kochi) were identified as these ports are direct competitors to Florida ports. Table 4 below shows the total amount of these commodities that arrive at the Port of Savannah and the Port of NY/NJ from India (Kochi) and also present the weighted average calculated by dividing the actual weight for each commodity by the total sum of weight and multiplying by the total weighted average weight.

Table 4. Top 10 commodities by weight considered for Mojo analysis from India (Kochi) to Port of Savannah and to port of NY/NJ

Commodity type by weight				
Commodities	Port of Savannah		Port of NY/NJ	
	Weight (%)	Weighted average weights (ktons)	Weight (%)	Weighted average weights (ktons)
Coal–not elsewhere classified	45.37	74,675.03	–	–
Gasoline	19.93	32,807.70	–	–
Textiles/leather	7.57	12,464.40	34.99	57,596.00
Fuel oils	1.93	3,171.62		
Machinery	6.16	10,134.07	4.75	7,822.40
Pharmaceuticals	0.72	1,191.09	–	
Articles-based metal	5.15	8,470.83	4.75	7,822.40
Plastics/rubber	4.84	7,961.59	9.29	15,289.80
Base metals	4.65	7,653.87	–	–
Basic chemicals	3.68	6,065.19	–	–
Nonmetal mineral products	–	–	18.62	30,643.70
Mixed Freight	–	–	8.92	14,681.60
Meat/Sea Food	–	–	7.37	12,130.10
Other food stuff	–	–	7.03	11,579.00
Other ag products	–	–	3.24	5,335.10
Electronics	–	–	2.90	4,777.70
Total	100.00	164,595.40	100.00	164595.40

Once these freight values are known, the research team then evaluated the major logistics efficiency and effectiveness gains for shippers, such as landed costs and landed delivery times, by analyzing these freight flows to/from competing ports and to/from Florida ports. More specifically, this paper includes the optimization of freight flows

using the Mercury Gate Mojo freight optimization tool/software for multiple scenarios in which specific freight routes have been analyzed using the Mercury Gate Mojo Optimization Tool.

3.2.1 Mercury Gate Mojo Optimization Tool

Mojo is a powerful freight transportation optimization solution that analyzes shipments, rates, and constraints to produce realistic load plans that reduce overall freight expenditure Using Mojo optimization software, the research team was able to accomplish certain objectives necessary for the project. The following processes were considered:

- *Sourcing*: Freight transportation routes were evaluated with the simulation of actual shipment data. The shipment rates were obtained through the synthesis of multiple sources.
- *Planning:* In addition to looking at optimal routes, mode shifting, cross docking, and pooling were evaluated. Moreover, the analysis of impact pool points and pool combinations in transportation costs was performed.
- *Execution:* As a result, optimized load plans across inbound and outbound shipments were generated and then consolidated across different states and ports (Monsreal-Barrera et al. 2019).
- *Sales:* Using the Mojo software, landed costs, CO2 emissions (Seman et al. 2019), and landed time were minimized. Finally, new freight transportation routes were evaluated using the existing route models.

4   Results

In this section, the Mojo optimization analysis scenarios calculated to provide landed cost and time savings for commodities shipped directly to Miami instead of Florida's competing ports from Port of Kochi, India are summarized.

First, the research team looked at the Port of NY/NJ and then the Port of Savannah. The comparison analysis for total optimized landed time and total optimized landed cost was calculated by Mojo and is provided in Table 5. The negative numbers represent cost savings when shipping to the Port of Miami instead of these two competing ports.

After the Mercury Gate Mojo Freight Optimization tool was run for all scenarios, as described in this paper, the research team identified specific scenarios and routes for which Florida ports were found to have an upper hand when compared to their competing ports in terms of either optimized landed costs or optimized landed times, or both.

Table 5. List of routes with optimized landed cost &/or landed time savings in Florida's favor

Origin	U.S. entry	Florida Port	Total optimized landed time (days)	Total optimized landed cost ($)	Total optimized landed time difference (days)	Total optimized cost difference ($)
Kochi	NY/NJ	Miami	33	$17,012,472	−3	−$6,759,838
Kochi	Miami	Miami	30	$10,252,634		
Kochi	NY/NJ	Jacksonville	33	$12,612,476	−2	$940,154
Kochi	Miami	Jacksonville	31	$13,552,630		
Kochi	NY/NJ	Tampa	33	$19,212,479	−2	−$6,759,848
Kochi	Miami	Tampa	31	$12,452,631		
Kochi	NY/NJ	Palm Beach	33	$15,912,473	−2	−$4,559,840
Kochi	Miami	Palm Beach	31	$11,352,633		
Kochi	Savannah	Miami	31	$17,382,074	−1	−$7,129,440
Kochi	Miami	Miami	30	$10,252,634		
Kochi	Savannah	Tampa	31	$12,982,079	0	−$529,448
Kochi	Miami	Tampa	31	$12,452,631		
Kochi	Savannah	Palm Beach	31	$16,282,076	0	−$4,929,443
Kochi	Miami	Palm Beach	31	$11,352,633		

5 Conclusion

The findings of this study highlight the fact that there is significant potential for Florida ports to attract direct shipments that are already destined to be consumed in Florida, but currently enter the U.S. through other competing ports. After the Mercury Gate Mojo Freight Optimization tool was run, as described in this paper, the research team identified specific scenarios and routes for which Florida ports were found to have an upper hand when compared to their competing eastern United States ports in terms of either optimized landed costs or optimized landed times, or both. Those scenarios, in which Florida ports were found to have an advantage on their competitor ports, have been highlighted in Table 5 under the results section. The reason these routes are currently not used as heavily may be due to the already established business relationships with existing ports of call.

As observed under the results section, Florida has a significant potential to attract freight business from its competing ports and can be the U.S. entry port for many freight shipments that are destined to Florida and beyond (e.g., Atlanta, Mobile, etc.). As these results suggest, this would directly benefit the shippers and carriers (improved landed times and landed costs). In addition, shippers and carriers would also be assured from risk mitigation and reliability standpoints due to Florida's ports not having major issues with union strikes or major natural disasters/disturbances compared to NY/NJ and other ports.

Also, to emphasize the importance of the results, it should be noted that these optimized landed time and landed cost estimates were calculated for commodities

destined to for only a one-year period. Therefore, these savings would be expected to grow/multiply in value over the many years with the growing Florida population (and demand for increased goods). This might prove to be a very convincing leverage point to earn more direct shipments to Florida (rather than losing this business to competing ports), which would in return help alleviate empty truck backhauling in Florida.

Overall, the analysis show that Florida has a significant potential to attract freight shipments from its competing ports and can be the U.S. port of entry for many freight shipments that are destined to Florida and beyond (e.g., Atlanta, Mobile, etc.).

References

Arizona Department of Transportation: Arizona State Freight Plan: Economic context of freight movement in Arizona (2014). https://www.azdot.gov/docs/default-source/planning/State-Freight-Plan/14325-arizona-state-freight-plan-phase-3-economic-context-report.pdf. Accessed 18 July 2016

Florida Department of Transportation, Florida Chamber Foundation: Florida: Florida trade and logistics study (2010). https://www.flchamber.com/wp-content/uploads/FloridaTradeandLogisticsStudy_December20102.pdf. Accessed 30 June 2016

Florida Department of Transportation: Florida transportation trends and conditions (2015a). http://www.dot.state.fl.us/planning/trends/tc-report/economy.pdf. Accessed 25 Sept 2016

Florida Department of Transportation: Florida Seaport System Plan (2015b). http://www.fdot.gov/seaport/pdfs/2015%20Florida%20Seaport%20System%20Plan_Final.pdf

Florida Department of Transportation: Florida fast facts, June 2016. http://www.dot.state.fl.us/planning/fastfacts.pdf. Accessed 25 Sept 2016

Florida Department of Transportation: SWOT analysis of TRANSEARCH and FAF data (2016a). http://www.dot.state.fl.us/planning/statistics/freight/SWOT.pdf. Accessed 25 Sept 2016

Florida Department of Transportation: Florida's transportation system by the numbers (2016b). http://www.dot.state.fl.us/Intermodal/system/. Accessed 25 Sept 2016

Florida Seaport Transportation and Economic Development Council: Analysis of global opportunities and challenges for Florida seaports (2014). http://static.flaports.org/GlobalOppsAnalysis2015web.pdf. Accessed 26 July 2016

Freight Efficiency Strategies Development Group (FESDG): Freight Efficiency Strategies: A White PaperSeries to Inform the California Sustainable Freight Action Plan. National Center for Sustainable Transportation, Davis, CA (2016)

Hanh, L.: Department of Civil and Environmental Engineering University of Southern California. Reengineering the logistics of empty cargo containers in the SCAG region (2003). https://www.metrans.org/sites/default/files/research-project/01-05_Final-Reengineering%20the%20Logistics%20of%20Empty%20Cargo%20Containers%20in%20the%20SCAG%20Region.pdf. Accessed 18 July 2016

House Committee on Transportation and Infrastructure: Improving the nation's freight transportation system (2013). http://transportation.house.gov/uploadedfiles/freightreportsmall.pdf. Accessed 25 Sept 2016

Islam, M., Ozkul, S.: Identifying fatality risk factors for the commercial vehicle driver population. Transp. Res. Rec., 0361198119843479 (2019)

ISO: Freight containers—Container Equipment Data Exchange (CEDEX) Part 1: General communication codes for general purpose containers (2014). http://www.iso.org/iso/catalogue_detail.htm?csnumber=53417. Accessed 3 Dec 2016

Kocatepe, A., Ozkul, S., Ozguven, E.E., Sobanjo, J.O., Moses, R.: The value of freight accessibility: a spatial analysis in the Tampa Bay region. Appl. Spat. Anal. Policy, 1–20 (2019)

Mardani, A., et al.: Application of decision making and fuzzy sets theory to evaluate the healthcare and medical problems: a review of three decades of research with recent developments. Expert Syst. Appl. **137**, 202–231 (2019a)

Mardani, A., et al.: A two-stage methodology based on ensemble adaptive neuro-fuzzy inference system to predict carbon dioxide emissions. J. Cleaner Prod. **231**, 446–461 (2019b)

Mercury Gate. Mojo transportation optimization. http://mercurygate.com/solutions/mojo-transportation-optimization/. Accessed 17 Mar 2017

Mercury Gate. Mojo transportation optimization. http://www.mercurygate.com/wp-content/uploads/2014/05/MercuryGate-TMS-Mojo-Brochure-2013-Web.pdf. Accessed 22 Mar 2017

Mercury Gate. Optimize loads with Mojo. http://mercurygate.com/2014/optimize-loads-with-mojo/. Accessed 22 Mar 2017

Michael Gallis and Associates: CH2M, and Cambridge Systematics: Nevada State Freight Plan, Appendix 2: Nevada's Freight Transportation System. Nevada Department of Transportation, Carson City, NV (2016)

Monsreal-Barrera, M.M., Cruz-Mejia, O., Ozkul, S., Saucedo-Martínez, J.A.: An optimization model for investment in technology and government regulation. Wireless Networks, 1–13 (2019)

Ropke, S., Pisinger, D.: A unified heuristic for a large class of vehicle routing problems with backhauls. Eur. J. Oper. Res. **171**(3), 750–775 (2006)

Salhi, S., Wassan, N., Hajarat, M.: The fleet size and mix vehicle routing problems with backhauls: formulation and set partitioning-based heuristics. Transp. Res. Part E Logist. Transp. Rev. **56**, 22–35 (2013)

Seman, N.A.A., et al.: The mediating effect of green innovation on the relationship between green supply chain management and environmental performance. J. Cleaner Prod. **229**, 115–127 (2019)

The PIERS Industry Blog: Imbalance in trade prompts clever solution for empty containers (n.d.d). https://pierstransportation.wordpress.com/2011/10/20/imbalance-in-trade-prompts-clever-solution-for-empty-containers-2/. Accessed 7 July 2016

Measuring Environmental Impact
of Collaborative Urban Transport Networks:
A Case Study

Andrés Muñoz-Villamizar[1] , Elyn L. Solano-Charris[1(✉)] ,
Javier Santos[2] , and Jairo R. Montoya-Torres[3]

[1] Operations and Supply Chain Management Research Group,
Escuela Internacional de Ciencias Económicas Y Administrativas,
Universidad de La Sabana, Chía, Colombia
erlyn.solano@unisabana.edu.co
[2] TECNUN Escuela de Ingenieros, Universidad de Navarra,
San Sebastian, Spain
[3] Grupo de Investigación En Sistemas Logísticos, Facultad de Ingeniería,
Universidad de La Sabana, Chía, Colombia

Abstract. Nowadays, urban freight transport networks must be environmentally sustainable. As a consequence, collaboration strategies have been implemented as one alternative to enhance the efficiency of the supply chain and to reduce its environmental burden. However, there is a lack of knowledge about how to correctly quantify the environmental impact of collaborative urban freight transport networks. To fill this gap, we present an optimization approach for evaluating the environmental performance of collaborative systems by applying the Overall Greenness Performance (OGP) tool. We illustrate our approach using real data from the city of Bogotá, Colombia. Our results provide insights for shifting towards sustainable freight transportation networks. For example, we quantify savings up to 11% of CO_2 emissions after applying collaboration in urban transportation contexts.

Keywords: Urban freight transport · Collaboration ·
Environmental performance · Optimization · Urban logistics · Case study

1 Introduction

The demand for urban freight transport activities has exponentially increased by the delivery of supplies and goods in urban areas (Eren Akyol and De Koster 2018). In 2017, around 55% of the population was living in cities, and by 2050 it is projected that the urban population will be about 70% (United Nations, 2012). A significant fact of the augmentation of urban freight transport is the generation of greenhouse gas (GHG) emissions and air pollution (Forsberg and Krook-Riekkola 2018), which accounts 25% of CO_2 emissions and 30–50% of other transport-related pollutants (ALICE, 2015).

This fact has attracted the interest of scholars, practitioners and government (Quak 2008) to align planning, investment and policy for sustainable development of transport

© Springer Nature Switzerland AG 2019
C. Paternina-Arboleda and S. Voß (Eds.): ICCL 2019, LNCS 11756, pp. 53–66, 2019.
https://doi.org/10.1007/978-3-030-31140-7_4

networks (Garg et al. 2017). Therefore, several initiatives in the urban transport context have emerged as a response to the negative impacts generated by goods transportation in urban areas, which proved to be successful in achieving the objectives set by all stakeholders (Muñoz-Villamizar et al. 2018). For example, one of the most popular solutions consists in using urban consolidation centers to consolidate and transfer loads in smaller vehicles (Rao et al. 2015; Muñoz-Villamizar et al. 2018). Another valuable alternative consists in a business agreement among firms operating at the same level of the supply chain to reduce shipping costs, provide a faster distribution service to customers and reduce the environmental impact of distribution activities, formally defined as horizontal collaboration (Nataraj et al. 2019).

Within the range of disciplines (i.e., economics, engineering, physics, chemistry, etc.) addressing the sustainable freight transportation, Operations Research (OR) has been an essential contributor to accomplishing this goal (Bektaş et al. 2019). According to Bektaş et al. (2019), there are generally two approaches addressing environmental issues in urban transportation within OR, i.e., optimizing the design of transportation networks and/or optimizing the operations of urban transportation services. However, implementation of these alternatives and others is often unsuccessful because of a lack of systematic assessment of short- and long-term effects (Buldeo Rai et al., 2017).

This paper aims to fulfill this gap and proposes a methodology for measuring and evaluating the environmental impact of collaborative urban freight transport networks. Our methodology uses the Overall Greenness Performance (OGP) tool (Muñoz-Villamizar et al. 2018) with a mixed-integer linear programming (MILP) model to assess the environmental impact of the activities and/or requirements of different stakeholders in an urban freight transportation system. Real data from the city of Bogotá, Colombia, are used to run numerical tests. This paper is organized as follows. Section 2 provides a literature review in sustainable and collaborative transportation systems. Section 3 presents the proposed methodology while Sect. 4 presents the case study. Finally, Sect. 5 presents the main conclusions of this study and the opportunities for further research.

2 Literature Review

Urban freight transportation networks are characterized by a set of vehicle routes that carry goods from producers' sites to consumers. This includes various stakeholders such as the providers of goods (i.e., suppliers), the consumers of goods (i.e., recipients) and the infrastructure providers (i.e., governing authorities) (Comi et al. 2018). However, there are other additional components of transportation networks, such as organizational (e.g., warehousing, inventory and routing strategies), technologies (e.g., type of vehicles, intelligent transport systems), public-private partnership schemes and regulations (i.e., restrictive policies, incentives or tariff-based regulations) (Gonzalez-Feliu 2017). The freight transport networks can, therefore, be defined as an essential growth component as it addresses all flows of goods and materials in a city (Kiba-Janiak 2017).

However, freight transportation is recognized as being one of the most significant sources of unsustainability in urban areas (Quak 2008). An augmentation of the population increases the demand for urban transportation, which in turn increases

congestion and environmental degradation (Kiba-Janiak 2017). When the negative external impacts of urban transport became more evident, academics, practitioners and policy-makers became increasingly interested in this issue (Quak 2008). According to Bektaş et al. (2019), the negative externalities of urban transportation fall within various categories: emissions, noise, land use and safety hazards. Therefore, reducing emissions has been the key element of international agreements, as well as a research priority (Bektaş et al. 2019).

In this context, urban transportation networks have received extensive attention in the OR community. The field of OR has developed various optimization models and solution techniques to reduce the environmental impacts of urban transportation (Bektaş et al. 2019). As Churchman (1968) outlined, an OR approach can characterize a problem, its resources, its decision-makers and its environment. Bektaş et al. (2019) argued that OR tools used within transportation have aimed to define efficient solutions, where efficiency is generally determined by measurable indicators. However, the several methods and indicators developed so far have been based on different meanings of environmental sustainability. This led Gonzalez-Feliu (2017) to point out that environmental performance is not perceived or measured in the same way by the different stakeholders.

Similarly, Comi et al. (2018) proposed that urban transportation research should be evaluated according to the stakeholders and elements associated with the supply chain. Therefore, as Gonzalez-Feliu (2017) argued, the best alternatives or solutions for urban transportation will be the result of evaluating and considering different stakeholders' perspectives. However, to assess their potential (i.e., ex-ante assessment) to improve the environmental sustainability, it is crucial to use coherent and robust methods (Gonzalez-Feliu 2017; Comi et al. 2018).

To the best of our knowledge, there is no standardized and/or systematic method for assessing the environmental impact of collaborative urban transport networks considering the different perspectives of stakeholders involved (Gonzalez-Feliu 2017; Comi et al. 2018). It is important to note these assessing methods must consider certain productivity factors, given that environmental performance should not be treated independently of productive performance (Comi et al. 2018; Muñoz-Villamizar et al. 2018). The next section presents the methodology proposed to fill this gap.

3 Proposed Methodology

This paper presents a methodology for measuring and evaluating the environmental impact of collaborative urban transport networks. Based on Muñoz-Villamizar et al. (2015a), this methodology adapts and integrates the OGP tool proposed in Muñoz-Villamizar et al. (2018) within an optimization framework for assessing the environmental impact according to the activities and/or requirements of stakeholders in the urban transportation network. A schematic representation of the proposed three-phase methodology is shown in Fig. 1. Using different versions of a MILP model, the idea is to categorize and quantify the externalities of the transportation network to be more effective at measuring the overall impact of the collaborative strategies.

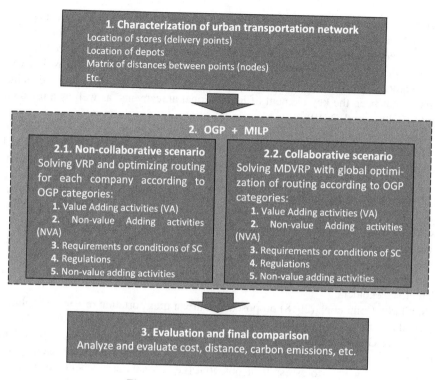

Fig. 1. Overall generic methodology.

3.1 Characterization of Urban Transportation Network

Transportation system characterization should include the location of delivery points (i.e., recipients) and depots (i.e., shippers), travel distances and travel speeds between nodes, time windows for delivery points and the regulated access defined by public authorities, demand at each delivery point, economic aspects (e.g., labor and fuel cost), etc. (Muñoz-Villamizar et al. 2019).

3.2 OGP and MILP Model

The second step consists of both to integrate of the OGP tool and the MILP model, and to define both non-collaborative and collaborative scenarios. On the one hand, the OGP tool allows the components and constraints of the transportation system to be categorized according to the generation of value. On the other hand, the mathematical model for Vehicle Routing Problem (VRP) variants allows the OGP components to be quantified in each scenario (i.e., non-collaborative vs. collaborative).

As already mentioned, this paper uses an adaptation of the OGP tool proposed in Muñoz-Villamizar et al. (2018) (see Fig. 2). The OGP is a hierarchy of metrics that considers the company's resource consumption and waste emissions (i.e., externalities) with its production level. Using the value-added concept, the OGP classifies a

company's consumption and waste processes according to the categories presented in Fig. 2. All these considerations about OGP categories, which are explained in the following sub-sections, are included in our methodology by using several variants for the VRP (i.e., Travelling Salesman Problem or TSP, Capacitated VRP or CVRP, VRP with time windows or VRPTW, distance-constrained VRPTW or DC-VRPTW). Regarding measuring environmental performance, the most commonly employed externalities are input-oriented (resource consumption) and output-oriented (emissions, toxic waste, oil and chemical spills, and discharges that are recovered, treated or recycled) (Molina-Azorín et al. 2009). In line with Bektas et al. (2019), in this study CO_2 emissions are the externality selected for evaluating the environmental impact of transportation systems. In estimating emissions, Ubeda CO_2 et al. (2011) pointed out that CO_2 computation can only represent an approximation because of the difficulty of quantifying certain variables such as speed, weather conditions, driving style, congestion, types of roads, etc. Therefore, to facilitate the implementation of the approach while also ensuring that valid managerial insights can be made, our methodology uses a distance-based emission factor to compute CO_2 emissions.

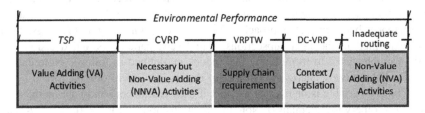

Fig. 2. OGP tool in transportation systems. Adapted from Muñoz-Villamizar et al. (2018)

According to the above, and terms of the OGP categories (see Fig. 2), the CO_2 emissions allocated to Value Adding (VA) activities correspond to the delivery of goods themselves in an 'ideal situation'. That ideal situation consists of the CO_2 emissions generated by delivering the goods without any constraint or condition (e.g., capacity, time windows, legislation, etc.) being placed on the system. This situation corresponds to the TSP. Then, CO_2 emissions generated for the extra trips due to vehicle capacity constraints are allocated to Necessary but Non-Value Adding (NNVA) activities (CVRP), while CO_2 emissions generated by the time window constraint are allocated to the Supply Chain category since this type of constraint is a client requirement (VRPTW). Similarly, CO_2 emissions generated by adhering to regulation initiatives (i.e., length constraints) are allocated to the Context category (DC-VRP). Finally, the CO_2 emissions in the NVA category correspond to system inefficiencies and mistakes commonly made in real-life operation, such as the CO_2 emissions generated by the inadequate selection of routing strategies.

Finally, it is necessary to define the difference between the non-collaborative and collaborative strategy. In the non-collaborative scenario, each company sets its delivery plan independently in which customers are never shared and each company is just trying to minimize its routing costs (see Fig. 1). Consequently, each company

individually solves its own VRP for local optimization. In the collaborative scenario, companies share trucks, routes and costumers to improve their individual turnovers. In our approach, this scenario is modeled as a multi-depot capacitated vehicle routing problem (MDCVRP) in which any company can deliver any of the stores for global optimization (Muñoz-Villamizar et al. 2015a).

The following subsections elaborate on the impact assessment of the five OGP categories. To quantify the OGP impact, each category is evaluated separately using its corresponding VRP variant and compared with the previous stage (excepting for first OGP category). That is, the model presented above is implemented each time using the variations proposed in its corresponding subsection, and then the E obtained in the previous category (i.e., total CO_2 emissions) is subtracted from the E value obtained in the current one. To facilitate the calculations, E_1 is defined as the emissions generated by VA activities; E_2 is defined as the emissions generated by NNVA activities; and so on.

VA Impact. The base MILP model employed to solve the different variants of the VRP is presented below. Note that calculating VA impact is the less difficult scenario since all the other OGP categories (i.e., NNVA, Supply Chain and Context) involve adding constraints in the transportation network. Therefore, it can be interpreted as the TSP. NVA impact, oppositely, corresponds to emissions from the real-life operation. The differences between each approach to measure each OGP category (see Figs. 1 and 2) are explained in the next subsections. Initial sets, parameters and decision (and binary) variables used in formulating this model are defined as follows. Note that for the non-collaborative scenario the number of depots is equal to one (i.e., $h=1$).

The binary variable is defined as $X_{ijk} = 1$ if both customers i and j are on the route of vehicle k; otherwise, it equals 0. Note that fictitious nodes i (or j) = 0, 1, ..., $(l-1)$ are used to represent the depots (l). In the non-collaborative scenario $l=1$, non-negative variables are defined as follows: Y_{ik} is the arrival time at node i of vehicle k. Other decision variables needed to calculate the selected key metrics are TD, the total travel distance, TC, the cost of the routing process, SL, the average of delivery times, and E, the total CO_2 emissions. It is important to note that the utilization rate is a non-linear expression as it corresponds to the ratio between orders shipped and vehicles used. Consequently, this metric should be computed later. Additionally, auxiliary variables u_i are used to prevent subtours. The following notation and model are used:

Sets:
h set of depots $\{1, ..., o\}$
i,j set of nodes $\{1,... , n\}$
k index for vehicles $\{1, ..., m\}$

Parameters:
q_i order for recipient i
Q_k maximum capacity of vehicle k
$c1_{hi}$ distance from depot h to node i
$c2_{ij}$ distance from node i to j
$c3_{ih}$ distance from node i to depot h
$v1_{hi}$ vehicle travel speed from depot h to node i

$v2_{ij}$ vehicle travel speed from node i to j
$v3_{ih}$ vehicle travel speed from node i to depot h
fc fuel cost
lc labor cost
vc fixed vehicle cost
ef emissions factor
us unloading speed
rl maximum route length
M a large number

Decision Variables:
Y_{ik} arrival time at node i of vehicle k
TD total travel distance
TT total travel time
TC total routing cost
TE total CO_2 emissions
u_i subtours auxiliary variables

Binary Variables:

$A_{hik} = \begin{cases} 1, \text{if vehicle } k \text{ goes from depot } h \text{ to node } i \\ 0, \text{otherwise} \end{cases}$

$X_{ijk} = \begin{cases} 1, \text{if vehicle } k \text{ goes from node } i \text{ to node } j) \\ 0, \text{otherwise} \end{cases}$

$B_{ihk} = \begin{cases} 1, \text{if vehicle } k \text{ goes from node } i \text{ to depot } h \\ 0, \text{otherwise} \end{cases}$

The mathematical model of the problem can be formulated as follows:

$$\sum_h \sum_k A_{hjk} + \sum_i \sum_k X_{ijk} = 1; \forall j \tag{1}$$

$$\sum_h \sum_k A_{hjk} \leq 1; \forall k \tag{2}$$

$$\sum_h A_{hjk} + \sum_i X_{ijk} = \sum_i X_{jik} + \sum_h B_{ihk}; \forall k, \forall j \tag{3}$$

$$\sum_i \sum_k A_{hik} = \sum_i \sum_k B_{ihk}; \forall h \tag{4}$$

$$\frac{c1_{hi}}{v1_{hi}} + q_i * us \leq Y_{ik} + M * (1 - A_{hik}); \forall k, \forall i, \forall h \tag{5}$$

$$Y_{ik} + \frac{c2_{ij}}{v2_{ij:}} + q_j * us \leq Y_{jk} + M * (1 - X_{ijk}); \forall k, \forall i, \forall j \tag{6}$$

$$u_i - u_j + n * X_{ijk} \leq n - 1; \forall k, \forall i, \forall j \qquad (7)$$

$$TD = \sum_h \sum_i \sum_k A_{hik} * c1_{hi} + \sum_i \sum_j \sum_k X_{ijk} * c2_{ij} + \sum_i \sum_h \sum_k B_{ihk} * c3_{ih} \qquad (8)$$

$$TT = \sum_h \sum_i \sum_k \frac{A_{hik} * c1_{hi}}{v1_{hi}} + \sum_i \sum_j \sum_k \frac{X_{ijk} * c2_{ij}}{v2_{ij}} + \sum_i \sum_h \sum_k \frac{B_{ihk} * c3_{ih}}{v3_{ih}} \qquad (9)$$

$$TC = TD * fc + TT * lc + \sum_h \sum_i \sum_k A_{hik} * vc \qquad (10)$$

$$TE = TD * ef \qquad (11)$$

$$Minimize\, TC \qquad (12)$$

Constraint (1) forces all recipients to be visited exactly once. Constraint (2) sets a limit of one route per vehicle. Routing continuity is represented by Constraint (3). Constraint (4) guarantees that each route begins and finishes at the depot. Constraint (5) computes arrival time for the first node visited according to travel and unloading time, while constraint (6) computes the arrival time for the remaining nodes. Constraint (7) forces the elimination of sub-sequences (i.e. subtours). Additionally, the computation of key metrics is presented in (8)–(11). Constraint (8) computes the total traveled distance. Constraint (9) calculates the total distance. Constraint (10) estimates the cost based on distance-fuel consumption, time-labor cost and fixed cost, respectively. Constraint (11) calculates CO_2 emissions. Finally, the objective function (12) minimizes the traditional routing cost in a VRP (i.e., fuel cost + labor cost + vehicle fixed cost) (Braekers et al. 2016).

NNVA Impact. Following the OGP definitions, the capacity constraint is considered a NNVA activity. This is because the vehicle capacity is an evident condition of urban transportation. However, this constraint does not add any value to the final customers. In order to model the CVRP, which is also known as the classic VRP (Braekers et al. 2016), we need to include constraint (13) in the model. Therefore, the E obtained after executing model (1)–(13), is used to compute the NNVA impact (i.e., $E_{2=}E-E_1$).

$$\sum_h \sum_i A_{hik} * q_i + \sum_i \sum_j X_{ijk} * q_i \leq Q_k; \forall k \qquad (13)$$

Supply Chain Impact. Using again the OGP definitions, the time window constraints is considered a supply chain requirement. The VRPTW is a popular extension of the VRP, in which delivery for each recipient must occur within a specified time interval (Braekers et al. 2016). To model this requirement, we include constraints (14) and (15) in the model. Similarly to NNVA, supply chain impact is computed by subtracting E_2 from the E obtained after executing (1)–(15) (i.e., $E_3=E-E_2$).

$$Y_{ik} \geq e_i * \left(\sum_h A_{hik} + \sum_j X_{jik} \right); \forall k, \forall i \tag{14}$$

$$Y_{ik} \leq l_i + M * \left(1 - \left[\sum_h A_{hik} + \sum_j X_{jik} \right] \right); \forall k, \forall i \tag{15}$$

Context Impact. In the same way, environmental legislation is considered to be a context constraint in the OGP. An environmental regulation could limit the routing length so that delivery occurs outside shopping hours (Quak 2008). This can be achieved by including constraint (16) in the model. Once more, the Context impact is computed by subtracting E_3 from the E obtained after executing (1)–(16) (i.e., $E_4=E-E_3$).

$$\sum_i \sum_h \frac{A_{hik} * c1_{hi}}{v1_{hi}} + \sum_i \sum_j \frac{X_{ijk} * c2_{ij}}{v2_{ij}} + \sum_i \sum_h \frac{B_{ihk} * c3_{ih}}{v3_{ih}} \leq rl; \forall k \tag{16}$$

NVA Impact. NVA impact, as mentioned earlier, corresponds to the CO_2 emissions generated by decision-makers' inappropriate selection of routing strategies. Thus, in order to obtain high levels of customer services, it is common for these decision-makers to seek to deliver the orders with the earliest due date first (Reyes et al. 2018). This inappropriate approach (i.e., minimizing lead times) could lead to not optimal solutions and create an inefficient usage of resources (i.e., NVA activities). Mathematically speaking, this strategy could be achieved by the objective function (17). That is, maximizing the difference between the latest due date and the delivery date. Therefore, NVA impact is computed by subtracting E_4 from E obtained after computing CO_2 emissions by executing (1)–(11) in conjunction with (13)-(17) (i.e., $E_5=E-E_4$).

$$\text{Maximize} \sum_i \sum_k l_i - Y_{ik} \tag{17}$$

Final Comparison. The last step of the proposed methodology consists of a performance analysis of different key metrics (i.e., travel distance, total costs, delivery times and CO_2 emissions). Please note that our approach is a simple-objective optimization (i.e., minimize cost) and then the other selected metrics are computed.

4   Case Study and Analysis of Results

This section applies the proposed methodology to a case study in Bogotá, Colombia. Bogotá is the capital of and largest city in Colombia. It is also the thirty-fourth largest city in the world and the fourth largest in Latin America (City Mayors 2017). As an emerging economy—and in contrast to developed economies— Colombia and its cities

are facing considerable sustainability challenges, boosted by their quick growth (Muñoz-Villamizar et al. 2019). As Gonzalez-Feliu (2017) pointed out, few studies on urban transportation refer to the complete methodology in scenario assessment and evaluation. Therefore, we compare non-collaborative vs. collaborative scenarios to assess the impact of aggregated demands, capacities into the routing policies. These analyses would allow a company to evaluate the convenience of increasing its current capacity, renegotiating new delivery dates with recipients or defining more effective legislation policies for urban transportation systems. Therefore, the urban transportation network in the case under study is characterized in Subsects 4.1 and 4.2 presents and analyzes the results of applying the methodology to the case under study.

The MILP models described below in this paper were implemented using GAMS commercial software version 24.1.3, with a time limit of 1000 s on an Intel(R) Core (TM) i5 personal computer with 1.4 GHz and 4 GB RAM.

4.1 Characterization of Urban Transportation Network

This study is based on real-world data from two of the major convenience stores (proximity shops) networks operating in Bogotá, Colombia. This work is an extension of the previous work from Muñoz-Villamizar et al. (2015b) in which the assessment of environmental impact in collaborative urban transport networks is added. For privacy reasons, the two companies will be called Company A and Company B. Company A owns 16 stores and Company B owns 10. The current situation for this transportation network is explained below.

The asymmetric origin-destination distance matrix for each company was obtained from the shortest paths derived from Google Maps™ mapping service. The vehicle selected was the Renault Kangoo Express Van with 650 kg of payload (Q) and 0.212 CO_2 kg/liters (Renault Colombia 2018). As mentioned before, GHG emissions, and CO_2 in particular have been traditionally evaluated within OR to quantify the externalities of transportation systems because there are models that estimate the emissions generated by using the energy required to move a vehicle from one point to another (Bektaş et al. 2019). Hence, CO_2 emissions are the externality selected in this study as the environmental performance key metric. However, note that other externalities (e.g., energy consumption) could be evaluated with our methodology, too.

Taking a gasoline price of 0.78 US$/liter in Colombia (GlobalPetrolPrices.com, 2018) and the energy consumption of the Kangoo Van (4.3 L/100 km), the assumed fuel cost is 0.034 US$/km. Besides, using a 5-year straight-line depreciation method over vehicle price (16,700 US$), a fixed vehicle cost of 13 US$/route was defined. A full month of weekly deliveries (i.e., four different sets of recipient orders) was randomly generated using the uniform distribution $q_i \sim U(0.1 * Q, 0.2 * Q)$ (i.e., 10% to 20% of the vehicle payload). Bogota's average traffic speed and its deviation are used to generate different weekly stochastic speeds between nodes i and j using the normal distribution: $v_{ij} \sim N(23.404, 3.986)$ (EL TIEMPO 2016). Other parameters considered in this case are the hourly salary of the workers involved in the freight transportation process (4.26 US$/h), an unloading speed of 0.018 min/kg, and the starting time (a_i) for a 3 h-long time window generated from a uniform distribution $a_i \sim U(0h, 7h)$ (Muñoz-Villamizar et al. 2019).

Consequently, the companies' current routing strategy responds to the earliest due date heuristic. That is, companies visit first the recipient with the smallest time window. It was also assumed that the companies' vehicle fleets are available to deliver every order within the time windows. Finally, Colombian legislation only allows urban freight delivery between 10 pm and 6 am (EL TIEMPO 2013). Hence, a routing length of 8 h was assumed. The full data are available upon request to researchers who would like to replicate the experiments.

4.2   Non-collaborative Scenario Vs. Collaborative Scenario

Figure 3 and Table 1 present the average results after applying the methodology in the full month generated instances (i.e., non-collaborative network vs. collaborative network). Several insights can be derived from these initial results. On the one hand, the inadequate selection of a routing strategy (i.e., NVA category) entails an overproduction of 18% in CO_2 in the collaborative scenario against the non-collaborative scenario. This difference is due to the increase in distance traveled (i.e., 441 km in the collaborative vs. 373.5 km in the non-collaborative). Conversely, NVA for the non-collaborative scenario has savings of 66% of the total costs. This statement could be an interesting trade-off between economic and environmental performance. However, as NVA corresponds to inefficiencies in the selection of the routing strategy, this trade-off is easily eliminated by removing the inadequate routing decision (i.e., delivering orders as soon as possible). On the other hand, if there are no NVA activities in the transportation network, the collaborative scenario represents savings up to 11% in CO_2 emissions. Total distance, total cost and average delivery time are also reduced by 10%, 65% and 94%, respectively. Those are interesting improvements by the implementation of the collaborative strategy.

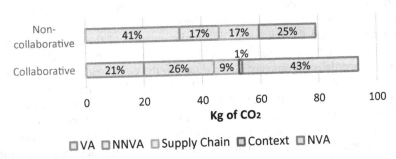

Fig. 3. OGP of non-collaborative network vs. collaborative network.

Regarding the impact of the Context Category, only the collaborative scenario is affected by the route length constraint. This impact is equivalent to 1% in CO_2 emissions. It is important to highlight the negative effect of legislation in the performance of the collaborative network. Furthermore, the collaborative scenario reduces the impact of Supply Chain constraints (i.e., time windows) from 17% to 9%. Besides, the collaborative scenario allows decreasing the VA category from 41% to 21% and the

Table 1. Results of non-collaborative network vs. collaborative network

Non-collaborative	VA	NNVA	Supply chain	Context	NVA
Travel distance (km)	153.7	216.8	281.5	281.5	373.5
Total cost (US$)	61.8	114.1	177.5	177.5	389.3
Average delivery time (h)	3.5	2.2	7.6	7.6	11.4
Total CO_2 emissions (Kg)	32.6	46.0	59.7	59.7	79.2
Collaborative	VA	NNVA	Supply chain	Context	NVA
Travel distance (km)	94.5	207.7	248.4	253.5	441.3
Total cost (US$)	16.5	59.7	61.2	61.4	133.4
Average delivery time (h)	0.2	0.3	0.4	0.4	0.7
Total CO_2 emissions (Kg)	20.0	44.0	52.7	53.7	93.6

reduction of the average delivery time. The collaborative transportation network can consolidate the orders and deliver them faster. However, the impact by vehicle capacity is increased in the collaborative scenario from 17% to 26%. Finally, in terms of cost and average delivery time (two of the most important criteria of companies), there is an average reduction of 64% in costs and 93% in the average delivery time, in all OGP categories.

5 Conclusions and Future Research

This study proposed a methodology for measuring and evaluating the environmental impact of collaborative urban transport networks. This methodology holistically characterizes and evaluates the environmental performance using different factors (i.e., value-added concept) of the transportation system. Therefore, our approach provides insights and highlights that are to be considered to shift towards a sustainable transportation network. On the one hand, the implementation of the collaborative transportation network entails savings up to 11% of CO_2 emissions. Total distance, total cost and average delivery time are also reduced in the collaborative scenario. On the other hand, there are negative impacts on CO_2 emissions by the incorrect selection of the routing strategy and by the current legislation that restricts the time of the routes.

The methodology has been validated in a practical case, demonstrating its potential for measuring environmental impact of collaborative urban transport networks to reduce environmental impact through OGP categorization. Besides, this article offers interesting results for decision-makers leading to savings up to 15% of CO_2. Future work should extend the proposed methodology to other contexts or alternatives (e.g., consolidation centers) and explore specific supporting tools and techniques (e.g., monitoring). Additionally, different environmental metrics could be tested according to the interest of decision-makers (e.g., energy and water consumption).

References

Bektaş, T., Ehmke, J., Psaraftis, H., Puchinger, J.: The role of operational research in green freight transportation. Eur. J. Oper. Res. **274**(3), 807–823 (2019)

Braekers, K., Ramaekers, K., Van Nieuwenhuyse, I.: The vehicle routing problem: state of the art classification and review. Comput. Ind. Eng. **99**, 300–313 (2016)

Buldeo Rai, H., van Lier, T., Meers, D., Macharis, C.: Improving urban freight transport sustainability: Policy assessment framework and case study. Res. Transp. Econ. **64**, 26–35 (2017)

Churchman, C.W.: The Systems Approach, 1st edn. Dell Publishing, New York (1968)

City Mayors: The Largest Cities in the World and Their Mayors (2017). http://www.citymayors.com/statistics/largest-cities-mayors-1.html. Accessed 02 May 2019

Comi, A., Schiraldi, M., Buttarazzi, B.: Smart urban freight transport: tools for planning and optimising delivery operations. Simul. Model. Pract. Theor. **88**, 48–61 (2018)

EL TIEMPO: El abecé de las restricciones a los vehículos de carga en Bogotá (2013). https://www.eltiempo.com/archivo/documento/CMS-13202956. Accessed 02 May 2019

EL TIEMPO: Así está la velocidad de 5 vías de Bogotá (2016). http://www.eltiempo.com/bogota/velocidad-en-vias-de-bogota/16560364. Accessed 02 May 2019

Eren Akyol, D., De Koster, R.: Determining time windows in urban freight transport: a city cooperative approach. Transp. Res. Part E **118**, 34–50 (2018)

Forsberg, J., Krook-Riekkola, A.: Supporting cities' emission mitigation strategies: modelling urban transport in a times energy system modelling framework. WIT Trans. Built Environ. **176**, 15–25 (2018)

Garg, C., Sharma, A., Goyal, G.: A hybrid decision model to evaluate critical factors for successful adoption of GSCM practices under fuzzy environment. Uncertain Supply Chain Manag. **5**(1), 59–70 (2017)

Gonzalez-Feliu, J.: Sustainability evaluation of green urban logistics systems: literature overview and proposed framework. In: Green Initiatives for Business Sustainability and Value Creation, pp. 103–134 (2017)

Kiba-Janiak, M.: Urban freight transport in city strategic planning. Res. Transp. Bus. Manag. **24**, 4–16 (2017)

Nataraj, S., Ferone, D., Quintero-Araujo, C., Juan, A., Festa, P.: Consolidation centers in city logistics: a cooperative approach based on the location routing problem. Int. J. Ind. Eng. Comput. **10**(3), 393–404 (2019)

Molina-Azorín, J., Claver-Cortés, E., López-Gamero, M., Tarí, J.: Green management and financial performance: a literature review. Manag. Decis. **47**(7), 1080–1100 (2009)

Muñoz-Villamizar, A., Montoya-Torres, J., Vega-Mejía, C.: Non-collaborative versus collaborative last-mile delivery in urban systems with stochastic demands. Procedia CIRP **30**, 263–268 (2015)

Muñoz-Villamizar, A., Montoya-Torres, J., Herazo-Padilla, N.: Mathematical programming modeling and resolution of the location-routing problem in urban logistics. Ingeniería y Universidad **18**(2), 271–289 (2014)

Muñoz-Villamizar, A., Santos, J., Montoya-Torres, J., Ormazábal, M.: Environmental assessment using a lean based tool. Serv. Orientation Holonic and Multi-Agent Manuf. **762**, 41–50 (2018)

Muñoz-Villamizar, A., Quintero-Araújo, C., Montoya-Torres, J., Faulin, J.: Short- and mid-term evaluation of the use of electric vehicles in urban freight transport collaborative networks: a case study. Int. J. Logistics Res. Appl. **22**(3), 229–252 (2019)

Quak, H.: Sustainability of urban freight transport: Retail distribution and local regulations in cities (PhD thesis). ERIM, Rotterdam (2008)

Renault Colombia (2018). https://www.renault.com.co. Accessed 02 May 2019

Reyes, D., Erera, A., Savelsbergh, M.: Complexity of routing problems with release dates and deadlines. Eur. J. Oper. Res. **266**(1), 29–34 (2018)

Rao, C., Goh, M., Zhao, Y., Zheng, J.: Location selection of city logistics centers under sustainability. Transp. Res. Part D Transp. Environ. **36**, 29–44 (2015)

Ubeda, S., Arcelus, F., Faulin, J.: Green logistics at Eroski: a case study. Int. J. Prod. Econ. **131**(1), 44–51 (2011)

Math-Heuristic for a Territory Design Problem

Mario A. Solana, Juan A. Díaz$^{(\boxtimes)}$ ⓘ, and Dolores E. Luna ⓘ

Universidad de las Américas Puebla, San Andrés Cholula Pue, 72810 Cholula, Mexico
{mario.solanavl,juana.diaz,dolorese.luna}@udlap.mx

Abstract. In this work, we study a territory design problem. This type of problem consists of dividing a geographic area into territories that meet certain characteristics or planning criteria. The problem studied in this paper considers the division of a geographical area into compact, contiguous and balanced territories, with respect to one or several measures of activity. To find feasible solutions for the problem, we propose a math-heuristic. The method first constructs solutions, using a minisum objective as measure of dispersion, that are balanced with respect to the different activity measures. Subsequently, the solutions are modified so that they satisfy the contiguity constraints. To test the performance of the proposed method, we use a set of test instances available in the literature. The results obtained are compared with the optimal solutions of the test instances. According to this comparison, the proposed method provides optimal solutions or solutions very close to the optimal solution with a reasonable computational effort compared to that required by an exact solution method.

Keywords: Territory design problem · p-median · GRASP · Math-heuristic

1 Introduction

Territory design problems can be seen as problems of grouping small geographic areas, called basic areas, into larger territories that satisfy a set of planning criteria. These planning criteria are defined according to the problem being studied. Among the most common, it is desired to have territories that are compact, contiguous and that are balanced (i.e. have similar sizes) with respect to one or several activity measures. Some examples of these activity measures are population size, sales potential, workload, etc. Although there is no rigorous definition of what a compact district is, it is said that a district is compact if it has a more or less round shape, has no holes and is not distorted. We consider that a district is contiguous, if we can travel between any two basic areas that compose it, without leaving the district [14].

Territory design problems have many applications. We find among the ones most studied in the literature, the geopolitical division of regions to ensure justice

© Springer Nature Switzerland AG 2019
C. Paternina-Arboleda and S. Voß (Eds.): ICCL 2019, LNCS 11756, pp. 67–82, 2019.
https://doi.org/10.1007/978-3-030-31140-7_5

in relation to the number of voters in each electoral district [8,9,11,12,15,17,22], the division of territories for the provision of services [1,3,4,13,21] and the design of sales districts in order to balance the workload and the potential sales of all territories [2,5,6,10,16,18,19]. In particular, the last two applications are very important to improve logistics both in providing public services and in the design of commercial territories. To find solutions and bounds for these problems, both exact algorithms and heuristic techniques have been used.

Hess et al. [11] are the first to study this problem to form constituencies. They propose to reconfigure districts using contiguity and compactness criteria. Garfinkel and Nemhauser [9] propose an implicit enumeration algorithm to reconfigure the electoral districts of the United States of America using criteria of contiguity and compactness. Hojati [12], studies this problem considering only the criterion of compactness. He uses a Lagrangian relaxation combined with a transportation technique to assign population units to the centers obtained with the Lagrangian relaxation and finally, he solves the splitting problem by solving a sequence of capacitated transport problems. Shirabe [22] recognizes that contiguity is a very important criterion in territory design. The paper introduces an integer-programming-based approach that enforced contiguity constraints independently of any other criteria that might be additionally imposed. Three experimental models are presented in which balance or compactness restrictions are relaxed. Although the proposal works well with small instances, the author recognizes the need to use heuristics to improve the results. Rincon-García et al. [17] propose three heuristic methods for the problem using contiguity, balance and compactness criteria: particle swarm optimization (PSO), artificial bee colony (ABC) and method of musical composition (DMMC). They compare their results with those used by the Federal Electoral Institute in Mexico and conclude that they are better than those currently being used.

Segal and Weinberger [21] study a territory design problem for the job of telephone repair-persons/installers. They use a heuristic that combines shortest path, minimum cost flow, and enumerative techniques. Blais et al. [1] present a case study in the area of health care services. They must partition a territory in a fixed number of districts by grouping territorial basic units. Five districting criteria must be respected: indivisibility of basic units, respect for borough boundaries, connectivity, visiting personnel mobility, and workload equilibrium. They solve the case study using tabu search. Carvajal et al. [4] use a territory design problem for proper forestry planning. They explain that there are three important practices in which the territory design makes a difference: (1) choose a maximum area for deforestation without causing erosion or aesthetic damage to the forests, (2) the conservation of mature forest areas that help to preserve the fauna and biodiversity, based on the fact that it has been demonstrated that the impact of these areas is even greater if there is contiguity; and (3) the correct selection of protected natural areas to guarantee the safety of endangered animals. They propose an integer programming model, capable of solving medium-sized forest problems, that involves the maximum area of deforestation as a measure of balance and the contiguity of the mature forest area. They conclude that the linear relaxations they use are good for achieving solutions

with small deviations from the optimal solutions. Hu et al. [13] use a mathematical model to locate shelters after an earthquake occurs. These shelters must be accessible, the population must know to which refuge they must go in case of an accident and the travel costs must be minimal. Therefore, contiguity and capacity restrictions must be satisfied. Camacho-Collado et al. [3] study a police districting problem. The problems consist of the efficient and effective design of patrol sectors in terms of performance attributes such as workload, response time, etc. They present a multi-criteria model that considers the attributes of area, risk, compactness, and mutual support. They find feasible solutions for the model using a heuristic that is empirically tested on a case study of the Central District of Madrid. Their solutions significantly improve the current ones.

Caballero-Hernández et al. [2] study a territory design problem motivated by a real-world application in a beverage distribution company in the city of Monterrey, Mexico. Their model seeks to balance the demand among the territories in order to delegate responsibility fairly. They propose a metaheuristic solution approach based on Greedy Randomized Adaptive Search Procedure (GRASP) and test the proposed metaheuristic on a set of randomly generated instances based on real-world data. Ríos-Mercado and Fernández [18] present a reactive GRASP approach to a commercial territory design problem very similar to the one studied in the previous work. Their model includes, as planning criteria, minimizing a measure of territory dispersion, balancing the different node activity measures among territories and territory contiguity. Fernández et al. [6] study a problem motivated by the recycling directive Waste Electrical and Electronic Equipment of the European Commission. However, this problem is different from those previously presented because the territories should be geographically as dispersed as possible to avoid that a company gains a monopoly in some region. They propose a solution based on a GRASP methodology. Elizondo-Amaya et al. [5] consider a Lagrangian relaxation scheme to obtain dual and primal bounds for the problem. This work takes into account the contiguity and compactness in the districts, as well as the balancing of two activity measures. The Lagrangian relaxation scheme allows finding a lower and upper bound for the solution by reducing the number of radii (maximum distance between a basic unit and the center of the territory to which it is assigned) that must be evaluated. This reduces significantly the computational time required to obtain the bounds. From there, the algorithm can use a branching scheme to obtain an optimal solution or use some other heuristic to provide a feasible solution of good quality with little computational effort.

Salazar-Aguilar et al. [20] study a territory design problem with balance constraints associated with two different activity measures and connectivity constraints. They consider two different objective functions to model territory compactness: a minisum objective and a minimax objective. They propose several formulations for the problems. These formulations include linear integer programming models and a quadratic integer programming formulation. They also introduce an exact solution framework, that is based on branch and bound and a cut generation strategy.

In this work we study the same problem studied in [20], but from a different perspective. In [20], two measures of dispersion are proposed, that when used as objective function, produce territories that satisfy the criterion of compactness. In this work, we only use one measure of dispersion, a minisum objective as the one used in the p-median problem. We propose a math-heuristic based on a greedy randomized adaptive procedure to find feasible solutions for the problem. We test the performance of the proposed math-heuristic with a set of instances from the literature. The obtained results show that the proposed method provides optimal or near-optimal solutions in reasonable computational time.

The rest of the paper is organized as follows. In Sect. 2, we present the problem formulation and the proposed math-heuristic to find feasible solutions for the problem. Section 3 presents the computational experience to evaluate the performance of the proposed math-heuristic. Finally, Sect. 4 presents some conclusions.

2   Problem Formulation and Methods

In this section, we first describe the territory design problem that we study in this work and show a mathematical programming model for it. Next, we describe in detail the proposed math-heuristic to find feasible solutions to the problem.

2.1   Integer Linear Programming Model

As we mentioned before, the territory design problem studied in this work is the one proposed by Salazar-Aguilar et al. [20], and considers compactness, connectivity and balance constraints. Although in that work several mathematical formulations are presented for the problem, we only present their integer linear programming formulation because it is the one that relates to our study.

Let $G = (V, E)$ be a graph with $V = \{1, \ldots, n\}$ a set of basic units (BUs) or nodes and E a set of edges representing adjacency between BUs. Also, let p be the number of territories in which we wish to divide the set of BUs, and it is required that each BU is assigned to only one territory. For each edge $\{i, j\}$, d_{ij} denotes the distance between basic units i and j. Additionally, let A be the set of activity measures, such as population, potential sales, workload, etc. For each $j \in V$ and $a \in A$, w_j^a, denotes the value associated to activity a in node j. Since it is not possible to have perfectly balanced territories with respect to each activity $a \in A$, a tolerance parameter τ^a for activity a is defined. This parameter measures the relative deviation from the average territory size with respect to activity $a \in A$. The average for activity a is given by

$$\mu^a = \sum_{j \in V} w_j^a / p$$

Therefore, an allocation of nodes to the selected median nodes will be feasible if, for each activity $a \in A$, the size of each cluster is between $(1 - \tau^a)\mu^a$ and

$(1 + \tau^a)\mu^a$. It is also required that for each $i, j \in V$ assigned to the same territory, there exists a path between them totally contained in the territory in order to achieve connectivity. Finally, to measure territory compactness a minisum objective function based on the p-median problem is used. Therefore, the problem studied in this work consists in finding a partition of V in p territories according to planning requirements of balancing and connectivity that minimizes the total sum of distances between the medians and the BUs allocated to them.

Now, we describe the integer linear model for the problem. Let us define N^i as the set of nodes adjacent to node i, this is:

$$N^i = \{j \in V : \{i, j\} \in E\}, i \in V$$

and let

$$x_{ij} = \begin{cases} 1, & \text{if basic unit } j \text{ is allocated to median } i, \\ 0, & \text{otherwise.} \end{cases}$$

for all $i, j \in V$ be the decision variables.
Therefore, the integer linear model can be formulated as:

$$(TDP) \qquad \min \sum_{i,j \in V} d_{ij} x_{ij} \tag{1}$$

$$\text{s. t.} \qquad \sum_{i \in V} x_{ii} = p \tag{2}$$

$$\sum_{i \in V} x_{ij} = 1 \qquad j \in V \tag{3}$$

$$\sum_{j \in V} w_j^a x_{ij} \geq (1 - \tau^a)\mu^a x_{ii} \qquad i \in V, a \in A \tag{4}$$

$$\sum_{j \in V} w_j^a x_{ij} \leq (1 + \tau^a)\mu^a x_{ii} \qquad i \in V, a \in A \tag{5}$$

$$\sum_{j \in \bigcup_{v \in S}(N^v \setminus S)} x_{ij} - \sum_{j \in S} x_{ij} \geq 1 - |S| \qquad i \in V, S \subset R(i) \tag{6}$$

$$x_{ij} \in \{0, 1\} \qquad i, j \in V \tag{7}$$

where $R(i) = \big(V \setminus (N^i \cup \{i\})\big)$.

The objective function (1) minimizes the sum of the distances between each BU and the location of the median it is allocated to. Constraint (2) ensures that a fixed number of p territories are created. Constraints (3) ensure that each basic unit is assigned to a single median. Constraints (4) and (5) guarantee territory balance with respect to each activity measure as they establish that the size of each territory must lie within the specified tolerance (measured by the tolerance parameter τ^a) around its average size μ^a. Finally, constraints (6) guarantee territory connectivity. They ensure that for any given subset S of BUs assigned to center i, not containing BU i, there must be an arc between S and the set containing i. These constraints are similar to the subtour elimination constraints in the Traveling Salesman Problem. It is important to note, that there is an exponential number of such constraints.

2.2　GRASP Math-Heuristic

The GRASP math-heuristic proposed in this paper is an iterative procedure that in each iteration builds a feasible solution through the following steps:

- *Selection of an initial set of medians*: Select a set of p BUs that will be the medians of each of the p territories.
- *Assignment of basic units to territories*: Assign each of the BUs to one of the medians selected in the previous step, in such a way that the sum of the distances between each basic unit and the median to which it is assigned is minimized while balancing constraints are satisfied.
- *Adjustment of territory medians*: For each territory, evaluate if the value of the objective function can be improved by replacing its median with another of the BUs of the territory. If one or more of the medians are replaced, re-execute the previous step to determine the optimal allocation of the BUs to the medians of the territories.
- *Secure connectivity constraints*: If the incumbent is improved, solve the separation problem to identify violated connectivity constraints.

Algorithm 1 depicts the pseudocode for the proposed GRASP math-heuristic. Next, we describe each step with more detail.

Selection of an Initial Set of Medians. It is desirable that the set of selected medians be geographically dispersed so that it is simpler to obtain territories that satisfy all the desired restrictions. The selection of the p medians is a greedy randomized adaptive procedure. Let M be the set of medians. The first median is selected randomly. While $|M| < p$, for each candidate BU, $j \in V \setminus M$, we calculate the greedy value

$$Dist_j = min_{i \in M}\{d_{ij}\}$$

denoting the smallest distance between the basic unit j and the previously selected medians. Let $Dist_{min} = min_{j \in V \setminus M}\{Dist_j\}$ and $Dist_{max} = max_{j \in V \setminus M}\{Dist_j\}$. Also, let $ThresholdValue = Dist_{max} - \alpha(Dist_{max} - Dist_{min})$ be a threshold value, where $\alpha \in (0,1)$ is a parameter that controls the degree of greediness or randomness of the greedy procedure. We randomly select a BU $j^\star$ from a restricted candidate list RCL that contains candidate BUs whose $Dist_j \geq ThresholdValue$. BU $j^\star$ is added to the set of medians M.

Assignment of Basic Units to Territories. Once we have the set M of p medians selected, we can obtain a feasible assignment of BUs to the p territories solving the following assignment problem.

$$(ASSIG(M)) \qquad \min \sum_{i,j \in V} d_{ij} x_{ij} \qquad (8)$$

$$\text{s. t.} \qquad \sum_{i \in V} x_{ij} = 1 \qquad j \in V \qquad (9)$$

Algorithm 1. GRASP Math-heuristic for the Territory Design Problem

Let $MaxIter$ be the number of iterations, δ be the increase of the α value at each iteration, and β be the maximum allowed value for the α parameter

$Best \leftarrow \infty$
$iter \leftarrow 0$
$\alpha \leftarrow 0$
while $(iter \leq MaxIter)$ **do**
    $M \leftarrow \emptyset$
    Select randomly a median r from the set V of basic units
    $M \leftarrow M \cup \{r\}$
    while $(|M| < p)$ **do**
        for all $(j \in V \setminus M)$ **do**
            $Dist_j \leftarrow \min_{i \in M}\{d_{ij}\}$
        end for
        $Dist_{min} \leftarrow \min_{j \in V \setminus M}\{Dist_j\}$
        $Dist_{max} \leftarrow \max_{j \in V \setminus M}\{Dist_j\}$
        $ThresholdValue \leftarrow Dist_{max} - \alpha(Dist_{max} - Dist_{min})$
        $RCL \leftarrow \{j \in V \setminus M : Dist_j \geq ThresholdValue\}$
        Select randomly j^* from RCL
        $M \leftarrow M \cup \{j^*\}$
    end while
    repeat
        Solve the assignment problem $ASIG_{Rel}(M)$
        for all $(j \in V : x_{mj} = 1$ in the optimal solution of $ASIG_{Rel}(M))$ **do**
            $a(j) \leftarrow m$
        end for
        $M' \leftarrow$ MEDIANSADJUSTMENT$((M))$
        if $(M' \neq M)$ **then**
            $M \leftarrow M'$
        end if
    until $(M' = M)$
    $SolValue \leftarrow \sum_{j \in V} d_{a(j),j}$
    if $(SolValue < Best)$ **then**
        repeat
            $CA \leftarrow$ VIOLATEDCONNECTSUBSET(M, a)
            if $(CA \neq \emptyset)$ **then**
                for all $(c \in CA)$ **do**
                    Add connectivity constraint associated to the
                    subset c in the optimal solution of $ASIG_{Rel}(M)$
                end for
                Solve the assignment problem $ASIG_{Rel}(M)$
            end if
        until $(CA = \emptyset)$
    end if
    $iter \leftarrow iter + 1$
    if $(\alpha < \beta)$ **then**
        $alpha \leftarrow \alpha + \delta$
    else
        $\alpha \leftarrow 0$
    end if
end while

$$\sum_{j \in V} w_j^a x_{ij} \geq (1 - \tau^a) \mu^a x_{ii} \qquad i \in M, a \in A \qquad (10)$$

$$\sum_{j \in V} w_j^a x_{ij} \leq (1 + \tau^a) \mu^a x_{ii} \qquad i \in M, a \in A \qquad (11)$$

$$\sum_{j \in \bigcup_{v \in S}(N^v \setminus S)} x_{ij} - \sum_{j \in S} x_{ij} \geq 1 - |S| \qquad i \in M, S \subset R(i) \qquad (12)$$

$$x_{ij} \in \{0, 1\} \qquad i \in M, j \in V \qquad (13)$$

It is worth noting that in this formulation there is an exponential number of connectivity constraints (6), which can be added as needed (those violated in the optimal solution). We call $ASSIG_{Rel}(M)$ to the formulation in which we have relaxed some or all of the connectivity constraints (6). This problem can be reduced to a Generalized Assignment Problem (GAP), if $|A| = 1$ and the lower bound of the balance constraints are set to zero, which is known to be NP-hard [7]. Therefore, to find the optimal allocation of BUs to the set of medians M, we solve the problem $ASSIG_{Rel}(M)$.

Algorithm 2. Adjustment of territory medians function

function MEDIANADJUSTMENT(M)
    $M' = M$
    for all $(m \in M)$ **do**
        $T_m \leftarrow \emptyset$
        for all $(j \in V : a(j) = m)$ **do**
            $T_m \leftarrow T_m \cup \{j\}$
        end for
        for all $(i \in T_m)$ **do**
            $SumDist_j \leftarrow \sum_{j \in T_m} d_{ij}$
        end for
        $j^* \leftarrow \arg\min_{j \in T_m} \{SumDist_j\}$
        if $(j^* \neq m)$ **then**
            $M' := M' \setminus \{m\} \cup \{j^*\}$
        end if
    end for
end function

Adjustment of Territory Medians. In this step, we seek to find the best median for each of the territories formed in the previous step. That is, let $T_1, T_2, \ldots, T_p$, be the territories associated with the set of medians M where $m_1, m_2, \ldots, m_p \in M$ are the medians of territories $T_1, T_2, \ldots, T_p$, respectively. Also, let $T_k = \{j \in V : x_{m_k j} = 1\}$, be the set of BUs assigned to territory T_k and $SumDist_i = \sum_{j \in T_k} d_{ij}$ for all $i \in T_k$. If $min_{i \in T_k}\{SumDist_i\} < \sum_{j \in T_k} d_{m_k j}$ then, there is some other BU in the territory T_k, for which the sum of distances with respect to the rest of the BUs of the territory is minimized, and therefore, if that BU is replaced for the BU m_k in the set M, the value of

the objective function is improved. In that case, the BU $i^\star \in T_k$ such that $i^\star \in \arg\min_{j \in T_k}\{SumDist_i\}$ replaces the median m_k and $M \leftarrow M \setminus \{m_k\} \cup \{i^\star\}$. Obviously, if there is any change in the set of medians, the assignment of BUs to territories may no longer be optimal and it is necessary to solve the problem $ASIG_{Rel}(M)$ again. Algorithm 2 depicts the pseudocode for this function.

Algorithm 3. Violated connectivity constraints identification function

```
function VIOLATEDCONNECTSUBSET(M, a)
    CA ← ∅
    for all (m ∈ M) do
        T_m ← ∅
        for all (j ∈ V : a(j) = m) do
            T_m ← T_m ∪ {j}
            CC ← CONNECTEDCOMPONENTS(m, T_m)
            for all (c ∈ CC) do
                if (m ∉ c) then
                    CA ← CA ∪ {c}
                end if
            end for
        end for
    end for
    return CA
end function
```

Secure Connectivity Constraints. In this step of the proposed algorithm, it is verified that the territories satisfy the connectivity constraints. This step is only executed when the incumbent is improved in some iteration of the algorithm. As mentioned above, a territory being contiguous means that there is a path between any pair of its BUs in which all the nodes belong to the territory. Therefore, if a territory is connected, the set of connected components of that territory must contain all its BUs. That is why in this step of the math-heuristic, the set of connected components for each territory is identified. In case the set does not contain all the BUs of the territory, the violated constraints are added for the next iteration of the algorithm. Algorithm 3 shows the pseudocode for the function to verify subsets of nodes for which the connectivity constraints (6) are violated, and algorithm 4 shows the pseudocode for the function to get all the connected components of a territory.

Algorithm 4. Connected components subset identification function

function CONNECTEDCOMPONENTS(m, T_m)
    $CC \leftarrow \emptyset$
    for all $(u \in T_m)$ **do**
        $visited(u) \leftarrow$ **false**
    end for
    $components \leftarrow 0$
    for all $(u \in T_m)$ **do**
        if (**not** $visited(u)$) **then**
            $CompConex \leftarrow \emptyset$
            Initialize list L with u
            $components \leftarrow components + 1$
            $visited(u) \leftarrow$ **true**
            while (L not empty) **do**
                Remove first element, l, from list L
                $CompConex \leftarrow \{l\}$
                for all $(v \in N^u : v \in T_m)$ **do**
                    if (**not** $visited(v)$) **then**
                        Add node v at the end of list L
                        $visited(v) \leftarrow$ **true**
                  end if
                end for
            end while
            $CC \leftarrow CC \cup \{CompConex\}$
        end if
    end for
    return CC
end function

3   Results and Discussion

In this section, we compare the results obtained with the GRASP math-heuristic proposed in this paper with the results of the exact method proposed by [20]. Both methods were implemented in FICO Xpress Optimization Suite ® and executed on a computer with an Intel(R) Xeon(R) E-2176M CPU @2.70 GHz processor and 32 GB of RAM.

Tests were made with 80 different instances of the problem from the literature, which are publicly available at http://yalma.fime.uanl.mx/~roger/ftp/tdp. The set of instances used contains:

- 20 instances of $n = 60$ nodes with $p = 4$.
- 20 instances of $n = 80$ nodes with $p = 5$.
- 20 instances of $n = 100$ nodes with $p = 6$.
- 10 instances of $n = 150$ nodes with $p = 8$.
- 10 instances of $n = 200$ nodes with $p = 11$.

For all instances, two activity measures are taken into account and a relative deviation value is considered with respect to the activity average of 5% to balance

the districts. That is, $|A| = 2$ and $\tau^{(a)} = 0.05$ for $a \in A$. Also, it is important to mention that for each of the test instances, the proposed method was executed ten times, since it has random elements.

The GRASP math-heuristic proposed in this paper has two parameters that affect its performance, the number of iterations $MaxIter$ and the value of the parameter α. To adjust the parameters, six experiments are carried out. In three of them the number of iterations is fixed to 50 for all instances and in the other three experiments it is varied according to the size of the instances by making $n/2$ iterations. The value of α is varied in all executions to have diverse solutions between 0 and 0.4. It always starts at zero and increases at each iteration until it reaches the value of 0.4. Once this value is reached, α is reset to zero and the process is repeated until the maximum number of iterations is reached. Three different values are used for these increments 0.01, 0.025 and 0.05. Table 1 summarizes the results of these experiments. The first column of the table shows the experiment number. The second column shows the number of iterations, while the third column shows the increment that is used for the value of α. Columns 4, 5 and 6 show the average percentage gap with respect to the optimal solution of the best solution obtained, the average of the solutions obtained and the worst solution obtained, respectively, for all the instances in the ten executions. Column 7 shows the average CPU time in seconds and column 8 shows the number of optimal solutions out of the 80 instances obtained by the GRASP-math-heuristic. As we can see, the results of the proposed algorithm are good in all cases. However, experiments 2 and 5 stand out. In experiment 2, we obtain 74 out of 80 optimal solutions in an average computational time of only 9.41 s. In experiment 5 is where we get the most optimal values and with an average percentage gap of only 0.06%.

Table 1. Parameter setting experiment results

Experiment number	Number of iterations	α increment	Min. % GAP	Ave. % GAP	Max. % GAP	Average CPU time (sec)	Number optimal
1	50	0.010	0.01%	0.09%	0.29%	9.22	72
2	50	0.025	0.01%	0.08%	0.24%	9.41	74
3	50	0.050	0.01%	0.06%	0.18%	9.78	72
4	$n/2$	0.010	0.00%	0.05%	0.20%	14.00	74
5	$n/2$	0.025	0.00%	0.06%	0.23%	14.40	75
6	$n/2$	0.050	0.00%	0.05%	0.20%	14.50	74

Tables 2 and 3 show the detailed results for experiment number 5. In both tables, the first column shows the name of the instance. The second column shows the value of the optimal solution for the problem. The next three columns show the best, the average and the worst value obtained in the 10 executions for each instance. Columns 6, 7 and 8 show the minimum, average and maximum percentage gap with respect to the optimal value obtained for each instance.

Column 9 shows the average CPU time in seconds required for the execution of the algorithm. Finally, the last column shows the execution time required by the algorithm proposed in [20]. As we can observe, only in 5 out of 80 instances the proposed GRASP math-heuristic does not find the optimal solution, but even in those cases the average percentage gap is only 0.076%. Additionally, the Wilcoxon signed-rank test is applied to compare the optimal solutions with the best ones obtained by the proposed method and with a 95% confidence level, there is no difference between the means. Figure 1 shows the optimal solution for the 2du100-05-1 instance. The BUs of each district are shown with different colors and symbols. The edges represent adjacencies between BUs.

Table 2. Results for instances with 60 and 80 nodes

Instance name	Optimal value	Minimum value	Average value	Maximum value	Minimum % GAP	Average % GAP	Maximum % GAP	Average CPU time	Optimal CPU time
2du60-05-1	5305.57	5305.57	5305.57	5305.57	0.00%	0.00%	0.00%	1.82	3.70
2du60-05-2	5451.68	5451.68	5451.68	5451.68	0.00%	0.00%	0.00%	1.43	3.30
2du60-05-3	5507.88	5507.88	5507.88	5507.88	0.00%	0.00%	0.00%	1.62	4.50
2du60-05-4	5935.67	5935.67	5935.67	5935.67	0.00%	0.00%	0.00%	1.50	2.60
2du60-05-5	5303.20	5303.20	5303.20	5303.20	0.00%	0.00%	0.00%	1.70	1.70
2du60-05-6	5253.94	5257.91	5257.91	5257.91	0.07%	0.07%	0.07%	2.10	4.30
2du60-05-7	5460.18	5460.18	5460.18	5460.18	0.00%	0.00%	0.00%	1.53	2.10
2du60-05-8	5309.96	5309.96	5309.96	5309.96	0.00%	0.00%	0.00%	1.76	1.70
2du60-05-9	5224.51	5224.51	5224.51	5224.51	0.00%	0.00%	0.00%	1.27	1.40
2du60-05-10	5350.16	5350.16	5350.16	5350.16	0.00%	0.00%	0.00%	1.47	2.00
2du60-05-11	5150.91	5150.91	5150.91	5150.91	0.00%	0.00%	0.00%	2.10	2.20
2du60-05-12	5597.50	5597.50	5602.83	5607.89	0.00%	0.09%	0.18%	1.45	5.80
2du60-05-13	5731.99	5731.99	5731.99	5731.99	0.00%	0.00%	0.00%	1.53	2.60
2du60-05-14	5462.96	5462.96	5462.96	5462.96	0.00%	0.00%	0.00%	1.41	2.70
2du60-05-15	5332.77	5332.77	5332.77	5332.77	0.00%	0.00%	0.00%	1.44	3.10
2du60-05-16	5399.55	5399.55	5399.55	5399.55	0.00%	0.00%	0.00%	1.46	10.90
2du60-05-17	5602.85	5602.85	5602.85	5602.85	0.00%	0.00%	0.00%	1.36	1.40
2du60-05-18	5773.96	5773.96	5773.96	5773.96	0.00%	0.00%	0.00%	1.44	3.00
2du60-05-19	5543.45	5543.45	5543.45	5543.45	0.00%	0.00%	0.00%	1.85	5.60
2du60-05-20	5767.54	5767.54	5767.54	5767.54	0.00%	0.00%	0.00%	1.42	2.60
2du80-05-1	6600.56	6600.56	6600.56	6600.56	0.00%	0.00%	0.00%	3.16	58.80
2du80-05-2	6408.82	6408.82	6408.82	6408.82	0.00%	0.00%	0.00%	4.51	6.40
2du80-05-3	6958.05	6958.05	6975.86	7047.09	0.00%	0.30%	1.50%	3.35	11.00
2du80-05-4	6900.16	6900.16	6902.90	6927.59	0.00%	0.05%	0.46%	3.20	63.00
2du80-05-5	6280.58	6280.58	6280.58	6280.58	0.00%	0.00%	0.00%	3.17	19.20
2du80-05-6	6521.08	6521.08	6521.08	6521.08	0.00%	0.00%	0.00%	4.34	17.30
2du80-05-7	6455.98	6455.98	6455.98	6455.98	0.00%	0.00%	0.00%	3.45	21.20
2du80-05-8	6680.29	6680.29	6680.29	6680.29	0.00%	0.00%	0.00%	4.01	20.60
2du80-05-9	6650.21	6650.21	6655.02	6698.32	0.00%	0.08%	0.81%	3.12	46.70
2du80-05-10	6534.77	6534.77	6535.90	6537.59	0.00%	0.02%	0.05%	3.97	32.30
2du80-05-11	6539.56	6539.56	6539.56	6539.56	0.00%	0.00%	0.00%	3.61	6.00
2du80-05-12	6704.00	6704.00	6704.00	6704.00	0.00%	0.00%	0.00%	3.59	7.20
2du80-05-13	6285.66	6287.23	6287.23	6287.23	0.03%	0.03%	0.03%	4.63	12.10
2du80-05-14	6615.80	6615.80	6615.80	6615.80	0.00%	0.00%	0.00%	3.25	23.50
2du80-05-15	6990.43	6990.43	6997.66	7004.89	0.00%	0.12%	0.24%	3.20	11.50
2du80-05-16	6391.67	6391.67	6391.67	6391.67	0.00%	0.00%	0.00%	3.65	12.20
2du80-05-17	6766.02	6766.02	6766.02	6766.02	0.00%	0.00%	0.00%	2.84	13.50
2du80-05-18	6808.45	6808.45	6808.45	6808.45	0.00%	0.00%	0.00%	4.16	13.10
2du80-05-19	6643.17	6643.17	6649.72	6708.69	0.00%	0.11%	1.10%	4.37	16.50
2du80-05-20	6873.62	6873.62	6873.62	6873.62	0.00%	0.00%	0.00%	3.24	26.80

Table 3. Results for instances with 100, 150 and 200 nodes

Instance Name	Optimal Value	Minimum Value	Average Value	Maximum Value	Minimum % GAP	Average % GAP	Maximum % GAP	Average CPU time	Optimal CPU time
2du100-05-1	7370.14	7370.14	7370.14	7370.14	0.00%	0.00%	0.00%	5.74	80.40
2du100-05-2	7278.46	7278.46	7278.46	7278.46	0.00%	0.00%	0.00%	5.92	55.60
2du100-05-3	7512.29	7512.29	7512.29	7512.29	0.00%	0.00%	0.00%	4.71	30.40
2du100-05-4	7581.57	7581.57	7584.09	7594.19	0.00%	0.04%	0.21%	5.83	10.60
2du100-05-5	7609.50	7609.50	7609.50	7609.50	0.00%	0.00%	0.00%	5.46	84.40
2du100-05-6	7243.00	7243.00	7249.19	7273.96	0.00%	0.10%	0.52%	5.34	31.30
2du100-05-7	7432.68	7432.68	7439.99	7469.24	0.00%	0.12%	0.62%	4.82	9.90
2du100-05-8	7052.89	7052.89	7058.38	7107.80	0.00%	0.09%	0.93%	5.23	8.90
2du100-05-9	7181.50	7181.50	7185.66	7223.12	0.00%	0.07%	0.70%	5.23	32.80
2du100-05-10	7432.89	7432.89	7438.17	7447.10	0.00%	0.09%	0.24%	4.50	34.20
2du100-05-11	6829.47	6829.47	6829.47	6829.47	0.00%	0.00%	0.00%	5.08	14.10
2du100-05-12	7461.20	7461.20	7465.44	7483.28	0.00%	0.07%	0.37%	5.38	13.30
2du100-05-13	7061.61	7061.61	7061.61	7061.61	0.00%	0.00%	0.00%	5.22	8.40
2du100-05-14	7825.62	7825.62	7827.64	7845.86	0.00%	0.03%	0.34%	5.34	11.00
2du100-05-15	7158.74	7158.74	7163.54	7172.65	0.00%	0.08%	0.23%	5.42	46.90
2du100-05-16	7653.16	7657.84	7669.37	7695.47	0.08%	0.27%	0.71%	6.45	69.20
2du100-05-17	6880.47	6880.47	6880.47	6880.47	0.00%	0.00%	0.00%	5.33	110.00
2du100-05-18	7438.51	7438.51	7438.51	7438.51	0.00%	0.00%	0.00%	5.97	50.40
2du100-05-19	7238.09	7238.09	7238.09	7238.09	0.00%	0.00%	0.00%	7.67	19.00
2du100-05-20	7590.10	7590.10	7590,10	7590.10	0.00%	0.00%	0.00%	5.32	72.60
du150-05-1	9511.76	9511.76	9530.04	9587.61	0.00%	0.31%	1.28%	21.38	145.10
du150-05-2	9400.51	9400.51	9401.43	9405.12	0.00%	0.02%	0.08%	26.19	258.40
du150-05-3	9134.57	9134.57	9134.57	9134.57	0.00%	0.00%	0.00%	22.81	71.10
du150-05-4	9359.00	9359.00	9359.00	9359.00	0.00%	0.00%	0.00%	22.53	96.70
du150-05-5	9506.58	9506.58	9506.58	9506.58	0.00%	0.00%	0.00%	20.09	89.50
du150-05-6	9039.06	9039.06	9039.06	9039.06	0.00%	0.00%	0.00%	25.87	37.60
du150-05-7	9854.70	9854.70	9869.10	9931.49	0.00%	0.24%	1.29%	19.99	829.40
du150-05-8	9199.29	9199.29	9199.29	9199.29	0.00%	0.00%	0.00%	46.19	3046.70
du150-05-9	9670.90	9670.90	9671.70	9673.58	0.00%	0.01%	0.05%	20.34	349.80
du150-05-10	9570.58	9570.58	9575.58	9611.38	0.00%	0.08%	0.69%	17.55	71.70
du200-05-1	10422.00	10422.00	10422.71	10429.10	0.00%	0.01%	0.12%	76.78	302.60
du200-05-2	10639.80	10639.80	10662.15	10684.90	0.00%	0.38%	0.76%	58.63	867.00
du200-05-3	10837.90	10837.90	10844.77	10868.80	0.00%	0.12%	0.52%	59.19	303.30
du200-05-4	11124.90	11124.90	11128.59	11139.40	0.00%	0.06%	0.24%	69.86	2323.30
du200-05-5	10874.50	10874.50	10885.96	10903.80	0.00%	0.19%	0.49%	69.62	2335.00
du200-05-6	10492.20	10492.20	10500.12	10512.50	0.00%	0.13%	0.34%	66.10	5574.10
du200-05-7	11020.90	11020.90	11034.41	11066.50	0.00%	0.23%	0.77%	77.35	2008.10
du200-05-8	10650.70	10652.70	10672.68	10707.10	0.03%	0.37%	0.95%	74.78	366.10
du200-05-9	11431.30	11441.40	11451.07	11485.50	0.17%	0.33%	0.91%	84.27	2293.90
du200-05-10	11039.50	11039.50	11048.80	11062.70	0.00%	0.16%	0.39%	57.41	352.40

The results obtained with the proposed method are compared with the results provided by the exact method proposed by [20]. Figure 2 describes the behavior of both algorithms in terms of CPU times. It is evident that in terms of the computational effort required, the exact method compares unfavorably with the proposed GRASP math-heuristic. The difference is not so noticeable for the 60 and 80 node instances, but as the number of nodes grows, the CPU times of the algorithm proposed in [20] grow in a non-linear fashion. In this figure it is easy to appreciate the advantages that the use of a math-heuristic provides as the size of the instances grows, which usually occurs in practical applications.

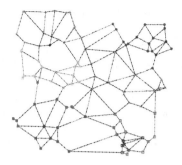

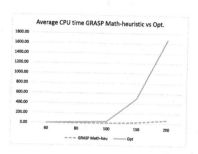

Fig. 1. Optimal solution example

Fig. 2. CPU time comparison

4   Conclusions

In this work a GRASP math-heuristic is proposed to solve the territorial design problem raised by [20]. The problem studied includes as criteria of territorial planning, contiguity, compactness and balance with respect to several measures of activity. To model the compactness, we use a minisum objective function as the one used in the p-median problem. The proposed algorithm is tested with a set of instances of the literature and to evaluate its performance is compared with the results obtained in [20].

The algorithm proposed is an iterative process. In each iteration a disperse set of medians is chosen using an adaptive randomized greedy procedure to avoid assignments of basic units that are very far from the medians of their territory, in order that the solutions satisfy the criterion of compactness. Subsequently, an assignment problem is solved to assign the BUs to the medians so that balance constraints are satisfied with respect to all activity measures. Next, we seek to reduce the objective function through an improvement procedure where medians are adjusted. Finally, a separation problem is solved to identify violated connectivity constraints (if any). The solution of the separation problem provides all the connected components of each of the territories. For each of the connected components that do not contain the median of the territory, a valid inequality is added, which prevents this connected component from repeating itself in the optimal solution of the assignment problem. This process is repeated until in the optimal solution of the assignment problem a single connected component is obtained for each territory.

According to the results obtained for the exact method proposed by [20] and the method proposed in this work, it is observed that the GRASP math-heuristic provides in all cases good quality solutions with reasonably small CPU times. This is more noticeable as the size of the instances increases, which is the case for real-life applications.

References

1. Blais, M., Lapierre, S.D., Laporte, G.: Solving a home-care districting problem in an urban setting. J. Oper. Res. Soc. **54**(11), 1141–1147 (2003)
2. Caballero-Hernández, S.I., Ríos-Mercado, R.Z., López, F., Schaeffer, S.E.: Empirical evaluation of a metaheuristic for commercial territory design with joint assignment constraints. In: Proceedings of the 12th Annual International Conference on Industrial Engineering Theory, Applications, and Practice (IJIE), pp. 422–427 (2007)
3. Camacho-Collados, M., Liberatore, F., Angulo, J.M.: A multi-criteria police districting problem for the efficient and effective design of patrol sector. Eur. J. Oper. Res. **246**(2), 674–684 (2015)
4. Carvajal, R., Constantino, M., Goycoolea, M., Vielma, J.P., Weintraub, A.: Imposing connectivity constraints in forest planning models. Oper. Res. **61**(4), 824–836 (2013)
5. Elizondo-Amaya, M.G., Ríos-Mercado, R.Z., Díaz, J.A.: A dual bounding scheme for a territory design problem. Comput. Oper. Res. **44**, 193–205 (2014)
6. Fernández, E., Kalcsics, J., Nickel, S., Ríos-Mercado, R.Z.: A novel maximum dispersion territory design model arising in the implementation of the WEEE-directive. J. Oper. Res. Soc. **61**(3), 503–514 (2010)
7. Fisher, M.L., Jaikumar, R., Van Wassenhove, L.N.: A multiplier adjustment method for the generalized assignment problem. Manag. Sci. **32**(9), 1095–1103 (1986)
8. Fryer Jr., R.G., Holden, R.: Measuring the compactness of political districting plans. J. Law Econ. **54**(3), 493–535 (2011)
9. Garfinkel, R.S., Nemhauser, G.L.: Optimal political districting by implicit enumeration techniques. Manag. Sci. **16**(8), B495–B508 (1970)
10. Hess, S.W., Samuels, S.A.: Experiences with a sales districting model: criteria and implementation. Manag. Sci. **18**(4–Part-II), P41–P54 (1971)
11. Hess, S.W., Weaver, J., Siegfeldt, H., Whelan, J., Zitlau, P.: Nonpartisan political redistricting by computer. Oper. Res. **13**(6), 998–1006 (1965)
12. Hojati, M.: Optimal political districting. Comput. Oper. Res. **23**(12), 1147–1161 (1996)
13. Hu, F., Yang, S., Xu, W.: A non-dominated sorting genetic algorithm for the location and districting planning of earthquake shelters. Int. J. Geogr. Inf. Sci. **28**(7), 1482–1501 (2014)
14. Kalcsics, J., Nickel, S., Schröder, M.: Towards a unified territorial design approach - applications, algorithms and GIS integration. Top **13**(1), 1–56 (2005)
15. Kaufman, A., King, G., Komisarchik, M.: How to measure legislative district compactness if you only know it when you see it. American Journal of Political Science (2017), to appear
16. Moreno, S., Pereira, J., Yushimito, W.: A hybrid K-means and integer programming method for commercial territory design: a case study in meat distribution. Annals of Operations Research, pp. 1–31. Springer, New York (2017). https://doi.org/10.1007/s10479-017-2742-6
17. Rincón-García, E.A., Gutiérrez-Andrade, M.Á., de-los Cobos-Silva, S.G., Mora-Gutiérrez, R.A., Ponsich, A., Lara-Velázquez, P.: A comparative study of population-based algorithms for a political districting problem. Kybernetes **46**(1), 172–190 (2017)

18. Ríos-Mercado, R.Z., Fernández, E.: A reactive GRASP for a commercial territory design problem with multiple balancing requirements. Comput. Oper. Res. **36**(3), 755–776 (2009)
19. Ronen, D.: Sales territory alignment for sparse accounts. Omega **11**(5), 501–505 (1983)
20. Salazar-Aguilar, M.A., Ríos-Mercado, R.Z., Cabrera-Ríos, M.: New models for commercial territory design. Netw. Spatial Econ. **11**(3), 487–507 (2011)
21. Segal, M., Weinberger, D.B.: Turfing. Oper. Res. **25**(3), 367–386 (1977)
22. Shirabe, T.: Districting modeling with exact contiguity constraints. Environ. Planning B: Planning Des. **36**(6), 1053–1066 (2009)

Maritime and Port Logistics

A Decomposed Fourier-Motzkin Elimination Framework to Derive Vessel Capacity Models

Mai L. Ajspur[1], Rune M. Jensen[1(✉)], and Kent H. Andersen[2]

[1] Department of Computer Science,
The IT University of Copenhagen, Copenhagen, Denmark
rmj@itu.dk
[2] Department of Mathematics, Aarhus University, Aarhus, Denmark

Abstract. Accurate Vessel Capacity Models (VCMs) expressing the trade-off between different container types that can be stowed on container vessels are required in core liner shipping functions such as uptake-, capacity-, and network management. Today, simple models based on volume, weight, and refrigerated container capacity are used for these tasks, which causes overestimations that hamper decision making. Though previous work on stowage planning optimization in principle provide fine-grained linear Vessel Stowage Models (VSMs), these are too complex to be used in the mentioned functions. As an alternative, this paper contributes a novel framework based on Fourier-Motzkin Elimination that automatically derives VCMs from VSMs by projecting unneeded variables. Our results show that the projected VCMs are reduced by an order of magnitude and can be solved 20–34 times faster than their corresponding VSMs with only a negligible loss in accuracy. Our framework is applicable to LP models in general, but are particularly effective on block-angular structured problems such as VSMs. We show similar results for a multi-commodity flow problem.

Keywords: Fourier-Motzkin elimination · Vessel capacity model · Liner shipping · Projection

1 Introduction

Container shipping is a central element in the clockwork of global trade. As part of this, container liner shipping companies operates a set of container vessels on services connecting major trade regions like Asia and Europe. Each company is focused on utilizing the cargo capacity in their service network, since unused capacity constitute a loss that can be fatal in a market with a profit margin of just a few percent. Thus, being able to estimate the residual capacity of a container vessel is central to the business. This is challenging in practice, however, since an empty slot may be impossible to utilize for a wide range of reasons, including the interaction of various stowage rules and seaworthiness requirements. As a result,

© Springer Nature Switzerland AG 2019
C. Paternina-Arboleda and S. Voß (Eds.): ICCL 2019, LNCS 11756, pp. 85–100, 2019.
https://doi.org/10.1007/978-3-030-31140-7_6

the free capacity of each container type is a complex function of the composition of cargo on board the vessel and the design of the vessel.

Usually, only the stowage planning team will be able to determine the residual capacity of a vessel accurately by spending hours of manual work. However, the knowledge is primarily needed in higher functions such as: *uptake management* that control the sale of cargo bookings to fill the vessels with profitable cargo; *capacity management* that route cargo through the service network; and *network management* that makes changes to the service network. Decision makers in these functions seldom have stowage insight or time to consult the stowage team. Instead, they usually boil down the free capacity of a vessel to its nominal volume, weight, and reefer (refrigerated containers) capacity minus total volume, weight and reefer number of containers already on board. This simple three dimensional capacity model is inherently optimistic, since it ignores stowage complications, and it has been shown that this can lead to revenue overestimates of more than 15% [5]. This can cause sub-optimal decisions that significantly harm business. Previous work has contributed frameworks for automated stowage planning (e.g., [2,13,16,20]), and recently, linear stowage planning models were shown to scale to large container vessels (e.g., [5]). These latter Vessel Stowage Models (VSMs) embed an accurate capacity model, but since they include positioning information about the containers, they are too large for use as capacity models in higher functions, since these tasks often require several hundred capacity models to be solved simultaneously.

In this paper, we introduce a novel method to calculate a Vessel Capacity Model (VCM) automatically from a VSM. Our basic idea is to derive the VCM by projecting out positioning variables from the VSM, such that the VCM only expresses the relationship between variables representing the total amount of each possible container type. Projection is an essential operation in polyhedral analysis with usage within e.g. program analysis and constraint (logic) programming (e.g., [4,7,14]), and Fourier-Motzkin Elimination (FME) is a projection method that has been successfully used in previous projection frameworks (e.g., [15,17,18]). Similar to our framework, these frameworks use simplifications, redundancy removal and approximation-procedures. The former uses the extreme-point method of [8], while the boundary-approximation of [15,17] involves a successive increase of the allowable deviation from the feasible area and a permissible maximal ratio of removed non-redundant inequalities. Seen from the perspective of capacity models, both frameworks are used on quite small systems. Furthermore, the systems in [18] are also sparse. Other methods exist for computing the projection of a feasible area of a constraint system, that are not based on FME. The method in [8] finds extreme points in the projection space incrementally. It can therefore also be used to approximate the projection as is done in [18]. It is recommended for dense systems. Another example is the method introduced in [11], which computes all facets of the projection iteratively using a face-lattice. This method is recommended by the authors for polytopes with a low facet count and a high vertex count.

Our main contribution is to improve the state-of-the-art of the above mentioned FME frameworks. In particular, we introduce a novel decomposition method that takes advantage of block-angular structured models such as VSMs to significantly speed-up the projection of the great number of unneeded variables in these models. To our knowledge, our FME-based framework is the first that can take advantage of block-angular structure in the system. Additionally, our removal of redundant constraints is parallellized, and the framework includes preprocessing of the constraint system including removal of less strict inequalities.

Our experimental evaluation of computing VCMs with this method shows that the number of constraints and non-zeros in the resulting VCMs typically are reduced by an order of magnitude compared to their corresponding VSMs. The decomposition reduces the size of the intermediary systems produced by FME, causing less time to be needed for removing redundant constraints, which speeds up the projection process significantly. In addition, for the models that include hydrostatic constraints, the resulting VCMs can be solved 20–34 times faster than their VSMs with only a negligible loss in accuracy. Although it can take several hours to derive a capacity model due to the clean-up of redundant constraints, this only has to be done one time for a vessel class, making the approach suitable for computing these models. Since the VCMs are linear and much faster to solve than their corresponding VSMs, they can be integrated in decision support systems for the higher functions in liner shipping. Multicommodity flow problems also have a block-angular structure and we found a speed-up and a reduction in final size similar to the ones seen for VCMs.

This paper is organized as follows. Section 2 introduces the required definitions and notation, and Sect. 3 briefly presents the VSMs that are projected. Then Sect. 4 outlines the methods used in our FME framework including how block-angular problems are decomposed. Our experimental results then follow in Sect. 5, before Sect. 6 concludes.

2   Definitions and Notation

A constraint system S is a set of linear equalities and inequalities over the same set of continuous variables, $VAR(S) = \{x_1, \ldots, x_n\}$. For the constraint $c \in S$, we say that c uses x if x's coefficient in c is nonzero. The set of points in $\mathbb{R}^n$ that satisfies all constraints in S is called the *feasible area* of S. A constraint $c \in S$ is *redundant* iff removing it from S does not influence the feasible area of S, otherwise it is called *non-redundant*.

For some variables $Y \subseteq VAR(S)$, we are not interested in their values in a feasible point - we just want to know that a satisfying value exists. This is captured by the *projection of S w.r.t. Y*, which is the largest set consisting of values for $VAR(S) \setminus Y$ that can be extended with values for Y such that all constraints in S are satisfied. The projection of a constraint system is a uniquely determined subset of $\mathbb{R}^{n-|Y|}$, but also the feasible region of another system S' (see e.g. [21]). We are mostly interested in the latter, since it is the relationship

between the values in the projection that is relevant to us, and we will allow ourselves to write that "S' is the projection of S w.r.t. Y" if the feasible area of S' equals the projection of S, though such a system is not uniquely determined. We note, that since we are dealing with subsets of multi-dimensional Euclidian spaces, the dimension and the order of the variables are important. However, in order to simplify the presentation, we do not explicitly specify these for every considered projection/constraint system S'. A more stringent exposition keeping track of the variable sets and ordering can be found in [1].

3   Vessel Stowage Model

A Vessel Stowage Model (VSM) is a constraint system defining the feasible stowage conditions of a vessel. For a stowage condition to be feasible, the vessel must among other things be seaworthy with proper transversal stability and stress forces within limits (hydrostatic constraints); capacities should not be exceeded; the container stacks must by physically possible; and there are separation rules for dangerous cargo to name a few. For an in-depth coverage of container vessel stowage, the reader is referred to a recent book on the topic [10].

Consider the container vessel shown in Fig. 1. Each *cell* on the vessel can hold two 20' containers or one 40', and some cells have power plugs, allowing refrigerated containers (*reefers*) to be stowed. Container stacks are arranged longitudinal in *bays* that can be further subdivided into *locations*, and each stack rests on sockets with maximum weight limits. To help achieve stability of the vessel, large water ballast tanks are placed on the vessel. The volume capacity of a vessel is measured in Twenty-foot Equivalent Units (TEUs) and can be more than 20 K.

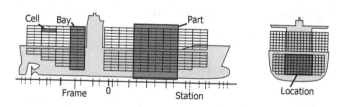

Fig. 1. Vessel structure and reference points

The VSMs used in this paper is based on previous work on stowage planning optimization (e.g., [5,16]). They considers 20' and 40' containers in three weight classes, and a container is either reefer or non-reefer. This gives a total of 12 container types T. For each container type and location on the vessel, a decision variable defines the number of containers of the type in the location. Due to the large number of containers, we ignore the integrality of these variables as in [16]. The models are based on industrial data from a large carrier and include a number of volume, weight and reefer capacity constraints for each location. The data specifies hydrostatic limits at *frame positions*, while input for hydrostatic calculations

are given at other *station positions* (see Fig. 1). A simplification of the representation of the hydrostatics is achieved by dividing the vessel into *parts* (P) spanning one or more succeeding bays or no bays at all as shown in Fig. 1. The weight of ballast water in each part is given by a continuous decision variable, and hydrostatic constraints are included that restrict the shear forces and bending moments at positions between parts. To this end, the model includes constraints defining the resulting force on each part using a linear approximation of its buoyancy and weight. The hydrostatic modeling approach of the VSM and the quality of the approximations are detailed in [9]. A Vessel Capacity Model (VCM) is derived from a VSM by adding auxiliary variables to equal the total of each container type on board the vessel and projecting all other variables out using FME. In this way the container positioning information is abstracted away.

Due to space limitations, we do not include a formal definition of the VSMs on which our experiments are carried out. However, it can be found in [1]. Instead, to explain our FME framework we briefly introduce the toy version VSM below. For this VSM, merely the structure is essential while the constraints themselves are of less importance.

$$
S_{\mathrm{g}} : \begin{cases} x_\tau = \sum_{p \in P} x_{p,\tau} & \forall \tau \in T & (1) \\[2ex] \sum_{p \in P} \sum_{\tau \in T} W_\tau x_{p,\tau} \leq D & & (2) \end{cases}
$$

$$
S_p \text{ for all parts } p \in P : \begin{cases} \sum_{\tau \in T^{20}} x_{p,\tau} \leq C_p^{20} & \sum_{\tau \in T^{40}} 2x_{p,\tau} \leq C_p^{40} & (3) \\[2ex] \sum_{\tau \in T^{20}} W_\tau x_{p,\tau} \leq C_p^{\mathrm{W20}} & \sum_{\tau \in T^{20}} 0.5 W_\tau x_{p,\tau} + \sum_{\tau \in T^{40}} W_\tau x_{p,\tau} \leq C_p^{\mathrm{W40}} & (4) \\[2ex] \sum_{\tau \in T^{\mathrm{R}}} x_{p,\tau} \leq C_p^{\mathrm{R}} & \sum_{\tau \in T^{20}} x_{p,\tau} + \sum_{\tau \in T^{40}} 2x_{p,\tau} \leq C_p^{\mathrm{TEU}} & (5) \end{cases}
$$

In this model, the only decision variables are the $x_{p,\tau} \in \mathbb{R}$, which denotes the number of containers in part $p \in P$ of type τ. Constraint (1) defines the auxiliary variables x_τ that specify the total of each container type (i.e., the only variables that will be left in the VCM), while (2) limits the total weight of cargo. (3) defines the volume capacity of part p of 20' and 40' containers, while (4) defines the weight capacity of 20' and 40' containers, respectively (W_τ is the weight of type τ). We notice that the weight limits of 40' containers includes half of the weight of 20' containers due to the arrangement of sockets, and that a 40' container counts two TEU. (5) defines reefer and total TEU capacity of part p, respectively.

This presentation clearly exposes the VSMs natural (primal) block-angular structure [19]: For each part p, S_p is a set of capacity constraints for part p that constitutes a local system, whose constraints only use variables that are not used in S_q for a different part q. The remaining constraints make up the global subsystem, S_{g}, where the constraints also use variables from several of the local subsystems.

4   FME-based Projection Framework

Our projection framework based on Fourier-Motzkin Elimination (FME) can be used for massive variable elimination in any linear inequality system but has been designed to take advantage of the block-angular structure often found in real-world models including the VSMs. The methods for projecting a constraint system are described in Sect. 4.1, while the decomposition used on block-angular structured problems is described in Sect. 4.2.

4.1   Projection Procedure

The projection procedure starts with a preprocessing of the constraint system S. Then we use the equalities in the reduced system to isolate variables from Y and substitute in the rest of the system (Gauss-eliminations). Subsequently, we successively eliminate one variable from Y at a time using FME and remove redundant inequalities. At the top-level, the pseudocode for our projection method is therefore as described in Algorithm 1. Each sub-procedure in this algorithm is detailed below.

PREPROCESS(S, Y): We reduce S by removing easily identifiable redundant constraints and assign necessary bounds and values to variables using well-known LP preprocessing steps (e.g., [3]). We further perform FME on variables in $x \in Y$ in easy cases where this only results in a deletion of a set of inequalities or a substitution of a variable with a value (when $|Pos_S(x)|$ or $|Neg_S(x)|$ is 0 or 1, see further below). We remove a redundant inequality when it is *linearly dependent* or *less strict* than another. This can be seen syntactically and happens in some of the cases when two constraint c and c' have the same coefficients (modulo a constant), or all coefficient in c are dominated by the coefficients in c' (modulo a constant). The steps are implemented with special care of equalities and working with the assumption that the system is feasible (details can be found in [1]).

GAUSS-ELIM(S, Y): An equality e can be used to isolate a variable $x \in Y$ which can then be substituted in all other constraints in S (a Gauss-elimination). This eliminates x from the system and does not cause the same combinatorial explosion of inequalities as FME may do. To avoid density, when the system S contains several equalities, we choose the variable x (used in any equality) that is used the fewest times in total in S and the equation e (among those using x)

Algorithm 1. Projection based on Fourier-Motzkin elimination

---

function PROJECT(System S, Variables Y)
    $(S, Y) \leftarrow$ PREPROCESS(S, Y)
    $(S, Y) \leftarrow$ GAUSS-ELIM(S, Y)
    while $Y \neq \emptyset$ **do**
        $(S, Y, New) \leftarrow$ FME-SINGLEVAR(S, Y)
        $S \leftarrow$ REMOVEREDUNDANCY(S, New)
    return S

---

that uses the fewest variables. This is repeated until there are no more equalities using variables in Y.

FME-SINGLEVAR(S, Y): FME is a classical algorithm for producing the projection of a set of variables from a system where all constraint are inequalities on the form $\mathbf{a} \cdot \mathbf{x} \leq b$. The method successively eliminates one variable $x \in Y$ until all required variables have been eliminated. To eliminate $x \in Y$, the constraints in S are first divided into three sets, $Pos_S(x)$, $Neg_S(x)$, and $Zero_S(x)$ depending on the sign of the coefficient of x. Here, bounds are treated as any other inequalities. A new system S' is then created, which is the projection of S w.r.t. $\{x\}$. It consists of $Zero_S(x)$, together with one inequality, $i_{p,n,x}$, for each pair $(p, n) \in Pos_S(x) \times Neg_S(x)$. $i_{p,n,x}$ is the addition of positive multiples of $p : \mathbf{a} \cdot \mathbf{x} \leq b$ and $n : \mathbf{a}' \cdot \mathbf{x} \leq b'$ such that the coefficient of x in the resulting inequality is 0. That is, $i_{p,n,x}$ equals $-a'_x \cdot \mathbf{a} \cdot \mathbf{x} + a_x \cdot \mathbf{a}' \cdot \mathbf{x} \leq -a'_x \cdot b + a_x \cdot b'$, where a_x and a'_x is the coefficient of x in c and c', respectively,

The order in which variables are eliminated naturally influences the size of the intermediary constraint systems. We have chosen to use the greedy heuristic that minimizes the number of new inequalities in the immediately next system [6], which is a commonly used heuristic and easily calculated from the current system. In the worst case scenario, the number of inequalities in the created system S' is $\frac{1}{4}|S|^2$, which implies that (both time and space) complexity is double-exponential. For a large, dense system, the growth will be substantial, which prohibits it from use for practical purposes *if* the added inequalities are non-redundant or the non-redundant inequalities are not removed (see e.g. [15]). It should, however, also be emphasized that not all inequalities in the succeeding system are necessarily non-redundant.

REMOVEREDUNDANCY(S, New): To detect redundancy, we examine each inequality $c : \mathbf{a} \cdot \mathbf{x} \leq b$ in turn and remove it from the system if $\max \mathbf{a} \cdot \mathbf{x}$ subject to $S \setminus \{c\}$ is less than or equal to b. The property can be checked using an LP solver. Equalities are not examined, since we want to keep these for use in Gauss-elimination. When removing redundancy after projecting x from S, we only check the newly added inequalities; if a constraint in $Zero_S(x)$ is non-redundant before the elimination, it will be non-redundant afterward as well. For large systems, checking all constraints for redundancy is time-consuming. We have therefore implemented a method for redundancy removal that uses several threads in parallel. Each thread checks one inequality at a time, while a manager takes care of the communication and keeps track of the redundant inequalities.

Several of the constants in the data used for our VSMs are results of various approximations and hence the boundary of the feasible area is not exact. Coarsening the boundary is therefore permissible, and we also remove inequalities that are *"almost redundant"*. An inequality $c : \mathbf{a} \cdot \mathbf{x} \leq b$ is almost redundant if $\max \mathbf{a} \cdot \mathbf{x}$ subject to $S \setminus \{c\}$ is less or equal to $b + \epsilon \cdot |b|$ for a small ϵ. Therefore, the manager also collects a set of almost redundant inequalities. After the parallel redundancy check, *one* thread is then used to go through all the almost redundant inequalities sequentially, and the ones that are still almost redundant are removed.

4.2   Decomposing a Block-Angular System

The toy VSM from Sect. 3 has a natural block-angular structure, where the set of capacity constraints for each part constitute a local subsystem (S_p for $p \in P$), and the remaining constraints form a global subsystem (S_g). To derive the VCM from this VSM, we want to eliminate all variables except those counting the number of containers of each type. However, when a variable $x_{p,\tau}$ is eliminated, the new inequalities constructed by FME uses variables from all subsystems, since $x_{p,\tau}$ is used in the global constraints that are combined with constraints from S_p. Continuing with FME, this result in an increasing number of global and more dense constraints, which again makes FME perform worse. To avoid the immediate "mix" of local subsystems, we will define and use auxiliary variables to ensure that we can project the local subsystems separately without producing global constraints. Afterward, we combine the projected subsystems and eliminate the auxiliary variables.

For the considered toy VSM, we first define a variable to hold the weight of cargo in each part p, w_p. We also define the variables $x'_{p,\tau} = x_{p,\tau}$, though this merely appears to be a renaming of variables. For each p, we then add the definition of w_p and $x'_{p,\tau}$ to S_p and rewrite the global constraints in terms of the new variables. That is, local subsystem p is now $S^0_p = S_p \cup \{w_p = \sum_{\tau \in T} W_\tau x_{p,\tau}\} \cup \bigcup_{\tau \in T} \{x'_{p,\tau} = x_{p,\tau}\}$, while the new system of global constraints is $S^0_g = \bigcup_{\tau \in T} \{x_\tau = \sum_{p \in P} x'_{p,\tau}\} \cup \{\sum_{p \in P} w_p \leq D\}$. Notice that due to the new variables, none of the variables $x_{p,\tau}$ for a given p is used in any constraints outside of S^0_p. Therefore, to project $\{x_{p,\tau} \mid \tau \in T\}$ for a given p from the whole system ($S^0_g \cup \bigcup_{q \in P} S^0_q$) we can start by just eliminating $\{x_{p,\tau} \mid \tau \in T\}$ from S^0_p. Then we can do so for the other parts. When all subsystems S^0_p have been projected, we can join these projections together with S^0_g and eliminate the remaining variables, $\{x'_{p,\tau} \mid \tau \in T, p \in P\} \cup \{w_p \mid p \in P\}$, from the resulting system.

More formally, for a block-angular structured system with local subsystems $S_1, \ldots, S_k$ (using the variables $X_1, \ldots, X_k$) and global subsystem S_g we do as follows for all subsystems S_i (detailed pseudocode and correctness proofs can be found in [1]).

- For each global constraint c using variables in S_i, we define an auxiliary variable $z^0_{c,i}$ that equals the variables in S_i's contribution to c. We add the equality defining $z^0_{c,i}$ to S_i, and we substitute it in c. We name the produced subsystem S^0_i.
- Then, we project S^0_i w.r.t. all variables from $Y \cap X_i$, resulting in the system S'^0_i. We *do keep* the auxiliary z^0-variables.

After projecting each S^0_i we combine the projections with S^0_g to create $\mathcal{S} \stackrel{\text{def.}}{=} S^0_g \cup S'^0_1 \cup \ldots \cup S'^0_k$. We then eliminate from $\mathcal{S}$ all the auxiliary z^0-variables, Z^0, plus any remaining variables in Y.

Comparing $\mathfrak{S} \stackrel{\text{def.}}{=} S^0_1 \cup \ldots \cup S^0_k \cup S^0_g$ with the original system S, all we have done is defining auxiliary variables and substituted them in the system. Thus, eliminating Y from S is equivalent to eliminating $Y \cup Z^0$ from $\mathfrak{S}$. When

eliminating $Y \cup Z^0$ from $\mathfrak{S}$, we can choose to first eliminate $X_1 \cap Y$, then $X_2 \cap Y$ up to $X_k \cap Y$, and finally $Z^0 \cup Y \setminus (X_1 \cup \ldots \cup X_k)$. Any variable in $X_1 \cap Y$ has a zero-coefficient in all constraints outside S_1^0, so $\mathfrak{S} \setminus S_1^0$ will not be changed by FME when $X_1 \cap Y$ is eliminated. It can therefore be put aside until that projection is done. Likewise, when eliminating $X_i \cap Y$, neither $S_{i+1}^0 \cup \ldots \cup S_k^0 \cup S_g^0$ nor the already projected systems contain any variables from $X_i \cap Y$ and can hence be put aside until the variables in $Z^0 \cup Y \setminus (X_1 \cup \ldots \cup X_k)$ are eliminated. Thus, the following holds.

Proposition 1. *The projection of S w.r.t. Y defines the same feasible area as the projection of S w.r.t. $Z^0 \cup Y \setminus (X_1 \cup \ldots \cup X_k)$.*

S has by construction a block-angular structure and instead of eliminating the remaining variables immediately, we can use the same approach as above to postpone "mixing" blocks, if for example the global constraints use too many variables. As an example, consider again the toy VSM that was decomposed into the subsystems S_g^0 and S_p^0 for $p \in P$ and assume that $P = \{1, 2, 3, 4\}$. Then we can group the subsystems into two groups, $\{S_1^0, S_2^0\}$ and $\{S_3^0, S_4^0\}$, and for each global constraint define a variable stating each new group's contribution to the global constraint. For example, for the global constraint $w_1 + w_2 + w_3 + w_4 \leq D$, we define a variable $w_{\{1,2\}}$ that is the weight of the containers in part 1 and 2. We similarly define $w_{\{3,4\}}$ and substitute with these in the global constraint. In total, we construct the following subsystems that can be arranged in a tree structure as shown in Fig. 2(a).

$$S_g^1 : \bigcup_{\tau \in T} \{x_\tau = x'_{\{1,2\},\tau} + x'_{\{3,4\},\tau}\} \cup \{w_{\{1,2\}} + w_{\{3,4\}} \leq D\},$$

$$S_1^1 : \bigcup_{\tau \in T} \{x'_{\{1,2\},\tau} = x'_{1,\tau} + x'_{2,\tau}\} \cup \{w_{\{1,2\}} = w_1 + w_2\},$$

$$S_2^1 : \bigcup_{\tau \in T} \{x'_{\{3,4\},\tau} = x'_{3,\tau} + x'_{4,\tau}\} \cup \{w_{\{3,4\}} = w_3 + w_4\}.$$

To obtain the projection of the original toy VSM, these systems are then projected recursively as shown in Fig. 2(b).

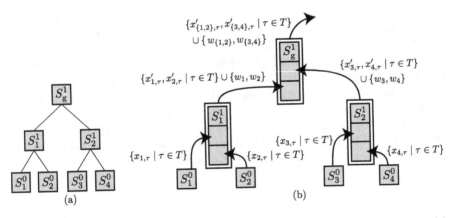

Fig. 2. (a) Tree-strucure of subsystems. (b) Projection using tree-structure from (a)

In more general terms, we divide all subsystems into k groups and do the following for $1 \leq j \leq k$.

- We define a system S_j^1. For each global constraint c using variables from systems in group j, we define a variable, $z_{c,j}^1$, that equals the contribution to c made by the variables in the systems in group j. We add the defining equality to S_j^1 and rephrase c using $z_{c,j}^1$.
- Then we project S_j^1, w.r.t. the previous, auxiliary z^0-variables, while we do keep the newly created z^1-variables.

Subsequently we can then join the projected S^1-systems with the new global constraints, and finally project the last auxiliary variables. Alternatively, we can repeat the steps above until the final projection can be done.

When we decompose a system as described above, we effectively create a tree structure of subsystems paired with a set of variables to be eliminated. An inequality system can then be projected by recursively projecting its children as shown in Fig. 2. The intuition why this works is as before; projecting all Y and Z-variables from the union of all the (unprojected) subsystems in the tree corresponds to projecting the Y variables from S, and because we can choose the elimination order of the variables, we only need to project the subsystems in the tree in the correct order (a rigorous proof can be found in [1]).

Proposition 2. *The projection of the system associated with the root of the tree constructed from S and Y w.r.t. the Y- and Z-variables as described corresponds to projecting S w.r.t. Y.*

Using our previously described projection method, we obtain the projection of S w.r.t. Y by calling PROJECTNODE(root of T), where T is the tree structure constructed from S and Y, and PROJECTNODE is described in Algorithm 2. We note that due to the elimination of almost redundant inequalities, this only approximates the projection. However, by setting $\epsilon = 0$, the algorithm indeed returns the correct projection.

Using the described decomposition, it is of course also possible to project nested block structured problems, i.e. systems that on the top-level can be divided into a global part and a number of local parts that in themselves can be further divided into local parts and a global part, and so on. Other block structured problems such as staircase problems can also be decomposed into a tree structure and projected using the described approach. Further, when the system S is decomposed into subsystems in a tree structure, the projection itself can be parallelized by maintaining a queue of not yet projected subsystems whose children have all been projected. The members of the queue (which initially are the leafs) are then projected independently in parallel.

Algorithm 2. Projecting a block-structured system via decomposition.

---

function PROJECTNODE(Node n)
$\quad$ $(S, Y) \leftarrow$ the system and variable set associated with n
$\quad$ **if** n is a leaf **then**
$\quad\quad$ **return** PROJECT(S, Y) $\qquad\qquad\qquad\qquad\qquad$ ▷ Algorithm 1
$\quad$ **else**
$\quad\quad$ **for all** children m of n **do**
$\quad\quad\quad$ $S \leftarrow S \cup$ PROJECTNODE(m)
$\quad$ **return** PROJECT(S, Y) $\qquad\qquad\qquad\qquad\qquad$ ▷ Algorithm 1

---

5 Results

We have constructed a number of different VSMs for a specific vessel, where the weight and hydrostatics are taken into account to various degrees. The first two VSMs do not model any hydrostatic constraints (similar to the toy-example but with more capacity constraints), and the first VSM has no total weight capacity imposed. The subsequent VSMs (referred to as complex VSMs) consider stress force constraints at the endpoint of 2, 4, 6 and 8 parts, respectively. Each VSM has been transformed into its corresponding VCM by eliminating all variables except the x_τ variables. Projections have been done in two different ways, *decomposed* and *flat*. For the decomposed projections, a tree structure has been used as described in Sect. 4.2, while the flat projections do not use any decomposition at all. CPU time is measured in both number of iterations and *ticks* calculated by the CPLEX Interactive Optimizer version 12.5.0.0. Our FME-based projection framework has been implemented in Java. The experiments were carried out on a computer with an Intel® Xeon® CPU with 8 cores and 32 GB RAM.

Table 1 shows the size reductions of the VCMs. It summarizes the size of the VSMs and the VCMs that are the result of the projection using decomposition. These sizes are given in terms of the number of inequalities (ineq), equalities (eq), variables (var), non-zero entries (nzs) and density (dens). The size of the VSMs are given both as they appear as input to our algorithm, and after it has been preprocessed by CPLEX. For comparison, the table includes a "Simple VCM" corresponding to the maximum volume, weight, and reefer capacity models used in liner shipping today. Since we project all but 12 variables, this naturally gives a large reduction in the number of variables. However, the complex VCMs also have 5.8–11.1 times fewer inequalities than even the presolved VSMs for complex VSMs (27.7 and 20.8 for the first two models). The VCMs also have fewer non-zero entries (3–6 times fewer for complex VSMs compared to the preseolved models, otherwise 18 and 24). The results reveal no apparent relationship between the size of the VSM and the size of its VCM.

Regarding the decomposition impact, Table 2 shows the time taken for the algorithm to do the projection, both decomposed and flat. For most VSMs, the flat projection timed out (TO) beyond 18 hours, in which case the variables left

Table 1. The size of the VSMs and corresponding VCMs

	VSM				VSM, presolved				VCM			
	ineq (eq)	var	nzs	dens	ineq	var	nzs	dens	ineq	var	nzs	dens
No weights	774 (12)	1142	6662	8.61	554	657	2784	5.03	20	12	155	7.75
No hydro.	806 (43)	1173	7854	9.74	555	657	3441	6.20	18	12	144	8.00
2 parts	810 (43)	1173	7860	9.70	556	661	3447	6.20	96	12	1113	11.59
4 parts	824 (49)	1179	7886	9.57	564	671	3471	6.15	64	12	731	11.42
6 parts	838 (55)	1185	7916	9.44	570	679	3496	6.13	80	12	888	11.10
8 parts	852 (61)	1191	7950	9.33	576	685	3522	6.11	52	12	582	11.19
Simple VCM	3	12	36	12.00	3	9	24	8.00				

Table 2. Projection time

	Decomposed		Flat	
	time	vars left	time	vars left
No weights	24.5m	-	2.5m	-
No hydro.	14.5m	-	1.8m	-
2 parts	7h 18m	-	(TO) 32h	551
4 parts	8h 4m	-	(TO) 61h	557
6 parts	3h 7m	-	(TO) 18h	577
8 parts	3h 19m	-	(TO) 65h	566

to be projected are given. Figure 3(a) shows the progression of the number of inequalities (ineq) and variables (var), respectively, as a function of time when the algorithm runs on the decomposed 8-part model. These numbers are the sum of all the inequalities and variables, respectively, in all the projected or unprojected subsystems in the decomposition at a given time. Likewise, Fig. 3(b) shows the progression for the flat projection of the same model; this figure includes the number of inequalities for the decomposed projection for comparison. Each graph shows the number of inequalities and variables after each step outlined in Sect. 4.1. The results in Table 2 shows that the decomposition has a substantial impact on the success of the projection of the complex VSMs. However, the non-complex VSMs are solved faster using a flat projection. This is probably because they are reasonably sparse without the hydrostatic constraints.

When considering each subsystem in a decomposition as a system in itself, in general, the number of inequalities after each call to FME-SingleVar in Algorithm 1 grows to begin with, as does the number of inequalities before this call. This continues until there are a few variables left, where both these numbers decrease. For the decomposed algorithm, though the number of inequalities grow after each FME-step, most of them are redundant or almost redundant. The same does not hold for the flat projection of the complex VSMs (at least not for the steps that are completed within the time limit). On the contrary, many of the produced inequalities are non-redundant, increasing the likelyhood that even more inequalities will be produced in the next elimination and that the redundancy removal will take longer time. We also note that the runtime, even for the decomposed projections, are not exactly small, and the main part of the execution time is spend doing redundancy removal. However, as mentioned in the introduction, these calculations only need to be done once per vessel class.

As a use-case example, the VSMs and their projected VCMs have been optimized for revenue. Each transported container yields a fixed revenue based on

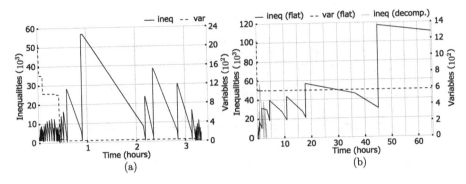

Fig. 3. Size progression for the 8-part model (a) with decomposition and (b) flat

its type. Table 3 shows the number of iterations (iter), the deterministic time in ticks (time) and the optimal objective value (obj) in 10M$. It likewise shows how many times faster, the projections are w.r.t. iterations and deterministic time, as well as the difference in objective value in percentage. For comparison, the number of iterations, deterministic time and objective value is shown for the simple VCM, too. As can be seen from the numbers in Table 3, in general, VCMs are much faster to solve than their corresponding VSMs. More specifically there are approx. 17–25 times fewer iterations and 20–34 times fewer ticks for the complex models. Meanwhile the difference in objective value is only modest; for the complex model, the difference is at most 0.5%. The differences (both in time and objective) is larger for the non-complex models. When comparing to the simple model, we see that this model of course is even faster, but the difference in objective is also between 72% and 76%. Hence, our results confirm the experiments by Delgado [5] showing a substantial revenue overestimation of capacity models used in liner shipping today.

Beside the VSM, we have studied another block-angular structured system, namely one describing a multi-commodity flow problem. In short, this problem considers a graph on which a number of commodities can flow on the edges. Each edge has a capacity (upper bound) for each commodity as well as a common total capacity. Demands and supply are modelled as variables, and we want to examine the relationship between the supply and demand of the commodities without having to care about how the items flow in the internal nodes. This can be done by eliminating all other variables than the ones denoting the demands and supply of each commodity. We have generated the flow graph shown in Fig. 4 with inspiration from the Chen.DSP collection [12]. It consists of seven "layers" with three nodes each, and there are two commodities. The capacity for each commodity and edge is 0 with a probability of 5% and otherwise drawn from a uniform distribution between 5 and 15, while the common capacity of the edge e is 0 with probability 25% and otherwise a number drawn from the uniform

Table 3. Iterations, time and objective values for the VSMs and VCMs.

	VCM			VSM			Difference		
	iter	time	obj	iter	time	obj	iter	time	obj
No weights	11	0.05	8.63	363	2.64	8.08	×33.0	×52.8	6.8%
No hydro.	9	0.04	7.87	188	5.48	6.22	×20.9	×137	26.5%
2 parts	14	0.29	6.09	251	5.88	6.07	×17.9	×20.3	0.196%
4 parts	13	0.18	6.17	228	4.95	6.16	×17.5	×27.5	0.153%
6 parts	9	0.20	6.17	227	5.02	6.18	×25.2	×25.1	0.202%
8 parts	12	0.14	6.21	233	4.79	6.18	×19.4	×34.2	0.490%
Simple VCM	4	0.02	10.7						

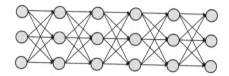

Fig. 4. A "layered" graph for a multi-commodity flow problem

distribution between $s - 10$ and s, where s is the sum of the individual capacities on e. A multi-commodity flow problem is naturally block-structured with a block for each commodity, but it contains usually many global constraints. Therefore, instead of using these blocks to decompose the system, we divide the graph into smaller subgraphs and use these as blocks. Similarly to Table 1, Table 4 shows the size of the original model and the projections resulting from a flat and decomposed projection, respectively. Figure 5 shows the progression over time of the number of inequalities and number of variables left to be projected, for both the flat and decomposed projection algorithm. Also here we see a reduction in the number of inequalities, variables and non-zero entries, of 3.5, 6.6, and 3.3/3.8 times, respectively (in both cases) compared to the presolved model. The density stays almost the same. This is not as large a reduction as for the VSMs, however, the unprojected models are also smaller to begin with. The decomposed projection is 8.2 times faster than the flat projection.

Table 4. Size of projection of a multi-commodity flow model.

	Size				Time
	ineq (eq) /	var /	nzs /	dens	
Original	204 (42) /	120 /	444 /	2.17	-
Presolved	59 /	79 /	201 /	3.41	-
Projected, decomposed	17 (2) /	12 /	61 /	3.59	2h 9m
Projected, flat	17 (2) /	12 /	53 /	3.12	17h 41m

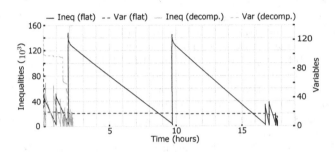

Fig. 5. Size progression during projection of a multi-commodity flow problem

6   Conclusion

This paper has introduced a novel FME projection framework that automatically translates a linear stowage model (VSM) into a smaller sized capacity model (VCM) by projecting unneeded variables. To our knowledge, our framework is the first to exploit a block-angular structure for projection and apply massive parallelization of computations. Our results show that the projected VCMs are reduced by an order of magnitude both in number of inequalities and number of non-zero entries. The VCMs including hydrostatic constraints are solved 20-34 times faster than their corresponding VSMs. Similar results are achieved for at multi-commodity flow problem. Future work includes further parallelization and approaches to automatically estimate the best way of decomposing a given system as well as testing the framework on further block-angular problems. From an application point of view, it could be interesting to test the VCMs in an actual setting for e.g. uptake management in a cargo flow graph, as well as adding more constraints to the VSM and/or investigate the limits for the decomposition framework.

Acknowledgements. We would like to thank Stefan Røpke, Thomas Stidsen, and David Pisinger for discussions on applications of the FME framework beyond container vessel capacity models. This research is supported by the Danish Maritime Fund, Grant No. 2016-064.

References

1. Ajspur, M.L., Jensen, R.M.: Using Fourier-Motzkin-elimination to derive capacity models of container vessels. Technical Report TR-2017-197, IT University of Copenhagen (2017)
2. Ambrosino, D., Sciomachen, A., Tanfani, E.: Stowing a containership: the master bay plan problem. Transp. Res. Part A: Policy Pract. **38**(2), 81–99 (2004)
3. Andersen, E.D., Andersen, K.D.: Presolving in linear programming. Math. Program. **71**(2), 221–245 (1995)

4. Benoy, F., King, A., Mesnard, F.: Computing convex hulls with a linear solver. Theory Pract. Logic Program. **5**(1–2), 259–271 (2005)
5. Delgado, A.: Models and Algorithms for Container Vessel Stowage Optimization. Ph.D. thesis, IT University of Copenhagen (2013)
6. Duffin, R.J.: On Fourier's analysis of linear inequality systems, pp. 71–95. Springer, Heidelberg (1974). https://doi.org/10.1007/BFb0121242
7. Fordan, A., Yap, R.H.C.: Early projection in CLP(R). In: Maher, M., Puget, J.-F. (eds.) CP 1998. LNCS, vol. 1520, pp. 177–191. Springer, Heidelberg (1998). https://doi.org/10.1007/3-540-49481-2_14
8. Huynh, T., Lassez, C., Lassez, J.L.: Practical issues on the projection of polyhedral sets. Ann. Math. Artif. Intell. **6**(4), 295–315 (1992)
9. Jensen, R.M., Ajspur, M.L.: The standard capacity model: towards a polyhedron representation of container vessel capacity. In: Cerulli, R., Raiconi, A., Voß, S. (eds.) ICCL 2018. LNCS, vol. 11184, pp. 175–190. Springer, Cham (2018). https://doi.org/10.1007/978-3-030-00898-7_11
10. Jensen, R.M., Pacino, D., Ajspur, M.L., Vesterdal, C.: Container Vessel Stowage Planning. Weilbach (2018)
11. Jones, C., Kerrigan, E.C., Maciejowski, J.: Equality set projection: a new algorithm for the projection of polytopes in halfspace representation. Technical report, Cambridge University Engineering Dept (2004)
12. Jones, K., Lustig, I., Farwolden, J., Powell, W.: Multicommodity network flows: the impact of formulation on decomposition. Math. Program. **62**, 95–117 (1993)
13. Kang, J.G., Kim, Y.D.: Stowage planning in maritime container transportation. J. Oper. Res. Soc. **53**(4), 415–426 (2002)
14. Lassez, J.L.: Querying constraints. In: Proceedings of the Ninth ACM SIGACT-SIGMOD-SIGART Symposium on Principles of Database Systems (PODS), pp. 288–298. ACM (1990)
15. Lukatskii, A.M., Shapot, D.V.: A constructive algorithm for folding large-scale systems of linear inequalities. Comput. Math. Math. Phys. **48**(7), 1100–1112 (2008)
16. Pacino, D., Delgado, A., Jensen, R.M., Bebbington, T.: Fast generation of near-optimal plans for eco-efficient stowage of large container vessels. In: Böse, J.W., Hu, H., Jahn, C., Shi, X., Stahlbock, R., Voß, S. (eds.) ICCL 2011. LNCS, vol. 6971, pp. 286–301. Springer, Heidelberg (2011). https://doi.org/10.1007/978-3-642-24264-9_22
17. Shapot, D.V., Lukatskii, A.M.: Solution building for arbitrary system of linear inequalities in an explicit form. Am. J. Comput. Math. **2**(01), 1 (2012)
18. Simon, A., King, A.: Exploiting sparsity in polyhedral analysis. In: Hankin, C., Siveroni, I. (eds.) SAS 2005. LNCS, vol. 3672, pp. 336–351. Springer, Heidelberg (2005). https://doi.org/10.1007/11547662_23
19. Williams, H.P.: Model Building in Mathematical Programming. Wiley, London (2007)
20. Wilson, I.D., Roach, P.A.: Container stowage planning: a methodology for generating computerised solutions. J. Oper. Res. Soc. **51**(11), 1248–1255 (2000)
21. Ziegler, G.M.: Lectures on Polytopes, Graduate Texts in Mathematics, vol. 152. Springer, New York (1995). https://doi.org/10.1007/978-1-4613-8431-1

A Note on Alternative Objectives
for the Blocks Relocation Problem

Stefan Voß[(⊠)] [iD] and Silvia Schwarze

Institute of Information Systems (IWI), University of Hamburg,
Von-Melle-Park 5, 20146 Hamburg, Germany
{stefan.voss,silvia.schwarze}@uni-hamburg.de

Abstract. The Blocks Relocation Problem (BRP) is an important problem arising at container terminals when containers have to be transshipped. Recently, we have seen the upcoming discussion whether some objectives are more appropriate than others. While most authors minimize the number of relocations as objective, some minimize the crane's working time. After adapting a given BRP formulation by including the second objective in two variants, a computational study is carried out on benchmark instances from literature applying some sensitivity analysis indicating to which extent optimal solutions change if the objective is replaced. If the number of relocations is minimized, often larger crane working times may be obtained. Conversely, if the crane working time is minimized, often solutions achieving also a minimum number of relocations are found. Based on our analysis, the consideration of alternative objectives like crane working time is recommended.

Keywords: Container relocation · Port operations ·
Integer programming · Crane working time · Blocks relocation problem

1 Introduction

The *blocks relocation problem (BRP)* describes the situation of outbound containers stored in a container yard until they are loaded to a vessel. Once a vessel arrives at the port, the containers have to be retrieved according to a given sequence. That is, each container has a priority number that indicates when it has to be loaded. As containers in the yard are stacked directly on top of each other, conflicts may appear if a container that has to be retrieved next, the *target container*, is buried below at least one other container. In this case, relocation activities have to be carried out, i.e., containers are moved within the stacking area in order to make the target container accessible. Given an initial layout of the container yard with N containers, the BRP aims at minimizing the number of relocation operations throughout the loading phase. In literature, some authors add the following assumption (A1): Only containers located above the current target container are allowed to be relocated (see, e.g. [9] and [18]). This condition allows to reduce the solution space and to decrease computational effort. However, it can be

© Springer Nature Switzerland AG 2019
C. Paternina-Arboleda and S. Voß (Eds.): ICCL 2019, LNCS 11756, pp. 101–121, 2019.
https://doi.org/10.1007/978-3-030-31140-7_7

shown that optimal solutions might get lost by adding (A1); see [4]. The stacking area of a BRP is 2-dimensional, i.e., it is defined by its width W (the number of stacks) and its height H (the number of rows). Such a 2-dimensional stacking area is called a *bay*. The dimension of the stacking area is increased in the *container relocation problem (CRP)*; see, e.g., [7]. For the CRP, containers are considered within a *block*, a 3-dimensional area composed of bays. Related to the BRP are remarshalling scenarios. Remarshalling can relate to blocks (*remarshalling problem (RMP)*) or bays (*premarshalling problem (PMP)*) and aims at transferring the given initial layout of the stacking area to a given final layout with a minimum number of relocation operations. The main difference to relocation problems like the BRP and the CRP is that in remarshalling problems, the number of containers in the stacking area is constant whereas in the relocation problems, the number of containers decreases as they are loaded to a vessel. Surveys on container rehandling are provided by [3,11].

A particular focus of this study is put on alternative objective functions for the BRP. In its initial version, the BRP follows the objective of minimizing the total number of relocations [9]. We refer to this criterion as f_1. For the subsequent discussion it is important to distinguish the notions *relocation* and *retrieval*. A relocation is a container movement within a bay or yard, usually carried out to resolve a conflict by relocating a blocking container. A retrieval, on the other hand, describes a container movement from the yard to the outside area. The number of retrievals is fixed and equals the number of containers N.

In recent publications, alternative objective functions are proposed. One class of objective functions aims at minimizing the number of conflicts in a bay and is therefore more related to the PMP. For our considerations on the BRP, we do not use these objective functions further; see [11] for a survey on related works. A second class of objectives considers a crane's working time and aims at minimizing this value. The working time of a gantry crane is determined by four basic activities: moving the trolley across the stacks, moving the gantry across the bays, picking up and placing down containers, and finally accelerating and decelerating the gantry, which is needed if bays are crossed. Depending on whether the time effort for pick-up/place-down is treated as a constant or depends on the number of tiers that are crossed in the vertical movement, two variants of measuring crane working time exist. Formal definitions of these variants, referred to as f_2 and $f_{2^{vert}}$, are given in Sect. 2.

For a particular solution, the value of $\hat{f}_1$ is determined by the number of all moves, i.e., relocations as well as retrievals are counted. That is, $f_1 = \hat{f}_1 - N$ and f_1 and $\hat{f}_1$ differ only by a constant, namely the number of retrievals N. In a combined approach, [10] develop a three-phase heuristic to minimize f_3, a combined objective being the weighted sum of $\hat{f}_1$ and f_2. The authors consider a 3-dimensional container yard, i.e. a block. Also with respect to a 3-dimensional yard, minimizing f_2, the crane working time with constant pick-up/place-down effort, is considered in [6]. A tree search heuristic is proposed and evaluated using the instances and parameter setting provided in [10]. For the 2-dimensional case, i.e. for a bay, [16] evaluate the objectives f_1 and f_2

separately. In a 2-dimensional setting, gantry operations (movement, accelera-tion, deceleration) can be neglected. An exact (branch and bound) approach is proposed together with a heuristic. The methods are tested using instances different from those given in [10]. Moreover, regarding the time consumption of the crane operations, the parameter setting of [16] gives a stronger weight to the pick-up/place-down operations, than it was chosen in [10]. The total distance traveled by a crane is considered in [5]. That is, objective $f_{2^{vert}}$, the crane work-ing time including tier-dependent pick-up/place-down, is minimized for a time effort of one time unit per traversed stack/tier in horizontal/vertical direction. The authors evaluate two exact approaches, one based on Dijkstra's shortest path algorithm and a second one implementing an A* search algorithm. Further, [1] study crane working time with constant pick-up/place-down effort. A heuristic based on the Good-Bad heuristic of [9] is proposed and evaluated. Related to stacking of slabs and coils in the steel industry, [13] consider the crane workload and apply f_2 in a 2-dimensional context. In an earlier work, [14] treat crane working time but neglect the distances between stacks during relocation, i.e., each relocation activity consumes a fixed amount of crane time.

The above mentioned references introduce crane working time as an alterna-tive objective and develop corresponding solution approaches. Moreover, those approaches are evaluated regarding their quality, i.e., computational times and optimality gaps are studied. In addition, some authors report the number of required relocations, i.e., refer to objective f_1. However, it is not discussed which influence the minimization of one objective has on the value of the alternative criteria. For instance, if one minimizes the number of relocations f_1, to which extent is the crane working time f_2 deviating from the optimal crane working time? This question leads to an analysis of the sensitivity with respect to the optimization of different criteria and gives insights regarding which trade-off one has to accept if the objective function is changed. A step into this direction is carried out in [12]. This work is mainly concerned with minimizing f_1, however, includes crane working times with tier-dependent pick-up/place-down effort into the design of a heuristic. In particular, when choosing a new stack for relocating a blocking container, a decision rule is applied that includes both, the priority numbers associated to the new stack as well as its location. The latter is used to penalize stack choices that lead to high crane working times. However, the method does not minimize $f_{2^{vert}}$ directly. The authors focus on a 3-dimensional yard and allow the consideration of multi-lift cranes. Moreover, they include a numerical study on the trade-off between $\hat{f}_1$ and $f_{2^{vert}}$. We refer to their results in our numerical study, see Sect. 3.

The goal of this paper is to provide an analysis and insights regarding the BRP under different objectives. After describing some implementation details, in particular regarding the modification of the mathematical model, simple but new complexity results for the BRP under f_2 and $f_{2^{vert}}$ are presented in Sect. 2. In Sect. 3, as the main focus of the paper, the sensitivity of the different objective functions is studied in numerical experiments for the case of a 2-dimensional stacking area providing as yet unknown insights. For this purpose, the objective

functions are evaluated using an exact approach from [18]. Note that we do not attempt to add towards theoretical issues in multicriteria decision making. Rather we argue towards (managerial) insights based on those issues which may lead towards an increased awareness and a rethinking of which objectives should be considered in realistic settings. Concluding remarks are provided in Sect. 4.

2   Implementation Details and Complexity

For solving the BRP under f_1, f_2, $f_{2^{vert}}$, and f_3, an exact approach is applied. For the BRP under f_1, mathematical formulations are available in literature and need not be (re-)developed. For these purposes the mathematical formulation (BRP-II-A) as proposed in [18] for f_1 is used and adapted to match the requirements of f_2, $f_{2^{vert}}$, and f_3. The formulation (BRP-II-A) includes the assumption (A1), i.e., only containers above the target container are allowed to be relocated. Thus, crane movements are only feasible to and from the target stack. We follow [10] in defining stack 0 as the outside area, e.g., referring to a truck or some other means of transport that transports a given container away from the bay.

We use the following notation and data. A bay is given with a width (number of stacks) W and a height of at most H tiers. N containers have to be retrieved in T periods. t_s indicates the time needed to cross one stack and t_{pp} gives the constant time effort for pick-up/place-down. In the case of tier-dependent pick-up and place-down effort, the time effort per tier for an empty spreader is given by t_{r0} and for a loaded spreader by t_{r1}. Later, $t_r = t_{r0} + t_{r1}$ is used.[1]

To compute the crane working time within (BRP-II-A), two variable classes are considered, namely x_{ijklnt} and y_{ijnt}. If $x_{ijklnt} = 1$, there is a relocation of container n from stack i, row j, to stack k, row l in period t. Period t ends when container t is retrieved. That is, the model includes one period per container and the number of periods is $T = N$. Moreover, $y_{ijnt} = 1$ indicates that container n is retrieved in period t from stack i, row j. Obviously, $y_{ijnt} = 1$ can only hold for $n = t$. Given these variables, the number of relocations is as follows:

$$f_1 = \sum_{i,k=1}^{W} \sum_{j,l=1}^{H} \sum_{n=1}^{N} \sum_{t=1}^{T} x_{ijklnt} \tag{1}$$

Moreover, the crane working time with constant pick-up/place-down effort is computed in (BRP-II-A) as:

$$f_2 = \sum_{i=1}^{W} \sum_{j=1}^{H} \sum_{t=1}^{T} (2t_s i + t_{pp}) y_{ijtt} + \sum_{i,k=1}^{W} \sum_{j,l=1}^{H} \sum_{n=1}^{N} \sum_{t=1}^{T} (2t_s |i-k| + t_{pp}) x_{ijklnt} \tag{2}$$

In (2) it is assumed that at the beginning the trolley is located at stack 0, i.e., in the outside area. For each period t, the trolley carries out the following

---

[1] Note that related to the 3-dimensional case, additional data t_b (time effort of the gantry) and t_{ad} (acceleration/deceleration of the gantry) is provided in the literature.

movements: It moves from stack 0 to the target stack, which is i if $y_{ijtt} = 1$ holds for some row j. In addition, after finishing some potential relocations, it moves back from stack i to stack 0 to retrieve the container t.[2] These two move activities consume $2t_s i + t_{pp}$ time units as they involve one pick-up/place-down activity as well as a movement across $2i$ stacks. Furthermore, in period t, relocation activities might take place, indicated by $x_{ijklnt} = 1$. If container n is relocated from stack i to stack k, then $(2t_s|i - k| + t_{pp})$ time units are consumed by the crane movement as in addition to the pick-up/place-down activity, the container crosses $2|i - k|$ stacks. Note that the trolley has to move back to stack i immediately after placing the container at stack k as relocations or retrievals are only possible from stack i according to assumption (A1). Note furthermore that it is not hard to get rid of the absolute values in (2) by determining the distance $|i - k|$ between stacks i and k for all combinations of i and k within a preprocessing step and providing these values as parameters.

On the other hand, if a tier-dependent pick-up/place-down effort is considered as part of the crane working time, additional parameters are required. Let h_{max} be that tier in which containers are moved horizontally, across the stacks. For practical issues, the tier above the uppermost stacking slot is chosen, i.e., $h_{max} = H + 1$. Even if a current stacking situation would allow movement of containers in lower tiers, we assume that a container is always elevated to tier $H + 1$, which maps the handling practices of container yards due to safety concerns. Moreover, h_{out} gives the height of the outside area, e.g., the truck on which the container is loaded. The crane working time with tier-dependent pick-up/place-down effort is computed as follows.

$$f_{2^{vert}} = \sum_{i=1}^{W}\sum_{j=1}^{H}\sum_{t=1}^{T}(2t_s i + (2h_{max} - j - h_{out})\,t_r)y_{ijtt}$$

$$+ \sum_{i,k=1}^{W}\sum_{j,l=1}^{H}\sum_{n=1}^{N}\sum_{t=1}^{T}(2t_s|i - k| + (2h_{max} - j - l)\,t_r)x_{ijklnt} \tag{3}$$

In (3), t_{pp} is replaced by tier-dependent expressions. A vertical spreader movement always starts at h_{max}. Picking up a container from tier j, it covers the distance $h_{max} - j$ twice, once empty and once loaded. Recall that $t_r = t_{r0} + t_{r1}$ sums the time effort of an empty and of a loaded spreader such that $(h_{max} - j)t_r$ is the time for a pick-up from tier j. Similarly, the place-down effort regarding tier l and regarding the outside area can be computed and $f_{2^{vert}}$ follows.

Finally, let $\bar{W} \geq 0$ be a weight for objective f_2, then the combined objective f_3 is given by:

$$f_3 = f_1 + \bar{W} f_2 \tag{4}$$

For the correct computation of f_2, $f_{2^{vert}}$, and consequently for f_3, it is necessary to add constraints to (BRP-II-A). For ease of exposition, the formulation

---

[2] This might be relaxed if we would define different BRP versions in the sense of having open versus closed BRP versions. This also opens up the discussion to relations of the BRP with older printed circuit board assembly research as, e.g., in [2].

from [18] is summarized in Appendix A, using the equation numbering therein. In (BRP-II-A), two constraints (see $(6a_z)$ and $(6b_z)$ in the Appendix) ensure a consistent relation between relocation, retrieval and position of containers. However, retrieval actions y_{ijTT} of the final period T are not covered by these constraints. For the evaluation of f_1, and thus for the (BRP-II-A), this is of no importance. However, for the computation of f_2 and f_{2vert} the following two constraints need to be included.

$$\sum_{i=1}^{W}\sum_{j=1}^{H} y_{ijTT} = 1 \tag{5}$$

$$y_{ijTT} \leq b_{ijTT} \qquad \forall i \in 1,\ldots,W, j \in 1,\ldots,H \tag{6}$$

Constraint (5) guarantees that container T is retrieved in period T. If (5) is not added to the constraints of (BRP-II-A) and if f_2 (or f_{2vert}) is minimized, then this final retrieval will be omitted due to cost reasons. Consequently, f_2 (f_{2vert}) will not be computed correctly. On the other hand, if f_1 is minimized, then variables y_{ijTT} do not account for the objective function.

Constraint (6) guarantees that in the final period T, a retrieval can only be carried out from a position that is not empty. If f_1 is minimized, then the values of y_{ijTT} are chosen arbitrarily in $\{0,1\}$ and the function value of f_2 (f_{2vert}) might not be computed correctly.

Summarizing, if the BRP is solved under minimization of f_2 or f_{2vert}, constraint (5) has to be added to the mathematical formulation. Moreover, to ensure a correct computation of f_2 (f_{2vert}) if f_1 is minimized, constraints (5) and (6) have to be added to the mathematical model.

For the BRP under f_1, [4] prove NP-hardness. This result directly carries over to the CRP in the 3-dimensional context, as the BRP is a particular case of the CRP. Let the *blocks relocation problem under crane working time* (BRP^{cwt}) be given as the BRP if the objective 'minimize the number of relocations', f_1, is substituted by the objective 'minimize the crane working time with constant pick-up/place-down effort' f_2. The complexity of the BRP^{cwt} is not yet clarified though the problem is treated in literature. We close this research gap in the following simple Lemma.

Lemma 1. *The BRP^{cwt} is NP-hard.*

Proof. Given a BRP instance, generate a BRP^{cwt} instance by fixing $t_s = 0$ and $t_{pp} = 1$. That is, regarding the crane time, each relocation and each retrieval accounts with value one, whereas move operations account with value zero. The crane time then equals the number of relocations plus the (fixed) number of retrievals, i.e., minimizing the crane time minimizes the number of relocations. Thus, any solution optimal for the BRP^{cwt} instance is optimal for the BRP instance, too. The NP-hardness of the BRP^{cwt} follows directly from the NP-hardness of the BRP [4].

Using the (BRP-II-A) notation, the relation between f_1 and f_2 described in the proof of Lemma 1 can be verified immediately. For $t_s = 0$ and $t_{pp} = 1$ one has that a solution optimal for f_1 minimizes also f_2, and vice versa:

$$f_2 = \sum_{i=1}^{W}\sum_{j=1}^{H}\sum_{t=1}^{T}(2t_s i + t_{pp})y_{ijtt} + \sum_{i,k=1}^{W}\sum_{j,l=1}^{H}\sum_{n=1}^{N}\sum_{t=1}^{T}(2t_s|i-k| + t_{pp})x_{ijklnt}$$

$$= \sum_{i=1}^{W}\sum_{j=1}^{H}\sum_{t=1}^{T}y_{ijtt} + \sum_{i,k=1}^{W}\sum_{j,l=1}^{H}\sum_{n=1}^{N}\sum_{t=1}^{T}x_{ijklnt}$$

$$= N + f_1 \ .$$

Analog to the definition of the BRP^{cwt}, let CRP^{cwp} be a modified CRP under the minimization of f_2. The NP-hardness of the CRP^{cwp} can be derived directly from Lemma 1 as BRP^{cwt} is a particular case of the CRP^{cwp}.

3 Computational Study

For investigating the sensitivity of BRP objective function values, the parameter setting is taken from [10] and [12]. [10] propose the following factors to model the time consumption of the four basic crane operations: Moving the trolley takes $t_s = 1.2$ s per crossed stack, moving the gantry takes $t_b = 3.5$ s per crossed bay, pick-up/place-down is considered as constant and takes $t_{pp} = 30$ s per rehandling activity (relocation or retrieval). Moreover, for tier-dependent pick-up/place-down effort, [12] define the time effort per tier of a spreader as $t_{r0} = 2.59$ if it is empty and $t_{r1} = 5.18$ if loaded. In addition, the height of a truck is specified as 0.5. As the ground level is given by tier 1 in our setting, we define the height of the outside area, which is determined by a truck standing on ground level, as $h_{out} = 1.5$. Acceleration and deceleration takes $t_{ad} = 40$ s and applies if the gantry is moved, i.e., if bays have to be crossed. As the BRP is based on a 2-dimensional stacking area, the parameters t_b and t_{ad} can be neglected for our purposes. Moreover, to evaluate the weighted sum of $\hat{f}_1$ and f_2, [10] apply the weight $\bar{W} = 1$ for f_2 and we adopt this setting to obtain $f_3 = f_1 + f_2$. Note that this objective differs by a constant from that of [10], who include $\hat{f}_1 = f_1 + N$ into their weighted objective $\hat{f}_3 = \hat{f}_1 + \bar{W}f_2$. As f_1 is the standard BRP objective, here we apply f_3 instead of $\hat{f}_3$. This is feasible, as constant values can be omitted for the optimization of an objective function.

The numerical tests are based on our implementation of the model (BRP-II-A) in [18] in IBM ILOG CPLEX 12.6. A time limit of 7200 s has been fixed. The preprocessing steps described in Sect. 4.2 of [18] have been included. However, the computation of upper bounds UB_t on the number of relocations has been neglected for our purpose. Thus, constraint (B) from [18] is set inactive. Computations have been carried out on a Linux Server with Intel Xeon Processor X5570 and 32 GB RAM. The tests are based on the instances provided by [4]. These instances are described by $H' - W$ for which W gives the number of stacks and H' the number of containers per stack. Each stack has two empty slots such that the height of the bay is $H = H' + 2$. Each instance set contains 40 instances which give different initial bay layouts of $N = W \cdot H'$ containers. Our experiments include 12 instance sets with a total of 480 instances.

Each instance is solved separately under objectives f_1, f_2, f_{2vert}, and f_3. Let f_1^*, f_2^*, f_{2vert}^*, and f_3^* be the respective optimal objective function values. Assume that the number of relocations f_1 is minimized. In this case, the crane times f_2 and f_{2vert} are evaluated and the deviation from the corresponding optimal objective function value is computed. Vice versa, the same is done when minimizing crane working times. Let $\bar{f}_k^m$ be the value of objective f_k if objective f_m is minimized. For instance, if f_1 is minimized, let $\bar{f}_2^1$, $\bar{f}_{2vert}^1$ and $\bar{f}_3^1$ be the function values obtained for f_2, f_{2vert} and f_3, respectively. Obviously, $\bar{f}_m^m = f_m^*$ holds for $m = 1, 2, 3$.

A first observation from the experiments is that for each instance, $\bar{f}_2^3 = f_2^*$ and $\bar{f}_3^2 = f_3^*$ is satisfied. That is, given the chosen parameter setting and instances, f_2 does not deviate from its optimal value if f_3 is optimized and vice versa. Due to this observation, the results for f_3 are omitted subsequently. It is important to note that this observation does not hold generally. Parameter settings or instances can be found where the objectives f_2 and f_3 lead to different optimal solutions, for instance, if the weight $\bar{W}$ is chosen very small.

The first set of experiments evaluates f_1 against f_2. From a total of 480 instances, only a single one has not been solved to optimality within the time limit of two hours for at least one objective function. This instance is excluded and the computational results are presented for the remaining 479 instances in Table 1.

Table 1. Optimal solution and deviation under alternative objective, f_1 vs. f_2.

Instance		f_1^*			$\bar{f}_1^2 - f_1^*$		f_2^*			$\bar{f}_2^1 - f_2^*$		
$H' - W$	solv.	avg	min	max	pos	max	avg	min	max	pos	avg	max
3–3	40	5.00	0	8	0	0	476.40	313.20	577.20	13	2.28	14.40
3–4	40	6.18	2	10	0	0	633.57	496.80	760.80	26	7.92	24.00
3–5	40	7.03	3	11	0	0	789.15	648.00	926.40	34	16.32	33.60
3–6	40	8.40	3	14	0	0	968.16	795.60	1178.40	39	22.44	62.40
3–7	40	9.28	6	14	0	0	1136.97	1016.40	1304.40	40	35.76	91.20
3–8	40	10.65	8	16	1	1	1331.25	1228.80	1512.00	40	43.29	115.20
4–4	40	10.20	5	14	0	0	911.88	730.80	1044.00	32	10.08	33.60
4–5	40	12.95	8	17	0	0	1170.18	1003.20	1326.00	36	19.80	48.00
4–6	40	14.03	8	18	0	0	1385.43	1180.80	1548.00	40	31.44	67.20
4–7	40	16.13	10	22	1	1	1647.66	1442.40	1840.80	40	53.01	139.20
5–4	40	15.43	7	21	0	0	1230.03	949.20	1422.00	37	16.08	33.60
5–5	39	18.69	12	24	0	0	1555.51	1323.60	1717.20	38	23.51	57.60
Total	479	11.15	0	24	2	1	1102.07	313.20	1840.80	415	23.49	139.20

The results are provided separately for each instance set and for the full set of instances. The first column gives the description of the instance sets denoted by '$H' - W$'. Column 'solv.' gives the number of instances that have been solved to optimality for each objective function within the given time limit of two hours per run. Columns 3–5 give for each instance set average, minimal, and maximal number of relocations in the optimal case, i.e., if the total number of relocations f_1 is minimized. Columns 6–7 present results for f_1 if f_2 is minimized. More detailed, $\bar{f}_1^2 - f_1^*$ gives the increase in the number of relocations (with respect

to the minimal number of relocations) if the crane working time is minimized. Column 6, 'pos', gives the number of instances for which an increase is reported; Column 7, 'max', gives the maximum increase. Note that the minimal increase of the number of relocations is 0 as there are instances without increase, see Column 6. Moreover, the average value of $\bar{f}_1^2 - f_1^*$ over all instances is 0.004 and is not reported separately for each instance set here. From Column 6 we have that minimizing crane times leads to an increased number of relocations for just two out of 479 instances. In these two cases, only a single extra relocation is required; see Column 7. That is, for the chosen parameter setting and instances, minimizing f_2 leads to optimal results regarding f_1 in almost all cases.

The relation between $\bar{f}_2^1$ and f_1^* is depicted in Fig. 1. Each point in the scatterplot relates to a single instance, shades of grey indicate the instance sets. Figure 2 gives a detailed view for the particular instance set 3–4. Both figures illustrate that the objectives f_1 and f_2 are pointing to some extent into the same direction. The same observation holds for the other instance sets.

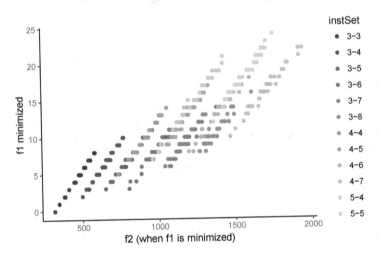

Fig. 1. Crane working time $\bar{f}_2^1$ vs. minimal number of relocations f_1^*, grouped by instance sets.

For each instance set columns 8–10 give the average, minimal, and maximal crane working time in the optimal case, i.e., if f_2 is minimized. Columns 11–13 present results for f_2 if f_1 is minimized. In more detail, $\bar{f}_2^1 - f_2^*$ gives the increase in crane working time (over the minimal crane working time) if the number of relocations is minimized. Column 11 gives the number of instances for which $\bar{f}_2^1 - f_2^*$ is positive, Columns 12 and 13 give average and maximum values. Note that the minimum value of $\bar{f}_2^1 - f_2^*$ over all instances is 0 as there are 64 instances without increase, see Column 11. Figure 3 gives boxplots of $\bar{f}_2^1 - f_2^*$, grouped by instance sets $H' - W$. Figure 3 illustrates that in these experiments, given a fixed H', the means of $\bar{f}_2^1 - f_2^*$ increase with increasing W. That is, the observed impact on the crane working times grows with increasing bay width.

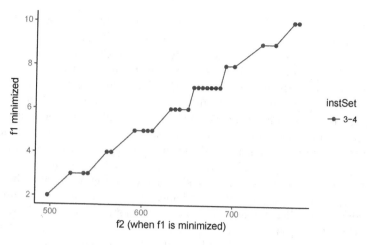

Fig. 2. Crane working time $\bar{f}_2^1$ vs. minimal number of relocations f_1^*, instance set 3–4.

From Column 11 it can be observed that in these experiments, for 415 out of 479 instances, an increased crane working time has to be accepted if the number of relocations is minimized. On average an increase of 23.5 s is observed and the maximal increase is 139.2 s. Thus, changing the objective function from min f_1 to min f_2 improves crane working times in the majority of cases. On the other hand, in these cases, the number of relocations are not affected strongly, see the observations regarding Columns 6 and 7 above.

Summarizing, for the chosen parameter setting [10], the reported deviation of one objective from its optimal value under minimization of the alternative objective is small. In the previous experiments, f_1 increases by at most 8.33% when f_2 is minimized and, vice versa, f_2 increases by at most 8.36% if f_1 is minimized. However, this does not hold in general. For alternative parameter settings results may be different; see an example under varied t_{pp} in Table 4.

Our results indicate that considering crane working times is recommended to exploit the potential improvement. In our tests f_2 can be improved without worsening f_1 in 86% of all instances. In these cases multiple optimal solutions exist for min f_1. This is illustrated in more detail in Tables 2 and 3. For these experiments, CPLEX is configured to provide all optimal solutions by using the populate() method. The related parameters have been fixed as follows.

– Limit the number of generated solutions and the size of the solution pool: populatelim=solnpoolcapacity=2,100,000,000;
– Guide the search effort: solnpoolintensity=4, i.e., generate all solutions;
– Absolute and relative tolerance for deviation from optimality: solnpoolagap=solnpoolgap=0.000001. The last parameters have been chosen slightly greater than zero in order to grasp also optimal solutions that are produced with rounding errors.

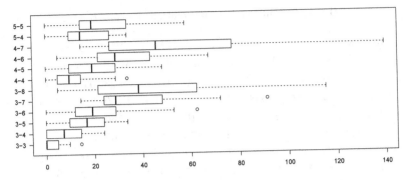

Fig. 3. Increase of crane working time if f_1 is minimized ($\bar{f}_2^1 - f_2^*$), grouped by instance sets.

Moreover, to avoid memory overrun, a time limit of 3600 s is fixed. Columns 1–3 provide for each instance set its name $H' - W$ and the number of containers N, as well as the number of instances for which the `populate()` method terminates within 3600 s. Columns 4–6 give average, minimum, and maximum values of the number of optimal solutions detected. Column 4 reveals that instances with unique optimal solutions exist. Moreover, already for small instances with nine containers, up to 40 optimal solutions are found. At most 3477 optimal solutions are detected for an instance with $H' = 3$ and $W = 6$. A second observation is that the average and maximum number of optimal solutions is not strictly increasing with the number of containers N; compare, e.g., the values for $N = 15$ and $N = 16$. Rather, for a fixed height H', the average and maximal number of optimal solutions grow with increasing bay width W. This effect is depicted in the boxplots in Fig. 4 and points out the impact of W. A similar observation regarding the increase of the crane working time when minimizing f_1 is derived from Fig. 3. Column 7 gives the average optimal crane working time f_2^*. Note that these values differ in some cases from the figures given in Table 1 as the set of instances varies for which the methods `populate()` or `solve()` terminated within the allotted time. Finally, Columns 8–10 give average, minimum and maximum of $f_2^{max} - f_2^*$, where f_2^{max} denotes the maximal value of f_2 that is obtained when searching all optimal solutions for f_1. That is, f_2^{max} gives for any instance the worst crane working time that might be realized when minimizing solely the number of relocations. Consequently, $f_2^{max} - f_2^*$ gives for any instance the maximal deviation of a crane working time from the optimal value when f_1 is minimized. This maximal deviation per instance amounts to 21.6 s on average and to at most 76.8 s. The minimal (over all instances) of the maximal deviation is given in Column 9. The 0 entries indicate that there are instances for which all solutions that are optimal for f_1 are also optimal for f_2. Values f_2^{min}, i.e., the shortest crane working time found when minimizing f_1, equal the optimal value f_2^* for all instances and are therefore not included in Table 2. That is, for each observed instance, there is an optimal solution that is optimal for both objective functions, f_1 and f_2.

Table 2. Multiple optimal solutions for min f_1.

Instance			# opt. sol.			f_2^*	$f_2^{max} - f_2^*$		
$H' - W$	N	solv.	Avg	Min	Max	avg	avg	min	max
3–3	9	40	6.10	1	40	476.40	5.28	0.00	14.40
3–4	12	40	53.08	1	580	633.57	15.00	0.00	38.40
3–5	15	40	314.95	4	1416	789.15	27.60	0.00	57.60
3–6	18	30	680.53	12	3477	938.80	43.68	9.60	76.80
4–4	16	40	144.23	1	1310	911.88	21.96	0.00	48.00
Total		190	216.58	1	3477	740.02	21.60	0.00	76.80

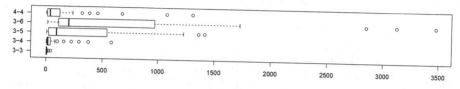

Fig. 4. Number of optimal solutions, grouped by instance sets, for min f_1.

Table 3 provides information on multiple optimal solutions found when minimizing f_2. Parameters in CPLEX are fixed as described for the experiments regarding Table 2. The structure of the columns in Table 3 resembles that of Table 2. Similarly, the average optimal number of relocations f_1^* in Column 7 may differ from those reported in Table 1. Columns 4–6 indicate that the number of optimal solutions when minimizing the crane working time is less than the number of optimal solutions when minimizing the number of relocations. Moreover, for each studied instance class there are instances with unique optimal solutions, indicated by the entries in Column 5. Figure 5 depicts boxplots of the number of optimal solutions (for min f_2), grouped by instance sets. There is one major difference in the structure of Tables 2 and 3 regarding the deviation in the non-optimized criteria. The values of $f_1^{max} - f_1^*$, i.e., the maximal deviation of the number of relocations from its optimal value when f_2 is minimized, is given only as maximum over all instances. These maxima equal 0 for all instance sets (and thus, averages and minima are 0, too). That is, for all studied instances, all solutions found when minimizing f_2 are optimal for f_1, too. Note that this is not true in general, as for larger instance sets there are three instances found in Table 1 for which the number of relocations increases when f_2 is minimized.

Note that the presented results for f_1 and f_2 do not hold generally. For an alternative parameter setting, different results may be obtained. In the current parameter setting, pick-up/place-down has a high time consumption ($t_{pp} = 30$ s) compared to trolley movement ($t_s = 1.2$ s). Thus, additional relocations do increase the crane working time a lot and it seems likely that minimizing the crane working time leads to good solutions regarding the number of relocations. For illustrating results under an alternative parameter setting, Table 1 is

Table 3. Multiple optimal solutions for min f_2.

Instance			# opt. sol.			f_1^*	$f_1^{max} - f_1^*$
$H' - W$	N	solv.	Avg	Min	Max	avg	max
3–3	9	40	2.18	1	8	5.00	0
3–4	12	40	4.40	1	26	6.18	0
3–5	15	40	12.45	1	126	7.03	0
3–6	18	40	25.23	1	650	8.40	0
4–4	16	40	3.43	1	12	10.20	0
Total		200	9.54	1	650	7.36	0

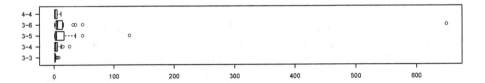

Fig. 5. Number of optimal solutions, grouped by instance sets, for min f_2.

Table 4. Optimal solution and deviation under alternative objective for $t_{pp} = 5$.

Instance		f_1^*			$\bar{f}_1^2 - f_1^*$		f_2^*			$\bar{f}_2^1 - f_2^*$		
$H' - W$	solv.	avg	min	max	pos	max	avg	min	max	pos	avg	max
3–3	40	5.00	0	8	0	0	126.40	88.20	152.20	13	2.28	14.40
3–4	40	6.18	2	10	6	2	178.39	146.80	211.00	31	8.73	24.00
3–5	40	7.03	3	11	15	1	236.56	198.00	271.80	36	18.29	38.20
3–6	40	8.40	3	14	18	3	303.24	261.20	355.00	39	27.37	62.40
3–7	40	9.28	6	14	20	2	375.16	341.40	429.40	40	40.70	91.20
3–8	40	10.65	8	16	28	3	455.12	428.80	502.60	40	53.17	115.20
4–4	40	10.20	5	14	10	2	255.50	205.80	294.00	36	11.46	38.20
4–5	40	12.95	8	17	17	2	342.95	303.20	382.20	39	23.28	48.00
4–6	40	14.03	8	18	24	3	428.49	380.80	469.60	40	37.76	76.40
4–7	40	16.13	10	22	30	4	532.77	473.40	590.80	40	64.78	143.80
5–4	40	15.43	7	21	8	2	343.50	274.20	406.40	37	16.99	33.60
5–5	37	18.51	12	24	23	3	455.46	398.60	496.60	36	29.68	57.60
Total	477	11.10	0	24	199	4	335.38	88.20	590.80	427	27.86	143.80

reproduced under a modified pick-up/place-down effort $t_{pp} = 5.0$. The trolley movement time per stack $t_s = 1.2$ remains unchanged. With this alternative parameter setting, the cost of extra relocations is reduced which leads to different properties of the optimal solutions. The results are given in Table 4.

First, optimal objective function values when minimizing the crane working time f_2^* are on average smaller than those reported in Table 1. Clearly, this is

related to the pick-up/place-down effort t_{pp} which is reduced by 25 s. Second, the columns reporting the deviation of the objective function values show different results for the alternative parameter setting. Regarding the number of relocations when minimizing crane working time ($\bar{f}_1^2 - f_1^*$) it can be observed that in 41.7% of all cases, the solutions are not optimal for f_1. Up to four additional relocations are reported, see Column 7, in relative figures, a maximal increase of f_1 by 25% is reported. Thus, as pick-up/place-down is not that time consuming anymore, minimizing f_2 does not necessarily have a positive impact on f_1. Concerning the deviation of f_2 from the optimum if f_1 is minimized, see Columns 11–13, a positive increase of crane working time is reported for a slightly larger number of instances. Also the average and maximum increase of crane working time is higher than that reported in Table 1. At most, a relative increase of 26.03% is reported for f_2.

To understand the impact of the tier-dependent pick-up/place-down effort, the experiments are carried out for the objective f_{2vert}. The structure of Table 5 is copied from Table 1. From 480 instances, 23 have not been solved to optimality within the time limit of two hours for at least one objective and, therefore, are excluded from the presentation.

Table 5. Optimal solution and deviation under alternative objectives, f_1 vs. f_{2vert}.

Instance		f_1^*			$\bar{f}_1^{2vert} - f_1^*$		f_{2vert}^*			$\bar{f}_{2vert}^1 - f_{2vert}^*$		
$H' - W$	solv.	avg	min	max	pos	max	avg	min	max	pos	avg	max
3–3	40	5.00	0	8	1	1	904.03	637.61	1060.85	23	16.35	68.10
3–4	40	6.18	2	10	7	2	1186.96	972.18	1406.16	36	40.20	102.84
3–5	40	7.03	3	11	10	4	1460.77	1254.08	1684.64	37	61.57	169.81
3–6	40	8.40	3	14	22	3	1758.37	1447.65	2081.37	39	89.80	196.08
3–7	40	9.28	6	14	17	3	2043.44	1846.79	2339.51	40	124.69	254.83
3–8	40	10.65	8	16	25	3	2362.62	2219.76	2584.50	40	147.25	300.30
4–4	40	10.20	5	14	9	2	1928.13	1664.40	2198.94	35	67.58	185.34
4–5	40	12.95	8	17	18	2	2424.61	2100.90	2640.24	39	131.57	272.64
4–6	37	13.76	8	17	24	4	2817.61	2495.58	3027.84	37	169.13	322.92
4–7	33	15.30	10	19	28	5	3295.69	3012.30	3418.86	33	214.37	396.96
5–4	39	15.28	7	20	13	3	2866.95	2345.88	3209.28	39	109.41	233.10
5–5	28	17.64	12	22	19	5	3494.28	3213.74	3705.31	28	177.71	320.40
Total	457	10.70	0	22	193	5	2156.27	637.61	3705.31	426	108.83	396.96

Table 5 indicates that the inclusion of tier-dependent pick-up/place-down time into the crane working times changes the relation of the objective functions. In 193 out of 457 solved instances, the number of relocations increases if the crane working time f_{2vert} is minimized. At most five extra relocations are required. Recall from Table 1 that for the constant pick-up/place-down effort, an extra relocation had been necessary only for three instances. Thus, minimization of crane working time with tier-dependent pick-up/place-down leads in the majority of cases to solutions that are optimal regarding the number of

relocations. However, the conclusion derived from Table 1, that minimizing the crane working time leads in almost all cases to an optimal scenario regarding the number of relocations, does not hold anymore for the inclusion of tier-dependent pick-up/place-down effort. One reason for this observation is that under f_{2vert}, relocations in the upper tiers are cheap, whereas relocations regarding the bottom tiers are expensive. Thus, when stacks are high, extra relocations can be meaningful if relocations in low stacks can be prevented.

Regarding the alternate case, the increase of crane working time if the number of relocations is minimized, both variants of crane working time behave similar. In Table 5, 426 out of 457 instances are found where the crane working time with tier-dependent pick-up/place-down effort is increasing if the number of relocations is minimized. In the previous experiments with constant pick-up/place-down effort, crane working time increased in 415 out of 479 instances; see Table 1. Comparing Tables 1 and 5, it can be noticed that the crane working time is on average higher if pick-up/place-down effort is tier-dependent. This difference stems from the choice of parameters where the constant time effort of $t_{pp} = 30$ is comparatively cheap, given $t_r = t_{r0} + t_{r1} = 7.77$ and stacking areas with a height of up to $H = H' + 2 = 7$.

A related study is provide in [12] where the trade-off between the number of relocations and the crane working time is investigated numerically. The presented model relates to a 3-dimensional yard and thus different instance sets are used. A heuristic is applied including penalties for relocation policies that increase the crane working time. More specifically, for each movement, a decision rule is applied including the number of traversed bays as well as the distance of the destination stack to the outside area. The results indicate that for some range of penalty factors, the number of relocations increases with increasing penalty, and, the total crane working time (including tier-dependent pick-up/place-down effort) decreases. However, for very large penalty values, both indicators, the number of relocations as well as crane working time, increase.

The result reported in [12] is not fully reflected by the outcome given in Table 1 where in almost all cases a decrease of crane time f_2 is possible without increasing f_1. However, the results for the alternative parameter setting, given in Table 4, and the results for crane working times including tier-dependent pick-up/place-down effort, f_{2vert}, see Table 5, show a similar trade-off between number of relocations and crane times. Note that, besides the impact of the problem setting, e.g., concerning the yard dimension, parameter setting, and the choice of a crane working time variant, the results of [12] and those of our experiments are comparable only to some extent. To address large instances, [12] apply a heuristic that considers crane times separately for each movement within a decision rule. Thus, the heuristic does not directly minimize f_2; rather it indirectly influences f_2 by penalizing long distances in container movement. On the other hand, in order to study the interaction between objective functions precisely, we consider an exact approach that directly addresses f_2, such that in detail, different outcomes are possible.

Summarizing, given the results presented in Table 1, for a constant pick-up/place-down effort and the parameter set taken from literature [10], a strategy could be to minimize only the crane working time as in most cases, good results for the number of relocations are obtained. However, for a changed parameter set or for the consideration of tier-dependent pick-up/place-down effort, one can not rely on such a basic strategy as the relation between the two objectives is not that simple anymore. Thus, minimizing the crane working time with a crosscheck regarding the number of relocations is recommended.

Moreover, the number of relocations may be of interest beyond its impact on the crane time, e.g., if frequent movement of items shall be prevented due to the fragility of goods. In this case, to include two objective functions into the considerations, methods from multicriteria optimization could be applied. For instance, lexicographic ordering allows to exploit the potential of both objectives by considering them in a hierarchical fashion.

4   Conclusions

In literature different objective functions have been considered for the blocks relocation problem. This work, to the best of our knowledge, is the first comprehensive and in depth study on the relationship of different objectives for the BRP once applied to the same problem instances. This is important as the literature on the BRP and related problems has been growing considerably in recent years without putting emphasis on this. That is, to understand the sensitivity of the BRP's objective functions, we raised the question to which extent the change of the objective function influences the function value of the initial objective. To that end, we added two variants of an alternative objective 'minimize crane working time' together with necessary constraints to a given mathematical formulation. We carried out computational experiments based on instance sets and parameter settings from literature. Furthermore, as a simple exercise we resolved the complexity status of the BRP if the crane working time with constant pick-up/place-down effort is minimized.

The computational experiments indicated that the number of relocations is not much affected by the change of the objective function if constant pick-up/place down effort is considered. In this case, minimizing the crane working time yields good results also with respect to the number of relocations. If the crane working time with constant pick-up/place down effort is minimized, for a majority of the instances, solutions turned out to be optimal for both objective functions. In particular, for almost all instances tested under the close-to-reality parameter setting of [10], there exist solutions that solve both criteria to optimality. Moreover, if an increase of the number of relocations was detected, this increase was small. However, this observation is not true in the opposite direction. If the number of relocations is minimized, very often solutions are found that are not optimal with respect to crane working time. While this could be rated as somewhat expected by specialists in multicriteria decision making, it is important for transportation research and operations research specialists

focussing on maritime shipping and container terminals. Consequently, as a lesson learned for the latter especially for real-world applications, our results indicate that crane working time should be included into the considerations as solutions optimal with respect to the number of relocations might have potential to improve the crane working time without increasing the number of relocations. Practically spoken, solutions that allow for a minimum number of relocations, for instance, might be further improved by choosing stacks for relocation that reduce the movement of cranes. While this may be to some extent intuitive for specialists in multicriteria decision making, it seems not for those decision makers dealing with problems in maritime shipping and stacking logistics.

Our results strongly recommend the inclusion of crane working times as they are relevant for improving operational decisions at ports from a practical point of view. Further, the results motivate to use alternative objectives in container rehandling as they may help to exploit potentials to improve given solutions further regarding different criteria. (However, we do not recommend to have dozens of papers copying existing BRP-algorithms for the modified objectives).

In the numerical study, smaller instances with up to 28 containers have been included. For future research, an interesting open question is whether the conclusions derived for these instances, beyond intuition, carry over to larger instance sizes (and we strongly expect that they do). Due to the NP-hardness of the BRP and the BRP^{cwt}, the instance size is limited when applying exact methods. Therefore, a related analysis based on heuristic solution methods is part of our future research [8]. While the BRP considers a single bay with given data, future research should incorporate the investigation of cases where the data is uncertain or may change over time (e.g., when the priority of a container changes on a short notice). One should also consider interference issues in cases where not only one crane is considered but when double or triple rail mounted gantry cranes are operating a large block with several bays. Future research should focus not only on the BRP but also on other types of problems like the premarshalling problem (for a recent solution method see [15]) and investigate to which extent different objectives make sense (see, e.g., the critical assessment of an earlier approach for this problem in [17]). Moreover, multiobjective methods could be applied to study the sensitivity of the objective functions in more detail, for instance, by producing the Pareto front.

A   Mathematical Model

To make this paper self-contained we repeat the mathematical model of [18] with modifications as described above. Whenever applicable, the numbering of constraints and equations follows the original appended by a small z for distinction.

Parameters

t_s	time consumed for traversing one stack with the trolley
t_{pp}	time consumed for pick up/place down of a container
$t_r = t_{r0} + t_{r1}$	cumulated time consumed for traversing one tier with the spreader, once loaded and once empty
h_{max}	tier in which container is moved horizontally to cross stacks
h_{out}	height of the outside area
M	Large number

Variables

$$b_{ijnt} = \begin{cases} 1 & \text{if block } n \text{ is at stack } i \text{ and row } j \text{ at the beginning of period } t, \\ 0 & \text{otherwise;} \end{cases}$$

$$\forall i = 1, \ldots, W, j = 1, \ldots, H, t = 1, \ldots, T, n = t, \ldots, N$$

$$x_{ijklnt} = \begin{cases} 1 & \text{if block } n \text{ is relocated from stack } i, \text{ row } j \text{ to stack } k, \text{ row } l \text{ in period } t, \\ 0 & \text{otherwise;} \end{cases}$$

$$\forall i = 1, \ldots, W, j = 1, \ldots, H, k = 1, \ldots, W, l = 1, \ldots, H, t = 1, \ldots, T-1, n = t+1, \ldots, N$$

$$y_{ijtt} = \begin{cases} 1 & \text{if block } n = t \text{ is retrieved from stack } i \text{ and row } j \text{ in period } t, \\ 0 & \text{otherwise;} \end{cases}$$

$$\forall i = 1, \ldots, W, j = 1, \ldots, H, t = 1, \ldots, T$$

Objectives

Objective 1: *Minimize number of relocations*

$$\min f_1 = \sum_{i,k=1}^{W} \sum_{j,l=1}^{H} \sum_{n=1}^{N} \sum_{t=1}^{N} x_{ijklnt}$$

Objective 2: *Minimize crane time*

$$\min f_2 = \sum_{i=1}^{W} \sum_{j=1}^{H} \sum_{t=1}^{T} (2t_s i + t_{pp}) y_{ijtt} + \sum_{i,k=1}^{W} \sum_{j,l=1}^{H} \sum_{n=1}^{N} \sum_{t=1}^{T} (2t_s |i - k| + t_{pp}) x_{ijklnt}$$

Objective 2^{vert}: *Minimize crane time with tier-dependent pick-up/place-down effort*

$$\min f_{2^{vert}} = \sum_{i=1}^{W} \sum_{j=1}^{H} \sum_{t=1}^{T} (2t_s i + (2h_{max} - j - h_{out}) t_r) y_{ijtt}$$

$$+ \sum_{i,k=1}^{W} \sum_{j,l=1}^{H} \sum_{n=1}^{N} \sum_{t=1}^{T} (2t_s |i - k| + (2h_{max} - j - l) t_r) x_{ijklnt}$$

Objective 3: *Minimize weighted sum of objectives 1 and 2*

$$\min f_1 + \bar{W} f_2$$

Constraints and Preprocessing
Constraints as given in [18] (they need to be appended by Constraints (5) and (6) from Sect. 2):

$$\sum_{n=t}^{N} b_{ijnt} \leq 1 \quad \forall i = 1, \ldots, W, j = 1, \ldots H, t = 1, \ldots, T-1 \tag{2_z}$$

$$\sum_{n=t}^{N} b_{ijnt} \geq \sum_{n=t}^{N} b_{ij+1nt} \quad \forall i = 1, \ldots, W, j = 1, \ldots H-1, t = 1, \ldots, T \tag{3_z}$$

$$b_{ijnt+1} = b_{ijnt} + \sum_{k=1}^{W}\sum_{l=2}^{H} x_{klijnt} - \sum_{k=1}^{W}\sum_{l=1}^{H} x_{ijklnt}$$
$$\forall i = 1, \ldots, W, j = 1, \ldots H, t = 1, \ldots, T-1, n = t+1, \ldots, N \tag{$6a_z$}$$

$$b_{ijnt} - y_{ijtt} = 0$$
$$\forall i = 1, \ldots, W, j = 1, \ldots H, t = 1, \ldots, T-1, n = t \tag{$6b_z$}$$

$$\sum_{i=1}^{W}\sum_{j=1}^{H} y_{ijtt} = 1 \quad \forall t = 1, \ldots, T-1 \tag{$7''_z$}$$

$$M \cdot \left(1 - \sum_{n=t+1}^{N} x_{ijklnt}\right) \geq \sum_{n=t+1}^{N}\sum_{j'=j+1}^{H}\sum_{l'=l+1}^{H} x_{ij'kl'nt}$$
$$\forall i = 1, \ldots, W, j = 2, \ldots H-1, k = 1, \ldots, W, l = 1, \ldots H-1, t = 1, \ldots, T-1 \tag{$8'_z$}$$

$$\sum_{j'=1}^{j-1} y_{ij'tt} \geq \sum_{k=1}^{W}\sum_{l=1}^{H}\sum_{n=t+1}^{N} x_{ijklnt} \quad \forall i = 1, \ldots, W, j = 2, \ldots H, t = 1, \ldots, T-1 \tag{A'_z}$$

(i_n, j_n) Position of container n in the initial bay

π_n Time period when n has to be relocated the first time, i.e.,
$\pi_n = \min n'$ such that n' is placed at any position $(i_n, 1)$ to (i_n, j_n)

$b_{i_n j_n nt} = 1$ $\forall n = 1, \ldots, N, t = 1, \ldots, \pi_n$

$b_{ijnt} = 0$ $\forall n = 1, \ldots, N, i \neq i_n, j \neq j_n, t = 1, \ldots, \pi_n$

$b_{i_n j_n n't} = 0$ $\forall n = 1, \ldots, N, n' \neq n, t = 1, \ldots, \pi_n$

$y_{i_n j_n tt} = 0$ $\forall n = 1, \ldots, N, t = 1, \ldots \pi_n - 1$

$x_{i_n j_n kln't} = 0$ $\forall n = 1, \ldots, N, n' \neq n, k = 1, \ldots, W, l = 1, \ldots, H, t = 1, \ldots, \pi_n$

$x_{iji_n j_n n n't} = 0$ $\forall n = 1, \ldots, N, n' \neq n, i = 1, \ldots, W, j = 1, \ldots, H, t = 1, \ldots, \pi_n$

$x_{ijklnt} = 0$ $\forall n = 1, \ldots, N, i, k = 1, \ldots, W, j, l = 1, \ldots, H, t = 1, \ldots, \pi_n - 1$

$x_{ijkln\pi_n} = 0$ $\forall n = 1, \ldots, N, i \neq i_n, j \neq j_n, k = 1, \ldots, W, l = 1, \ldots, H$

$x_{ijklnt} = 0$ $\forall n = \pi_n, i, k = 1, \ldots, W, j, l = 1, \ldots, H, t = 1, \ldots, n - 1$

$y_{ijnn} = 0$ $\forall n = \pi_n, i \neq i_n, j \neq j_n$

$x_{ijkln'n} = 0$ $\forall n = \pi_n, i \neq i_n, j, l = 1, \ldots, H, k = 1, \ldots, W, n' = 1, \ldots, N$

$$x_{i_n jkln'n} = 0 \qquad \forall n = \pi_n, j = 1, \ldots, j_n, k = 1, \ldots, W, l = 1, \ldots, H, n' = 1, \ldots, N$$

$$b_{i_n jn'n+1} = 0 \qquad \forall n = \pi_n : n < T, j = j_n, \ldots, H, n' = 1 \ldots, N$$

$$y_{i_n jn+1n+1} = 0 \qquad \forall n = \pi_n : n < T, j = j_n, \ldots, H$$

$$x_{i_n jkln'n+1} = 0 \qquad \forall n = \pi_n : n < T, j = j_n, \ldots, H, k = 1, \ldots, W, l = 1, \ldots, H, n' = 1, \ldots, N$$

$$b_{ijnt} = 0 \qquad \forall t = 1, \ldots, T, i = 1, \ldots, W, j = N_t + 1, \ldots, H, n = 1, \ldots, N$$

$$y_{ijt} = 0 \qquad \forall t = 1, \ldots, T, i = 1, \ldots, W, j = N_t + 1, \ldots, H$$

$$x_{ijklnt} = 0 \qquad \forall t = 1, \ldots, T, i, k = 1, \ldots, W, j = N_t + 1, \ldots, H, l = 1, \ldots, H, n = 1, \ldots, N$$

$$x_{ijklnt} = 0 \qquad \forall t = 1, \ldots, T, i, k = 1, \ldots, W, j = 1, \ldots, H, l = N_t, \ldots, H, n = 1, \ldots, N$$

$$x_{ijilnt} = 0 \qquad \forall i = 1, \ldots, W, j = 1, \ldots, H, l = 1, \ldots, H, n = 1, \ldots, N, t = 1, \ldots, T$$

References

1. Azari, E., Eskandari, H., Nourmohammadi, A.: Decreasing the crane working time in retrieving the containers from a bay. Sci. Iranica Trans. E Ind. Eng. **24**(1), 309–318 (2017)
2. Bard, J., Clayton, R., Feo, T.: Machine setup and component placement in printed circuit board assembly. Int. J. Flex. Manuf. Syst. **6**(1), 5–31 (1994)
3. Caserta, M., Schwarze, S., Voß, S.: Container rehandling at maritime container terminals. In: Böse, J.W. (ed.) Handbook of Terminal Planning. Operations Research/Computer Science Interfaces Series, pp. 247–269. Springer, New York (2011). https://doi.org/10.1007/978-1-4419-8408-1_13
4. Caserta, M., Schwarze, S., Voß, S.: A mathematical formulation and complexity considerations for the blocks relocation problem. Eur. J. Oper. Res. **219**, 96–104 (2012)
5. Firmino, A., de Abreu Silva, R.M., Times, V.C.: An exact approach for the container retrieval problem to reduce crane's trajectory. In: IEEE 19th International Conference on ITS (ITSC), pp. 933–938 (2016)
6. Forster, F., Bortfeldt, A.: A tree search heuristic for the container retrieval problem. In: Klatte, D., Lüthi, H.J., Schmedders, K. (eds.) Operations Research Proceedings 2011. Springer, Heidelberg (2012). https://doi.org/10.1007/978-3-642-29210-1_41
7. Jin, B., Zhu, W., Lim, A.: Solving the container relocation problem by an improved greedy look-ahead heuristic. Eur. J. Oper. Res. **240**, 837–847 (2015)
8. Jovanovic, R., Tuba, M., Voß, S.: An efficient ant colony optimization algorithm for the blocks relocation problem. Eur. J. Oper. Res. **274**, 78–90 (2019)
9. Kim, K.H., Hong, G.P.: A heuristic rule for relocating blocks. Comput. Oper. Res. **33**, 940–954 (2006)
10. Lee, Y., Lee, Y.J.: A heuristic for retrieving containers from a yard. Comput. Oper. Res. **37**, 1139–1147 (2010)
11. Lehnfeld, J., Knust, S.: Loading, unloading and premarshalling of stacks in storage areas: survey and classification. Eur. J. Oper. Res. **239**, 297–312 (2014)
12. Lin, D.Y., Lee, Y.J., Lee, Y.: The container retrieval problem with respect to relocation. Transp. Res. Part C **52**, 132–143 (2015)
13. Tang, L., Zhao, R., Liu, J.: Models and algorithms for shuffling problems in steel plants. Nav. Res. Logist. **59**(7), 502–524 (2012)
14. Tang, L., Ren, H.: Modelling and a segmented dynamic programming-based heuristic approach for the slab stack shuffling problem. Comput. Oper. Res. **37**, 368–375 (2010)

15. Tierney, K., Pacino, D., Voß, S.: Solving the pre-marshalling problem to optimality with A* and IDA*. Flex. Serv. Manuf. J. **29**, 223–259 (2017)
16. Ünlüyurt, T., Aydin, C.: Improved rehandling strategies for the container retrieval process. J. Adv. Transp. **46**, 378–393 (2012)
17. Voß, S.: Extended mis-overlay calculation for pre-marshalling containers. In: Hu, H., Shi, X., Stahlbock, R., Voß, S. (eds.) ICCL 2012. LNCS, vol. 7555, pp. 86–91. Springer, Heidelberg (2012). https://doi.org/10.1007/978-3-642-33587-7_6
18. Zehendner, E., Caserta, M., Feillet, D., Schwarze, S., Voß, S.: An improved mathematical formulation for the blocks relocation problem. Eur. J. Oper. Res. **245**, 415–422 (2015)

Estimating Discharge Time of Cargo Units – A Case of Ro-Ro Shipping

Beizhen Jia[1]([✉]), Niels Gorm Malý Rytter[1], Line Blander Reinhardt[3], Gauvain Haulot[2], and Mads Bentzen Billesø[2]

[1] Aalborg University, A. C. Meyers Vænge 15, 2450 Copenhagen SV, Denmark
bj@m-tech.aau.dk
[2] DFDS, Sundkrogsgade 11, 2100 Copenhagen Ø, Denmark
[3] Roskilde University, Universitetsvej 1, 4000 Roskilde, Denmark

Abstract. Ro-Ro shipping is a dominant form of short sea freight transport. Ro-Ro ship operators are today unable to provide customers with precise information about when trailers are available for pick-up by customers on the terminal despite vessel arrival times being well known in due time. This results in reduced truck utilization, longer waiting time for drivers, less efficient yard space utilization, potential terminal congestion and dissatisfied customers. In this paper the cargo unit discharge time estimation problem of Ro-Ro shipping is solved in collaboration with a European short-sea Ro-Ro shipping company. A module-based framework using statistical analysis for estimating the discharge time is proposed and tested. The initial framework is able to estimate the earliest pick-up time of each individual truck or trailer within 1 h accuracy for up to 70% of all cargo. The results of the study show potential for improving performance and accuracy. Further investigation and testing is currently ongoing by the case company based on the results from this study.

Keywords: Cargo discharge time estimation · Short sea shipping · Terminal operations · Integrated logistics chain · Industry implementation

1 Introduction

Roll-on/Roll-off (Ro-Ro) shipping is a large part of the maritime freight transport of coastal communities and also deep sea due to the versatility of most Ro-Ro vessels. For over 1.8 billion tonnes of goods transported through short-sea shipping (SSS) in European Union in 2017, Ro-Ro units accounted for 13.6% with only 1% less than cargo transported through containers [1]. In Europe the Ro-Ro shipping is very dominant, due to the extensive coastal line compared to the landmass of Northern, Western and Southern Europe. The fact that this landmass consists of a large number of peninsulas, makes the short-sea Ro-Ro shipping an attractive alternative to land-based and container transport and in some cases such as the British Isles there does not exist a land-based alternative. Ro-Ro vessels consist of two major types: deep-sea going Ro-Ro vessels which are commonly car carriers traveling across continents, and short-sea Ro-Ro vessels that transport mostly trailers and heterogeneous cargo, sometimes with a mixture of passengers as well. The short-sea vessels are in Europe strongly present

© Springer Nature Switzerland AG 2019
C. Paternina-Arboleda and S. Voß (Eds.): ICCL 2019, LNCS 11756, pp. 122–135, 2019.
https://doi.org/10.1007/978-3-030-31140-7_8

between countries separated by sea but located closer to each other such as the North Sea, Baltic Sea and Mediterranean areas. Ro-Ro SSS like short-sea container transport generally operates with fixed schedules servicing often just two ports although in occasions the routes can include from 3 to 10 port calls in a round trip even though longer routes are more common in short-sea container transport. Although Ro-Ro vessels have a much smaller capacity than container vessels, the Ro-Ro vessels have the advantage of a larger choice of ports due to container vessels crane requirement.

The main competitors for Ro-Ro vessels are road transportation and short-sea container shipping and it is important to remain competitive which implies offering clients a short transit time and reliable schedules. However, speeding up the vessels increases the bunker consumption significantly. Increasing the bunker consumption is both costly and also not applicable regarding the International Maritime Organization (IMO) announced goals for reducing CO_2 emission in maritime transportation by 50% until 2050.

The European Commission has set ambitions for enhancing the further development of SSS through three actions, one of which is improved integration of SSS in full logistics chains [2]. The integration includes among others the loading and discharge of Ro-Ro vessels. These processes can take up to several hours depending on the vessel size thus leaving a large time interval for the first trailer available for pick-up to the last one available at a given destination port. Lack of information can result in customers' trucks waiting around at the terminal for hours for a trailer or terminal congestion caused by trailers taking up the limited terminal space longer than actually needed.

2 Background

Today, the information about trailer availability for pick-up at the yard is often released after the discharge of all the cargo from the Ro-Ro vessel. In order to increase customer satisfaction without increasing operational costs, one option is to provide customers with the planned discharge time for their trailers or general cargo so that they can avoid waiting for the discharge of all the cargo before retrieving it. Being able to provide customers with information about availability of individual trailers for pick up at yard in due time, e.g. several hours before vessel ETA, can enable customers to increase the utilization of their logistics assets and resources. Moreover, it can reduce the congestion at the gate and the surrounding road network as all customers are not arriving to the terminal to pick up their trailers at the same time. Meanwhile, reduced 'turnaround' time of trailers in the terminal means a better utilization of the yard with more throughput. However, despite extensive effort spent on stowage planning and execution, Ro-Ro shipping companies are today unable to produce and deliver this information to their clients.

Researchers have previously investigated the challenge of terminal congestion in relation to truck arrival, however most of the research so far has been focused on the segment of container shipping rather than the Ro-Ro sector. Moreover, the focus of the research has been on investigating problems of terminal congestion due the unpredictable arrival time of trucks for pickup of import cargo, which impacts resource allocation at terminals and implies inefficiencies for ports and haulage companies.

For container terminals, research on improving efficiency of landside drayage operations has proposed implementing a Truck Appointment System (TAS), extended gate hours and pricing policies to control truck arrival rates to handle these challenges [3]. For example,, the impact of TAS on truck-related port emissions, turn-around time, congestion and air pollution has been studied extensively [4, 5]. Furthermore, there are some studies investigating the use of optimization methods to support TAS [6–8]. For example, Phan and Kim have proposed a solution for negotiations of truck arrival time among trucking companies and terminal [9]. Reinhardt et al. applied several optimization techniques to solving the bottleneck of the inland transport of containers connecting customers and terminals for more efficient liner shipping operations [10, 11].

If we zoom out and consider the overall flow of logistics operations at terminals, it is interesting to observe that most research is focusing on TAS and truck arrivals. Thus investigating options for predicting discharge time of individual cargo units (containers) at terminals as a way forward attempting to improve terminal operations and customers' processes has been overlooked. In a situation where a ship operator or terminal is able to predict the available pick-up time of the individual cargo units, a TAS with better accuracy and reliability could be developed which would result in reduced terminal congestion. For container shipping, the challenge of predicting cargo availability for pickup might be difficult to embrace due to variability of stowage situations from ship to ship; however, for Ro-Ro shipping this issue might be more addressable as loading and discharge procedures across decks; lane sections etc. can be assumed more regular and stable across voyages. In general, but in particular from the perspective of Ro-Ro shipping we consider estimating the discharge time of cargo units as an overlooked topic when solving terminal congestion problems and logistics efficiency problems. Quality estimation will enable TAS and truck arrival management systems to perform much better. It is also an issue so far not studied for Ro-Ro terminals, where we mainly identified a few studies focused on simulation and decision support for terminal capacity planning and operational execution [12–17].

In this paper, we have in collaboration with a European short-sea Ro-Ro shipping company identified the discharge time estimation problem for Ro-Ro shipping, developed a module-based framework for estimating the time available for customers' pick-up of individual trailers. We have completed a subsequent evaluation of accuracy of the methods on data collected from actual discharge cases, and compared the results with different time windows.

The remainder of this paper continues with defining the discharge time problem for Ro-Ro vessels in Sect. 3, followed by a description of the framework structure in Sect. 4. In Sect. 5, we present a case study on its application and discuss the results. Finally, we conclude the paper and point out directions for future research.

3   Problem Description

The Ro-Ro cargo unit discharge time problem is a challenge involving various stakeholders of the cargo logistics chain, as shown in Fig. 1. A cargo unit can be either unaccompanied or accompanied depending on if there is a truck and driver travelling with the cargo. Unaccompanied cargo requires tugs in order to be placed on/off board.

All cargo is loaded under the instruction of a dispatcher (or foreman), who manually creates an overall stowage plan and controls cargo flows in an import/export terminal. When a vessel arrives at a terminal, a local dispatcher plans the discharge of the vessel for both types of cargo. Once all cargo is discharged from the vessel and onto the terminal, import customers are able to pick up their unaccompanied cargo and complete the rest of the logistics chain. One of the pain points for both terminals and customers is that the import customers do not have information of the available pick-up time for their cargo in advance as this is assumed difficult to be provided by the Ro-Ro vessel operators for multiple reasons.

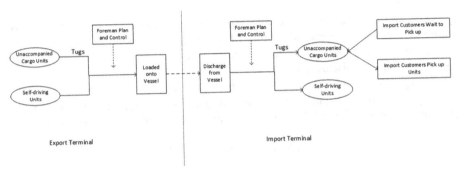

Fig. 1. Ro-Ro cargo logistics chain at terminal

First, different unit types require a different amount of time to be fastened to or be released from the vessel. For example, it's faster to lock/unlock the trestles attached to standard trailers, whereas mafis and cassettes require longer time due to their special operational requirements (heavy weight, gooseneck, translifter, lashing etc.)

Besides this, general cargo, hazardous cargo, refrigerated cargo, livestock, bulk can also be transported. They have dedicated zones or warehouses where they are supposed to be discharged to inside the terminal. Refrigerated cargo must be plugged in, therefore the area where these trailers are stored in the terminal is usually on the edge or furthest away. Same goes for bulk cargo, like steel and wood. Hence the cycle time is much longer for the above-mentioned cargo, compared to standard trailers.

Where the unit is loaded onboard a vessel also influences the discharge time as a unit can be discharged only when all units which stand in front of it are discharged in order to make a path out. Moreover, it requires more time for tugs to travel to the weather deck, which is the top deck of a vessel, than the time required to pick up a trailer on the main deck. Therefore, it is the relative position of a unit on a deck and the deck that determines the discharge time.

When a vessel arrives at a terminal, it also takes some time to set up the ramp and arrange tug masters before discharging the first unit. If the vessel is early or late according to the schedule, it will have an impact on the exact time when units are discharged, and in the case where multiple vessels arrive in one time slot it will also have an impact on the schedule of the tug usage.

Tug availability is one of the most important factors determining the discharge speed of a vessel and hence the discharge time. The more tug masters are assigned to the vessels, the faster the vessel gets discharged. However, depending on the day of the week and the number of vessels arriving, the tug availability fluctuates throughout the discharging process. Day of week is an external factor that has an impact on the number of tugs to be used. It indirectly influences the discharge speed by directly influencing the number of tugs scheduled for the discharge process. Weekends and weekdays with more vessel arrivals will have less tugs scheduled for each vessel's discharge, hence lowering down the speed. The tug availability is not a fixed number of workers as illness and other issues may affect the number of tugs available, thus making it difficult to model and plan.

Moreover, extreme weather requires extra lashing of the units for safety reasons during sailing. When it comes to the time of discharge, bad weather can slow down the tug masters' driving speed, and it requires extra time to release the lashing on the units before they can get discharged.

Having captured the influence of these factors or variables enables us to model the discharge time of each unit as a function of unit type or type group, cargo type, position, vessel arrival condition, tug availability, day of week, and weather condition.

$$\text{EDT}_{unit} = F(t, c, p, v, n, d, w)$$

EDT_{unit} − discharge time per unit
t − unit type
c − cargo type
p − position on board
v − vessel arrival condition
n − tug availability
d − day of week
w − weather condition

The factors influencing the discharge time of the unit are at the same time the challenges affecting the model of estimated discharge time (EDT). The challenges are of different risk types, as shown in the risk matrix in Fig. 2, depending on the availability of knowledge and the ease of control of the factors. As can be seen, the factors fall into two major quadrants by the time of vessel departure – known but uncontrollable; unknown but controllable.

Some information is known but uncontrollable, like day of week, cargo type, unit type, and weather condition. Regarding weather condition, one could argue that it is known through weather forecasts but it can also be considered slightly unknown due to inaccuracy or uncertainty of weather forecasts in general. For this study, it is considered as a piece of known information as operational efficiency is not sensitive to slight weather changes, and that weather forecast is sufficient to catch significant weather shifts. Whereas already by the time of loading, some of the factors are unknown; however, still controllable which means that the information could be captured with a certain degree of human intervention. This includes the position on board, tug availability, and vessel arrival condition. These three factors have the highest influence on

the discharge time of a unit. However, the challenges in estimating the discharge time are, to the authors' knowledge, lack of traceability where the unit is loaded on board; shifting tug usage; and an uncertain discharge sequence deck-wise but also position-wise within a deck. Furthermore, the challenges when implementing solutions to control these factors are the standardization of loading and discharge processes across routes and voyages while considering human participation and business complications stemming from customer requirements.

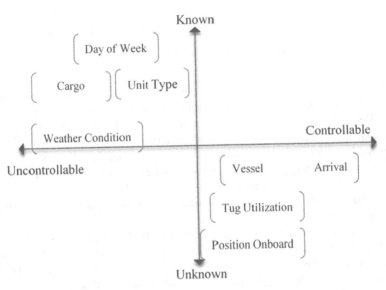

Fig. 2. Categorization of potential variables for cargo unit EDT

4 Framework

To estimate individual discharge times of the cargoes from the loading information, this paper proposes a modular framework for the Ro-Ro cargo discharge time estimation problem (Fig. 3). The framework consists of basic statistical methods and logics combined in different modules to form the framework for delivering a good discharge time estimation. The framework consists of three modules:

Module 1: The loading position is estimated from loading information such as loading timestamps, standardized loading sequence and its position (first in last out).

Module 2: Estimates the discharge sequence from the estimated loading position provided by Module 1 (furthest in last out).

Module 3: Estimates the discharge time based on the discharge sequence generated in Module 2 with certain discharge speed.

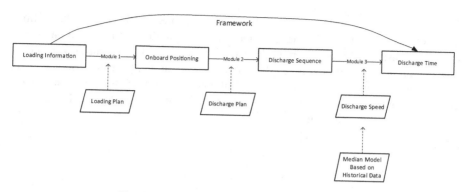

Fig. 3. The modular discharge time framework

The combination of the three modules constitutes the Ro-Ro cargo discharge time estimation framework, and the overall accuracy depends on the performance of each module. Depending on what information is available in the operation, the discharge time estimation can be constructed with only one or two modules. For example, if the company makes a detailed stowage plan and executes accordingly, the first module will be omitted as real loading positions of cargoes will be available as input to module 2. In this paper, we are more interested in the cases where loading positions are not recorded and thus unknown when vessels depart, which is also close to situations experienced in real-life operations.

For the first two modules, a fixed loading and discharge plan is assumed, which means that the vessel loads and discharges in a specific sequence, however, a limited number of usually minor shifts in position in the plan is possible in reality. The third module estimates the discharge time based on the estimated discharge sequence and discharge speed which arises as a sub-problem.

4.1 Discharge Speed

As discussed in Sect. 3, the discharge speed is influenced by various different factors. To find the discharge time in Module 3 we have constructed a model which we call a situational median model to estimate discharge speed for different discharge situations. A situation is a combination of various factors that have a significant influence on the discharge speed, such as unit type, week day, tug availability and deck loaded. An example of a situation is illustrated in Fig. 4, and it is a situation where the discharge happens on a Wednesday, for trailers on the weather deck with four tugs working simultaneously.

Each situation is connected to a discharge speed based on historical data, assuming no significant changes of processes, equipment or systems in the relevant time horizon. Let S be the set of situations, and V be the set of discharge speed, where v_i in V is the discharge speed of situation i in S. The Binary variable x_{ni} equals to 1 if i is the situation of the nth discharged unit, and 0 otherwise. Discharge time for one unit is defined as the time interval between the current discharging unit and the previously discharged unit. It is formulated as below:

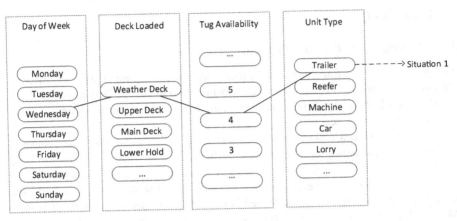

Fig. 4. Example of a discharge situation

$$\nabla^i_{EDT(n)} = DT^i_n - DT^{i'}_{n-1} \qquad\qquad i, i' \in S \qquad (1)$$

DT^i_n is the discharge timestamp of the nth discharged unit, and $n - 1$ is the previous unit in the discharge sequence. The situation i of $\nabla^i_{EDT(n)}$ is determined by the situation of the discharging unit n such as unit type, deck, weekday, and tug availability and is thus independent of the situation of unit $n - 1$. This means that each discharge unit has its independent speed calculated from the situation i of the unit.

The situational median discharge speed equation is the median of discharge time intervals categorized by different situations from historical voyages. The discharge speed of situation i is irrelevant to the unit's discharge sequence n. Thus we can define the situational median speed v_i as:

$$v_i = median(\cup \nabla^i_{EDT}) \qquad\qquad i \in S \qquad (2)$$

The estimated discharge time of the nth unit is the sum of the time needed to discharge the individual unit from the first in the discharge sequence up until the nth, based on the unit's situation. And it is formulated as:

$$EDT_n = \sum_{m=1}^{n} \sum_{i=1}^{|S|} v_i x_{mi} \qquad (3)$$

The framework is configured with more details from the industry case which is tested and evaluated with real data in the next section.

5   Case Study

5.1   Description of the Case Problem

The problem and the framework are further researched in a case study with a Ro-Ro shipping company that operates short-sea transportation in Europe. The chosen route of the study is a 15-h voyage from Vlaardingen, the Netherlands to Immingham, England, with two identical vessels servicing a daily schedule.

A three-week data collection was conducted in collaboration with the company. Loading and discharging operations were instructed by foremen, based on the standardized sequence plans per deck. For Module 1, an example of the loading sequence of main deck drawn by a foreman is given in Fig. 5. The first trailer loaded is estimated to be in position 1 and the last one loaded in position 63. If this was a discharge plan in Module 2, position 63 would be estimated to be the first discharged etc. Exact loading positions have been captured for framework validation. In addition, nine-month historical data starting from January 2018 was retrieved from the company's database for the situational median discharge speed model. No significant changes in the process were made throughout the selected nine-month and three-week period.

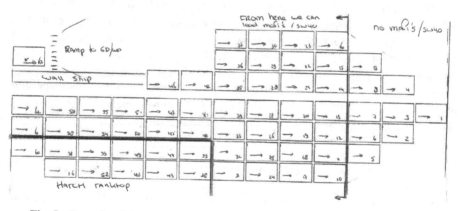

Fig. 5. Example of the loading plan of the main deck. Source: DFDS Vlaardingen

The majority of the data is automatically logged through booking and terminal management systems. For each unit, information on the time of loading, the time of discharging, deck loaded and unit type etc. is available. However, due to changes in tug availability, it has been very difficult to determine the number of tugs available per deck at a certain time. Therefore, this information will not be considered and included in the framework for the present, and we assume the constant availability of tugs per deck every day. The unit type, as discussed above in the problem formulation, has an impact on the speed of discharge as well. However, based on analysis, the discharge process appears stable and units are evenly scattered over time, indicating that the unit type is not a significant influencing factor, therefore it is not considered in this case. Lastly, vessel arrival conditions and weather are not included in the case study.

According to interviews with the company, foremen, and managers, among others, the study of the discharge speed is delimitated by the focus on loaded deck and day of week. This, however, also indicates the level of terminal activity and thus indirectly indicating the average number of tugs used. A diagram with data input and output for each module in the case study is illustrated in Fig. 6. Initial data input to the first module of the framework is the timestamp at which a unit was loaded onto the vessel and the deck the unit was loaded onto. Based on the actual sequence of loading the standardized loading sequence plan, the output of Module 1 will be the estimated loaded position for each unit. In the second stage, the output of Module 1 is fed into Module 2 in order to estimate the discharge sequence based on the standardized discharge sequence plan. Lastly, the overall deliverable of the framework, which is the individual discharge time of a unit is estimated based on the discharge sequence and discharge speed, calculated as in Eq. (3).

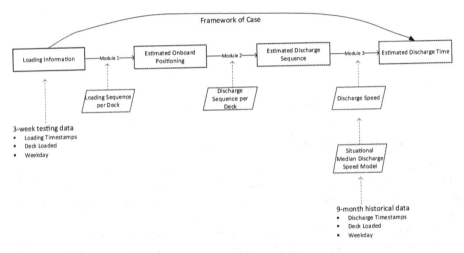

Fig. 6. Framework configuration of case study

5.2 Framework Evaluation

For a module-based framework, it is important to separate the individual module performance to understand the overall framework accuracy and to improve the performance if possible. Therefore, it is important to look at individual module performance as well as combined performance. To achieve this, we have conducted a three-week data collection where the company, terminal and crew were actively involved. Among other things, we have collected the onboard positioning of cargoes, actual discharge sequence and the actual discharge timestamp.

Individual module performance tells how well a module estimates given the input to the framework is real data instead of estimated. Illustrated in Fig. 7, the error of Module 1 is the difference between estimated position and actual position; if the actual onboard positioning is known, the discharge sequence estimated from Module 2 compared with actual discharge sequence is the individual performance of Module 2,

and the same logic applies to the individual performance of Module 3. Combined module performance is the result of a combination of two or more modules. A combined performance of all three modules makes the accuracy of the overall framework. By comparing combined performance to individual performance, we are able to tell how well modules can be integrated into one framework and what the accuracy loss is by predicting in a modular way. It also makes it possible for the company to see where the discharge time estimations would improve the most with real-life data.

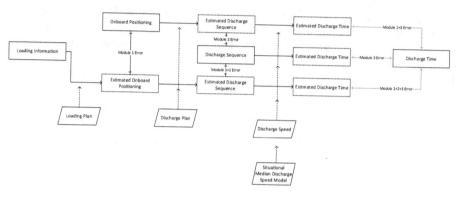

Fig. 7. Framework evaluation map

5.3   Computational Results

The discharge speed was calculated in MySQL and fed into the overall framework, which was coded in excel. Table 1 presents the results for individual modules, combined modules and the overall framework, with a 15-min, 30-min and 60-min time window.

As mentioned in the previous section, individual performance for Module 1 represents how well the module estimates loading positions from loading timestamps of units; for Modules 2 and 3, it is based on actual unit position on board and actual unit discharge sequence, respectively. Actual data was gathered during the three-week data collection. Because of the nature of the data input and output in Modules 1 and 2, the errors are measured by the differences in the sequences. In order for the results to be comparable, we converted it into to a time estimate in minutes by multiplying errors in position by discharge speed. Combined performance of Modules 1 and 2 presents an integrated result when the input of Module 2 is not actual data but predicted data from the output of Module 1.

Table 1 shows the computational results of each individual module of the framework and two different combinations of the modules. The overall performance shows an undesirable accuracy of 32.5% with a 15-min delay window, 43.2% and 65.8% for 30-min and 60-min delay window, respectively. When we compare the combined and overall results of modules to individual module performance, the difference in accuracy is relatively small. This means that the three modules have little influence among each other and prove the robustness of the modular EDT framework.

Table 1. Results of framework performance evaluation

	% units <15 min late			% units <30 min late			% units <60 min late		
	Individual performance	Module 1 + 2	Overall performance	Individual performance	Module 1 + 2	Overall performance	Individual performance	Module 1 + 2	Overall performance
Module 1	93.0%	90.3%	32.5%	95.0%	94.8%	43.2%	96.8%	98.3%	65.8%
Module 2	91.8%			95.0%			98.8%		
Module 3	32.4%	–		45.1%	–		67.2%	–	

From loading information to loading sequence (Module 1), and from loading sequence to discharge sequence (Module 2), we could predict the loading and discharge sequence with an accuracy of more than 90%. Furthermore, the combined result of Module 1 and 2 does not show a significant drop in the accuracy. The robustness of the modules relies on the standardization of the loading and discharge procedures. From experience and practices, there already exist patterns of loading and discharge Ro-Ro vessels. Standardization of patterns is a challenge however, as the result shows, it is not impossible to overcome and acquire robust outcome out of it.

Module 3 has the lowest accuracy – 32.4%, 45.1% and 67.2% predicted within 15, 30 and 60 min late, respectively. This is the bottleneck of the framework since the overall accuracy follows closely the accuracy of Module 3 with little difference. However, this result is expected without pulling in tug availability and other factors discussed in Sect. 3.

From a business perspective, almost 70% of the units can be estimated its discharge time with an hour time window. This means 70% of the customers get correct available pick-up time for their trailers, instead of hours after the ship's arrival and they can therefore avoid traffic jams around the terminal and time waste in general.

6   Concluding Remarks

This paper describes a currently unsolved problem for the Ro-Ro shipping industry – the estimation of discharge time for individual cargo units before vessel arrival - and proposes a data-driven module-based approach for the problem. The motivation behind predicting cargo unit discharge times was that it enables ship and terminal operators to deliver a more efficient cargo supply chain for customers, a better utilization of the Ro-Ro terminal as well as a better service product from the shipping company.

The main idea of the proposed solution method is to approach the discharge time from loading information step by step, on a modular basis. With the input of an instructed loading sequence plan, by ranking the timestamps when units are loaded, their positions on board are estimated. Based on the discharge sequence plan and position on board, a discharge sequence of all units is estimated. Then the discharge time of the individual unit can be estimated by incorporating the discharge speed, which was solved as a sub-problem where we introduced a situational median approach to find the discharge speed suitable for each unit.

The weak part of the framework is Module 3, which was expected due to framework simplifications and limited data availability for tug usage, unit types, vessel

arrival conditions and weather for the case study. Nevertheless, the overall results achieved with data obtained from real Ro-Ro cargo operations seem to verify the relevance and robustness of a modular and quantitative approach. Compared to individual performance, the combined and overall performance of modules deteriorates only to a trivial degree. The framework is widely applicable and customizable to different routes, ships and companies by tuning individual modules and adjusting the set of situations based on various influencing factors in Ro-Ro shipping. As for container terminals, it provides the framework and inspiration to potential research on discharge time of containers as input to TAS.

Further work could be focused on improving current solutions to calculating discharge speed or modelling discharge time against discharge sequence to improve the accuracy in Module 3. Machine learning could also be an interesting investigation compared to the modular framework method, provided there is sufficient historical data. Another focus could be the problems related to cargo operations, for example, Ro-Ro stowage automation and optimization problems to be incorporated in Module 1; dual cycling of loading and discharge operations, tugs planning and scheduling, etc. which have a significant impact on discharge speed.

Acknowledgements. We would like to thank our industrial partner DFDS for its openness on problems and data sharing, and all the help and support from DFDS in Copenhagen, Vlaardingen and Immingham. The research is funded by EU Baltic Sea Interreg project – ECOPRODIGI, aiming at improving eco-efficiency in shipping industry processes through digitalization.

References

1. Eurostat: Short sea shipping - country level - gross weight of goods transported to/from main ports, by type of cargo (2019). https://appsso.eurostat.ec.europa.eu/nui/show.do?dataset=mar_sg_am_cwk&lang=en. Accessed 30 July 2019
2. COMM/TREN: Short sea shipping - Mobility and Transport - European Commission (2016). https://ec.europa.eu/transport/modes/maritime/short_sea_shipping_en. Accessed 12 Oct 2018
3. Huynh, N.N.: Methodologies for reducing truck turn time at marine container terminals (2005)
4. Giuliano, G., O'Brien, T.: Reducing port-related truck emissions: the terminal gate appointment system at the ports of los angeles and long beach. Transp. Res. Part D Transp. Environ. (2007). https://doi.org/10.1016/j.trd.2007.06.004
5. Huynh, N., Walton, C.M.: Improving efficiency of drayage operations at seaport container terminals through the use of an appointment system. In: Böse, J. (ed.) Handbook of Terminal Planning. Operations Research/Computer Science Interfaces Series, vol. 49, pp. 323–344. Springer, New York (2011). https://doi.org/10.1007/978-1-4419-8408-1_16
6. Huynh, N., Walton, C.M.: Robust scheduling of truck arrivals at marine container Terminals. J. Transp. Eng. (2008). https://doi.org/10.1061/(ASCE)0733-947X(2008)134:8(347)
7. Zehendner, E., Feillet, D.: Benefits of a truck appointment system on the service quality of inland transport modes at a multimodal container terminal. Eur. J. Oper. Res. (2014). https://doi.org/10.1016/j.ejor.2013.07.005

8. Chen, G., Govindan, K., Yang, Z.: Managing truck arrivals with time windows to alleviate gate congestion at container terminals. Int. J. Prod. Econ. (2013). https://doi.org/10.1016/j.ijpe.2012.03.033
9. Phan, M.-H., Kim, K.H.: Negotiating truck arrival times among trucking companies and a container terminal. Transp. Res. Part E Logist. Transp. Rev. (2015). https://doi.org/10.1016/j.tre.2015.01.004
10. Reinhardt, L.B., Pisinger, D., Spoorendonk, S., Sigurd, M.M.: Optimization of the drayage problem using exact methods. INFOR Inf. Syst. Oper. Res. (2016). https://doi.org/10.1080/03155986.2016.1149919
11. Reinhardt, L.B., Spoorendonk, S., Pisinger, D.: Solving vehicle routing with full container load and time windows. In: Hu, H., Shi, X., Stahlbock, R., Voß, S. (eds.) ICCL 2012. LNCS, vol. 7555, pp. 120–128. Springer, Heidelberg (2012). https://doi.org/10.1007/978-3-642-33587-7_9
12. Tang, G.L., Guo, Z.J., Yu, X.H., Song, X.Q., Wang, W.Y., Zhang, Y.H.: Simulation and modelling of roll-on/roll-off terminal operation. In: Chan, K., Yeh, J. (eds.) International conference on Electrical, Automation and Mechanical Engineering (EAME 2015), Phuket, Thailand, 26–27 July 2015, 7/26/2015–7/27/2015. Atlantis Press, Amsterdam (2015). https://doi.org/10.2991/eame-15.2015.201
13. Vadlamudi, J.C.: How a discrete event simulation model can relieve congestion at a RORO terminal gate system: case study: RORO port terminal in the Port of Karlshamn
14. Özkan, E.D., Nas, S., Güler, N.: Capacity analysis of ro-ro terminals by using simulation modeling method. Asian J. Shipp. Logist. (2016). https://doi.org/10.1016/j.ajsl.2016.09.002
15. Keceli, Y., Aksoy, S., Aydogdu, Y.V.: A simulation model for decision support in Ro-Ro terminal operations. IJLSM (2013). https://doi.org/10.1504/IJLSM.2013.054896
16. Iannone, R., Miranda, S., Prisco, L., Riemma, S., Sarno, D.: Proposal for a flexible discrete event simulation model for assessing the daily operation decisions in a Ro–Ro terminal. Simul. Model. Pract. Theory (2016). https://doi.org/10.1016/j.simpat.2015.11.005
17. Balaban, M., Mastaglio, T.: RoPax/RoRo: exploring the use of simulation as decision support system (2013)

Port Community System Adoption: Game Theoretic Framework for an Emerging Economy Case Study

Adriana Moros-Daza[1,2]([✉]), René Amaya-Mier[2], Guisselle Garcia-Llinas[2], and Stefan Voß[1]

[1] Institute of Information Systems (IWI), University of Hamburg, Hamburg, Germany
{adriana.moros.daza,stefan.voss}@uni-hamburg.de
[2] Department of Industrial Engineering, Universidad del Norte, Barranquilla, Colombia
{ramaya,gagarcia}@uninorte.edu.co

Abstract. Port community systems (PCS), as platforms for connecting the stakeholders of a port community (PC), build a basis for improving information exchange. These systems support automation and visibility along global supply chains, thereby helping to enhance the overall PC performance in both local and foreign trade activities. However, successful real-world PCS adoption often involves multiple collaborating parties interacting with at least two different types of PC stakeholders. This study assesses the payoff of the alliances between three types of stakeholders to ensure the adoption of a PCS. In order to accomplish the former, it also identifies the different possible collaborations at the inception of a PCS. We use different simulated scenarios to obtain the expected payoff of each stakeholder, in terms of lead time savings achieved, and based on the Shapley value method allocate profits to each coalition. Also, we present a sensitivity analysis to test each coalition with different variation of the market while adopting a PCS. The main result is that profits gained from the adoption of a PCS funded by the port community grand coalition are higher than those from pairs or individual members. Also, the profits gained from the coalition of two cooperative members are higher than any individual effort. Therefore, the stakeholders' individual incentive to join a PCS coalition should be relatively high coming from a steady coalition.

Keywords: Port community system · Game theory · Shapley value · Profit allocation

1 Introduction

Given the increase in foreign trade activities and volume of cargo, ports are now globally facing challenges to remain competitive [27, 32]. This is specially true

C. Paternina-Arboleda and S. Voß (Eds.): ICCL 2019, LNCS 11756, pp. 136–153, 2019.
https://doi.org/10.1007/978-3-030-31140-7_9

for ports from emerging economies [32,48], which are on the way of developing towards a fundamental part of global markets. The use of innovative technologies to transform the different services at ports is becoming an essential strategy to improve the competitiveness and productivity of them [2,32]. Currently, those innovative technologies are known as smart port initiatives, related to the introduction and adoption of information technology platforms, mobile devices and apps, cloud-based services, sensors and Internet of Things technologies, augmented reality, autonomous transportation, big data, blockchain technology, and others. The development of smart initiatives is becoming a critical and essential part of the port industry, because of their growing necessity of performing operations in a beneficial, fast, transparent, reliable, and safe way [12] to maintain competitive and at the forefront of ports across the globe.

However, not every smart port initiative is appropriate for every port worldwide; [27] exposed that the adoption of different smart port initiatives depends mainly on the type of port and its focus. [27] explained that the focus for emerging ports is the ease of doing business, making reference to the coordination of cargo flows with information flows in order to increase the efficiency of the processes and gain competitiveness. In order to achieve this, one of the applicable smart port initiatives is the development of Port Community Systems (PCS) [13,27] or, on a smaller and more restricted scale, a Single Window (SW) [11,20,43]. A PCS can be defined as an inter-organizational system for promoting commercial services and information exchange between the port and its customers and a variety of stakeholders, such as forwarders, carriers, importers, exporters, customs, among others. In this sense, a PCS is a tool that enhances the information flow between the stakeholders by electronically integrating heterogeneous compositions of public and private repositories, technologies, systems, processes, and standards within a port community [11,13,20,43]. On the other hand, a SW can be described as a single point of data entry on which different stakeholders can submit and share information and documents in order to comply with regulatory requirements that involve import, export and transit [42]. This definition, which to some extent restricts the use of a SW to customs procedures, makes the adoption of a PCS a supplementary asset, because it builds a basis for implementing SW and also improves additional commercial services for the port community like traceability, tracking, customs clearance and other services [11,13].

For the adoption of a PCS the collaboration of different stakeholders of the port community is necessary, which has been signaled in previous studies [22,23] as a significant barrier for the adoption of PCS in the context of emerging economies. A possible reason to this is given by the uneven payoff allocation in the alliances (port community stakeholder coalitions) [50], meaning by this, the unfair share of benefits derived from a PCS collaboration, which may be uncorrelated to risk and/or effort incurred by any given party. This is supported by the fact that every stakeholder of the port community is an individual economic entity, not willing to sacrifice its own benefit for other's payoff on behalf of the alliance. Still, achieving the maximum payoff is the most powerful motivation for

stakeholders to join the alliance. Therefore, the key for adopting and maintaining a PCS lies in whether there is a reasonable "profit sharing and risk sharing" system among the allied port community partners [50], which constitutes a strategic research topic for improving PCS adoption.

Despite the recognized benefits of PCS worldwide, it is challenging to find documentation on overcoming PCS adoption barriers. The most relevant literature about PCS emphasizes its benefits and exposes successful examples around the world [4–6,40,45]. Yet, there is scarce literature on developing and/or implementing PCS. Notwithstanding, none of those studies have investigated the way to overcome barriers like collaboration resistance for adopting PCS in emerging economies. One way to overcome collaboration resistance for adoption is to prove the greater benefit of coalitions [50], and in order to quantify it, many methods and theories have been used. Game theory [10,36,50], which includes the analysis of non-cooperative and cooperative scenarios, is very common [10] for analyzing the port communities' behavior and to propose future actions related to the adoption of technology within the port industry.

Considering the previous discussion, this work aims to evaluate the feasibility, as well as stakeholders coalitions, regarding the adoption of a PCS in emerging economies. We use the Port of Barranquilla (Colombia) as a case study to illustrate the former, using a DES simulation model as a means for estimating the payoff of a variety of collaborating scenarios considering different stakeholders colluding as active members of the port community when a PCS is adopted. The simulation model is based on two different scenarios, an as-is scenario (current situation without a PCS) and a to-be scenario (expected situation with PCS adoption). Moreover, the payoff in the to-be scenario is defined in terms of saved process time from the as-is scenario. We proposed the use of the Shapley value function for estimating the benefits of colluding with PCS adoption for different stakeholders of the port community.

The remainder of this paper is structured as follows: Sect. 2 provides a description of the current works related with the adoption of smart technologies and the use of game theory in port communities. Section 3 exposes the theory regarding the allocation of profits with PCS stakeholder alliances and makes references to the use of the Shapley value function. Section 4 introduces a case study and evaluates the different possible coalitions. Also, Sect. 4 presents a sensitivity analysis based on possible market capture with the adoption of a PCS. Finally, Sect. 5 presents the main conclusions and some further research directions.

2   Related Work

In the port industry, the use of game theory approaches to analyze the port communities' behavior and to propose future actions related to the adoption of new technologies is very common [36]. Most studies focus on the behavior of each stakeholder of the community. For example, [39] studied cooperative scenarios (possible alliances) between worldwide renown shipping lines finding

good results, i.e., it was found that individual companies that have come together in alliances are serving many more ports than before. An example is the merger of Cosco and Orient Overseas Container Line (OOCL), which is leading Cosco to become the second-biggest mover of United States imports with an 11.8% market share [41].

Some studies have found [38,39] that some liners achieve great success by cooperation and may also show good economic performance, compared with those that have failed from it. But this is not the case for all types of stakeholders within the port communities. This means that in some cases the port communities can be also approached as non-cooperative games and an excellent example for non-cooperative stakeholders in port communities are Port Terminals [15]. It is common to find studies about their behavior with solutions for their problems [18]. The main difference between cooperative and non-cooperative games is that the former focus on the distribution of payoffs among stakeholders, while the latter studies what the stakeholder should do according to the strategies of other stakeholders and its own situation [3].

Also, game theory can be used to measure the impact of collaboration on technology adoption in different industries [50]. In the port industry, game theory can be used to develop and adopt information technology tools and methods in seaports and for addressing new economic issues and problems [10]. An example is the use of game theory to solve the problem of forming cooperations on the inter-organizational level, while allocating costs efficiently to adopt a digital platform [10].

Table 1 summarizes some research related to the use of game theory in the port industry. It shows that a considerable amount of the research focuses on oligopolies in port communities, as well as game theory to overcome barriers in port communities. A few studies relate to the use of game theory for technology adoption coalitions.

Table 1. Game theory in port communities

GT[a] in PC[b]	References
Oligopolies in PC	[9, 15, 17–19, 30, 31, 34–37, 39, 51, 52]
GT as a way to overcome barriers	[7, 17, 19, 28–31]
GT for technology adoption	[10, 50]

[a]GT = game theory, [b]PC = port community

Taking into account the studies presented in Table 1, we could identify that it is not common to find studies about the port industry analyzing stakeholders' interactions within a port community altogether. However, we could identify and represent the specific idea towards common interaction. Among those references in Table 1, 87% of the game theory approaches in the port industry are related to individual efforts and only 13% assess pairs of stakeholders. To our best

knowledge, there are no game theory approaches related to the port industry in the literature, evaluating three or more different types of stakeholders.

Upon checking those references it seems more common to see investigations focused on one single stakeholder of the port community, like Port Terminals or Local Carriers, or at best interactions between pairs of players. Complementing the former, Fig. 1 shows the common combination of stakeholders' interactions. Correspondingly, Fig. 1 shows that 59% of those papers about game theory in the port communities are based only on one single type of player, i.e. Port Authorities, Freight Forwarders, Terminal Operators, Carriers or Shipping Companies. The rest is divided between the interaction of two types of players, the most common interaction is Terminal Operators - Carriers 35% of the total percentage, followed by the interaction between Freight Forwarders - Carriers or Freight Forwarders - Logistics operators with 3% each.

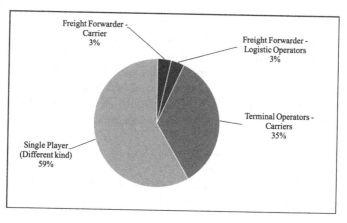

Fig. 1. Reported combinations of stakeholders' interactions (regarding references from Table 1)

Following the above, we can conclude that there is a lack of investigation about games or mathematical models using the combination of all the important stakeholders of a port community. Moreover, there is no published research related to the coalition of more than two different types of stakeholders that measure the impact of the adoption of an information technology tool (technology alliances) in port communities. Technology alliances can be analyzed through two different strategies, cost allocation [10] or profit allocation [50]. This paper focuses on profit allocation and proposes a contribution to the extant literature by using the Shapley value to ensure information technology alliances between three different types of stakeholders of a port community.

3 Profit Allocation for Information Technology Alliances in Port Communities Using the Shapley Value

Taking into account the Shapley value foundations[1], let us assume a set of n players and $I = \{1, 2, ..., n\}$, where $v(S)$ is a characteristic function corresponding to any subset S in I, which represents the maximum payoff of the alliance S, given that it satisfies the following conditions,

Efficiency:

Let x_i be a payoff gained by member i from the maximum cooperative payoff $v(I)$, the payoff allocation of cooperative game can be written as $x = (x_1, x_2, ..., x_n)$. The following are the essential conditions for successful cooperation and alliance,

$$\sum_{i=1}^{n} x_i = v(I), \tag{1}$$

$$x_i \geq v(i); i = 1, 2, ..., n \tag{2}$$

Symmetry:

If 1 and 2 are two players who are equivalent in the sense that

$$v(S \cup \{1\}) = v(S \cup \{2\}) \tag{3}$$

for every subset S of I which contains neither 1 nor 2, then $\varphi_1(v) = \varphi_2(v)$, where $\varphi_i(v)$ means the rewards of player i acquired in the cooperative game.

Linearity:

If two coalition games v and w are combined, then the combined gains should correspond to the gains derived from v and the gains derived from w:

$$\varphi_i(v + w) = \varphi_i(v) + \varphi_i(w) \tag{4}$$

for every i in I. Also, for any real number a, $\varphi_i(av) = a\varphi_i(v)$ for every i in I.

Null player:
$$v(\emptyset) = 0, \tag{5}$$

Stand-alone test (if v is a super-additive set function):

$$v(s_1 \cup s_2) \geq v(s_1) + v(s_2), s_1 \cap s_2 = \emptyset, \tag{6}$$

where $s_1 \subseteq I, s_2 \subseteq I$, then $[I, v]$ is a n-person cooperative game, meaning that, if the cooperation has positive externalities, all players (weakly) gain.

---

[1] For general game theoretical issues and some foundation of the Shapley value see, e.g., [33, 50].

The goal of multiple people cooperative gambling which has many solutions is to seek a unique most reasonable solution. $\Phi(v) = (\varphi_1(v), \varphi_2(v), ..., \varphi_n(v))$ is an allocation formula,

$$\varphi_i(v) = \sum_{s \in s_i} w(\mid s \mid)[v(s) - v(s \setminus i)], i = 1, 2, ..., n \qquad (7)$$

$$w(\mid s \mid) = \frac{(n - \mid s \mid)(\mid s \mid - 1)!}{n!}, \qquad (8)$$

where s_i denotes all subsets including player i in set I, $\mid s \mid$ is the element number in the subset, $w(\mid s \mid)$ is a weight factor. $v(s \setminus i)$ is the payoff if player i is removed from set S.

4 Case Study Port Community System

There are about 47 PCS established around the world [14, 26], and depending on their geographical location they can be operating in a country with a developed, emerging, frontier, or standalone market [25]. Each country can be indexed in any of the above mentioned categories based on their economic development and market accessibility. By this, a developed market is associated to a country with an advanced economy, where predominant characteristics include highly developed capital markets with high levels of liquidity, meaningful regulatory bodies, large market capitalization, and high levels of per capita income. Developed markets are mostly in Western Europe, North America, and Australasia. Then, an emerging market is associated with a country in the process of rapid growth and development with lower per capita incomes and less mature capital markets than developed countries. On the other hand, a frontier market is a subset of the emerging market category, characterized by little market liquidity, marginally developed capital markets, and lower per capita incomes. It is important to highlight that different firms differ on the definition as to what constitutes a developed market, and consequently the indexing of a country in each category can vary from one firm to another. For our research, we follow the indexing and the last measurements by MSCI Inc.[2] [24]. In addition to these three categories, MSCI Inc. proposes the standalone markets category [25] which makes reference to countries with severe deterioration in market accessibility or size and liquidity for that market. Also, this category includes newly eligible markets that can demonstrate a relative openness and accessibility for foreign investors and are not undergoing a period of extreme economic or political instability.

According to the International Port Community Systems Association (IPCSA) [13], a PCS can increase the competitiveness of the port communities by focusing on the improvement of four areas: electronic data interchange and interoperability, governance and institutionalism for logistical facilitation,

---

[2] MSCI Inc. is a company, considered as a global provider of equity, fixed income, hedge fund stock market indexes, and multi-asset portfolio analysis tools. It publishes the MSCI BRIC, MSCI World and MSCI EAFE Indexes [25].

integral operation of the port logistics chain, and assurance of quality and efficiency in logistics services. For example, in Mauritius the introduction of a PCS helped to decrease the average customs clearance time of goods from about 4 hours to around 15 min for non-litigious declarations, with estimated savings of around 1% of GDP[3] [44]. By 2014, the Competitiveness Rank in Brazil averaged 56 points out of 100 in the 2014 edition of the Global Competitiveness Report[4] published by the World Economic Forum [49]. After the adoption of the PCS in 2015, Brazil reached an all-time high of 75 in 2016 [49]. Another successful example is India, which with the adoption of a PCS achieved to link 9 national banks for e-payment and also manages to have an effective tracking of goods movement and exchange of messages to all ports in the community [8]. A United Nations calculation (UN/Cefact) [44] stipulates that the savings obtained with this type of platform, is around 7% of the total costs of international trade operations.

Within the scope of the LogPort project,[5] a simulation of the adoption of a PCS in the Colombian Caribbean Coast was conducted [21], taking into account the implementation of features like BPM (Business Process Management), EDI (Electronic Data Interchange) and biometric technology [11,13,21]. The simulation focuses on ports as central nodes for import activities. The model begins with the arrival of the ships, considering a ship as an entity entering the system. Then a decision regarding resource capacity associated with the available bays of the port is then made; if available, the vessel can enter for later attention. At the same time, a documentation entity enters and joins the entity of the ship, and proceeds with the process when the document preparation time is completed. When the two entities meet, depending on the type of vessel and the type of merchandise, there is a decision to be made regarding what type of unloading to perform. Each of these imports passes to an inspection decision which can be physical, documentary or simply not having inspection. After this, the merchandise is ready for dispatch. From this point on, it follows a series of processes associated with the coordination of the inland transport. Once completed, the cargo is loaded onto trucks and follows a series of processes. Followed by this a decision is made, if the merchandise was loaded in its entirety it is proceeded with the dispatch and the process is finished, if the merchandise has not been dispatched it is loaded in trucks until it is complete. Based on the BPM simu-

---

[3] By its acronym in English: Gross Domestic Product.

[4] The most recent 2018 edition of Global Competitiveness Report assesses 140 economies. In 2018, the World Economic Forum introduced a new methodology emphasizing the role of human capital, innovation, resilience and agility, as not only drivers but also defining features of economic success in the 4th Industrial Revolution. As a result, the Global Competitiveness Index (GCI) scale changed to 1 to 100 from 1 to 7, with higher average score meaning higher degree of competitiveness [49].

[5] LogPort was an officially launched Colombian project on May 16, 2014, through an alliance between universities, companies and the Colombian Government, which aimed to improve port logistics measurements in Colombia. In general, it was a platform for scientific, technological and innovation management to improve the operational efficiency of the national logistics system, with broad impact in the Colombian Caribbean region.

lation of the above mentioned model it was possible to determine the reduction in time of the foreign trade macro processes, while adopting a PCS.

Figure 2 represents the current (as-is) macro processes from the time of an import activity in the Colombian Caribbean Coast and the simulation based on the adoption of a PCS (to-be). The macro processes are defined after a set of indicators from the *World Bank's Doing Business report*[6], which stipulate four macro processes for foreign trade activities: *Inland transport and handling, Customs clearance and inspections, Port terminal handling*, and *Document preparation* [46].

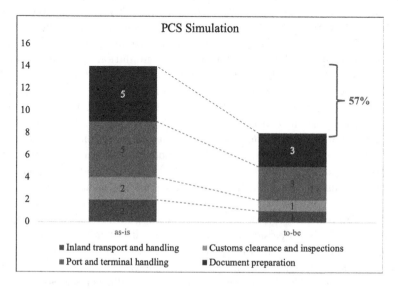

Fig. 2. PCS simulation results in the port of Barranquilla [21]

The adoption of a PCS has an impact on the entire chain and the processes associated with foreign trade activities [1,21]. Taking into account the features of a PCS, some specific processes can be affected. In this case, the main feature of the adoption of a PCS is the elimination of waiting times associated with the document preparation and the customs clearance. Moreover, the simulation made in the previous study [21] showed that a PCS can improve significantly the time to complete a foreign trade activity in the country. Figure 2 shows that with an adoption of a PCS, the total time of the import activities in the Colombian Caribbean Coast can be reduced on average by 57%. The most significant

---

[6] Doing Business is a set of indicators from the World Bank that measures aspects of business regulation for domestic firms through an objective lens. Based on standardized case studies around the world, Doing Business presents quantitative indicators on the regulations that apply to firms at different stages of their life cycle. The results for each economy can be compared with those for 189 other economies and over time [47].

impacts are in *Inland transport and handling and document preparation* macro processes. However, the span of both processes and benefits brought by a PCS implementation exceeds a port terminal, in terms of the firms involved in foreign trade supply chains and overall port community. Thus, how can a PCS be successfully adopted by a heterogeneous community? We propose in this paper to use the Shapley value to assess information technology alliances of two or more port community stakeholders by using a PCS.

Based on the previous simulations [21], we identify the most relevant stakeholders for the adoption of a PCS; in terms of information sharing, they are: *Port/Terminal, Customs,* and *Local Carrier.* Each player has a significant role within the process (see Fig. 3), and taking into account that the reduction in processing times can be associated with economical gains [16], an expected payoff for each stakeholder can be obtained. This is, more agile and lean processes can render two non-exclusive effects conducing to an increased payoff: cost reductions and/or increased revenue by charging premium rates. A PCS adoption implies reduction on processing times through the migration of traditional exchange of information (paper-based) to a digital platform. Such migration can increase the productivity of all the stakeholders, considering that this software also enables the integration of different smart technologies[7] [10,21]. Therefore, the payoff for the *Port/Terminal* is due to the increased efficiency of their services, making this stakeholder more attractive for the global market [44]. The payoff for *Customs* and *Local Carrier* is similar to the *Port/Terminal,* since an improvement in the system means more goods declaration for the *Customs* (e.g., more taxes) and more trips for the *Local Carrier* (increase in productivity, equal greater cash flow), which is directly associated with an increase of economic gains.

Therefore, when the set of players is $I = \{Port/Terminal, Local Carrier, Customs\} = \{PT, LC, C\}$ the expected payoff v in different alliance combinations is as shown in Table 2.

Table 2. Payoff allocations

$v(PT)$	22
$v(LC)$	7
$v(C)$	7
$v(PT \cup LC)$	29
$v(PT \cup C)$	29
$v(C \cup LC)$	14
$v(PT \cup LC \cup C)$	57

---

[7] PCS are able to integrate different smart technologies, i.e., Single Window, Vessel Traffic services, Port River information systems, Terminal Operating System (TOS), Port Road and Traffic control system, Intelligent transport system, Port Hinterland Intermodal IS, and Port Security systems. For general issues and foundation related to those smart technologies see, e.g., [10,21].

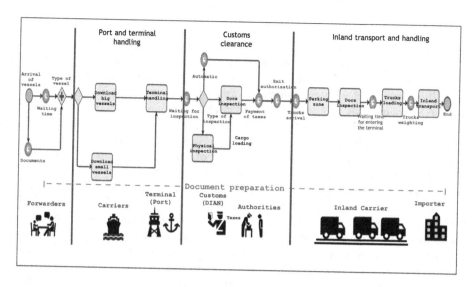

Fig. 3. Stakeholders interaction within the PCS

Taking into account the Shapley value method, the subsets containing the Customs are $\{C\}$, $\{C \cup PT\}$, $\{C \cup LC\}$, and $\{C \cup PT \cup LC\}$, and the payoff allocation for the Customs are shown in Table 3.

Table 3. Payoff allocation of the Custom

S	C	$C \cup PT$	$C \cup LC$	$C \cup PT \cup LC$
$v(s)$	7	29	14	57
$v(s \setminus C)$	0	22	7	29
$v(s) - v(s \setminus C)$	7	7	7	28
$\lvert s \rvert$	1	2	2	3
$w(\lvert s \rvert)$	0,333333	0,16666667	0,166667	0,333333333
$w(\lvert s \rvert)[v(s) - v(s \setminus C)]$	2,33	1,17	1,17	9,33

Subsequently, the subsets containing the Port/Terminal are $\{PT\}$, $\{PT \cup C\}$, $\{PT \cup LC\}$, and $\{PT \cup C \cup LC\}$, and the subsets containing the Local Carrier are $\{LC\}$, $\{LC \cup C\}$, $\{LC \cup PT\}$, and $\{LC \cup C \cup PT\}$. The payoff allocation of the Port/Terminal and the Local Carrier are shown in Tables 4 and 5, respectively.

As can be seen from the tables and using Eq. (7), the payoff of Customs C is $\varphi_C(v) = 14$. We can get $\varphi_{PT}(v) = 29$, $\varphi_{LC}(v) = 14$, with the same procedure. It is easy to show that $\varphi_C(v) + \varphi_{PT}(v) + \varphi_{LC}(v) = 57$, and $\varphi_C(v) \geq 7$, $\varphi_{PT}(v) \geq 22$, and $\varphi_{LC}(v) \geq 7$. Further, $\varphi_C(v) + \varphi_{PT}(v) \geq 29$, $\varphi_C(v) + \varphi_{LC}(v) \geq 14$, $\varphi_{PT}(v) + \varphi_{LC}(v) \geq 29$. Based on the previous results, the profits gained from the

Table 4. Payoff allocation of the Port/Terminal

S	PT	$PT \cup C$	$PT \cup LC$	$C \cup PT \cup LC$
$v(s)$	22	29	29	57
$v(s \setminus PT)$	0	7	7	14
$v(s) - v(s \setminus PT)$	22	22	22	43
$\mid s \mid$	1	2	2	3
$w(\mid s \mid)$	0,333333	0,16666667	0,166667	0,333333333
$w(\mid s \mid)[v(s) - v(s \setminus PT)]$	7,33	3,67	3,67	14,33

Table 5. Payoff allocation of the Local Carrier

S	LC	$LC \cup C$	$LC \cup PT$	$C \cup PT \cup LC$
$v(s)$	7	14	29	57
$v(s \setminus LC)$	0	7	22	29
$v(s) - v(s \setminus LC)$	7	7	7	28
$\mid s \mid$	1	2	2	3
$w(\mid s \mid)$	0,333333	0,16666667	0,166667	0,333333333
$w(\mid s \mid)[v(s) - v(s \setminus LC)]$	2,33	1,17	1,17	9,33

adoption of a PCS funded by the port community coalition are higher than those from individual or two cooperative members. Also, it is important to highlight that the profits gained from the coalition of two cooperative members are higher than those from the individual joining the coalition. Therefore, the incentive of the studied PC stakeholders to join the smart port technology alliance should be relatively high and with a remarkable coalition stability.

Sensitivity analysis

This section introduces a sensitivity analysis taking into account three additional different scenarios based on the simulation mentioned above [21]. Each scenario represents different variations of the market while adopting a PCS. Scenario 1 (S1) is the base expected to-be scenario previously analyzed. Scenario 2 (S2) represents an increase of 10% of the market in general cargo. Scenario 3 (S3) presents an increase of 20% of the market in general cargo. Finally, Scenario 4 (S4) represents an increase of 10% of containers.

Figure 4 shows that, in the best case scenario, the total time of the import process can be reduced by 35% with a PCS with an increased 10% of the imports of general cargo. It is important to clarify that the port of the case study is specialized in general cargo. Because of that, with an increase of 10% of containers the total time can only be reduced by 23%. Still, there is a time improvement conducent to the possibility of opening new markets in the region.

Taking into account the results from Fig. 4 it was possible to allocate the expected payoff of each stakeholder for each scenario, as shown in Table 6.

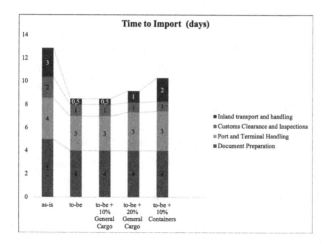

Fig. 4. PCS simulation scenarios results in the port of Barranquilla [21]

Table 6. Payoff allocation for scenarios of the sensitivity analysis

	S1	S2	S3	S4
$v(PT)$	22	21	21	21
$v(LC)$	7	4	7	17
$v(C)$	7	7	7	7
$v(PT \cup LC)$	29	25	28	35
$v(PT \cup C)$	29	28	28	28
$v(C \cup LC)$	14	11	14	24
$v(PT \cup LC \cup C)$	57	57	64	71

Following the same procedure and considering the Shapley value method, the results of each scenario are exposed in Table 7. The results show and confirm that the payoff gained from the adoption of a PCS funded by the port community coalition are higher in every scenario than those from individual or two-fold cooperative members. Also, Table 7 shows that each scenario satisfies the conditions of efficiency, symmetry, and linearity, and the highest overall payoffs come from the scenario S4, i.e. the scenario with the increase of 10% of containers. The main difference between the scenarios is in its individual payoffs; even if the global payoff of the grand coalitions are the same in some cases, depending on the scenario, each stakeholder could gain differently.

Table 7. Payoff allocation for scenarios of the sensitivity analysis

	S1	S2	S3	S4
$\varphi_{PT}(v)$	29	29	32	31
$\varphi_{LC}(v)$	14	13	15	23
$\varphi_C(v)$	14	15	17	17
$\varphi_{PT}(v) + \varphi_{LC}(v) + \varphi_C(v)$	57	57	64	71
Efficiency	✓	✓	✓	✓
Linearity	✓	✓	✓	✓
Symmetry	✓	✓	✓	✓

5 Conclusion

Due to the increase of foreign trade activities around the world, ports are facing challenges to remain competitive. This implies having a good infrastructure, technology, and interoperability through the integration of port services that can render more efficient operations. A common solution around the world to integrate and to assure interoperability through the chain is the implementation of a PCS [13]. However, successful real-world PCS adoption often involves multiple collaborating parties interacting, at least two different types of PC stakeholders for a start. Therefore, the aim of this study was to identify the different possible collaborations within the PC for the adoption of a PCS.

Taking into account various simulated scenarios for Barranquilla, Colombia [21], we obtained the payoff of PCS adoption for three different types of stakeholders to ensure the adoption of such a tool. For this study, the selected stakeholders were: *Port/Terminal, Customs,* and *Local Carrier*. As mentioned above, each player (stakeholder) has a significant role within the process (without any of them, an import/export activity cannot be completed). Basically, each of the above-mentioned stakeholders has a key role in one or more of the four main processes of which the import and export activities are composed. Furthermore, taking into account previous simulations [21], we identify that a future adoption of a PCS means a possible reduction on process' time. Moreover, the reduction on time can be associated with gains [16], which means that an expected payoff for each stakeholder can be obtained. It is important to highlight that the payoff was defined in terms of saved process time from the as-is scenario (current scenario). We evaluated the expected payoff of each stakeholder and, based on the Shapley value method, allocated the profit to each coalition.

The main result was that profits gained from the adoption of a PCS funded by the port community grand coalition are higher than the individual efforts or the pairs of members. Also, the profits gained from the coalition of two cooperative members are higher than any individual effort. Therefore, the stakeholders' individual incentive to join a PCS coalition should be relatively high with a stable coalition. This was also proved while testing different scenarios in a sensitivity analysis. Each scenario represented a different variation of the market

while adopting a PCS. Still, the main difference between all the scenarios was lying in the individual payoffs; even if the global payoff of the coalition of all the stakeholders are the same in some cases, depending on the scenario, each stakeholder will gain differently.

For future works we recommend to take into account the costs of PCS adoption, in order to evaluate in a more complete way the coalitions between the different stakeholders of the port community. While the elaboration is somewhat simplified it may be used to exemplify managerial benefits and implications in real cases. Furthermore, we propose as future research the development of a cost allocation rule for each stakeholder within different coalitions; taking into account that the interpretation of the payoff could be also related to the savings that each stakeholder obtains by changing the current exchange of information (paper-based) to a digital platform (paperless). Also, we consider as future work the evaluation of a game where more than one player from the same type of stakeholder is considered for the cost allocation of a PCS.

References

1. Amaya, R., et al.: Intervención sobre Prácticas Integrativas en el Clúster de Logística del Atlántico Cadenas Logísticas De Comercio Exterior. Ediciones Uninorte (2017)
2. Arvis, J.F., et al.: Connecting to compete 2018: trade logistics in the global economy. World Bank (2018). https://doi.org/10.1596/29971
3. Aumann, R.J.: Rationality and bounded rationality. In: Hart, S., Mas-Colell, A. (eds.) Cooperation: Game-Theoretic Approaches. NATO ASI Series (Series F: Computer and Systems Sciences), vol. 155, pp. 219–231. Springer, Heidelberg (1997). https://doi.org/10.1007/978-3-642-60454-6_15
4. Carlan, V., Sys, C., Vanelslander, T.: How port community systems can contribute to port competitiveness: developing a cost-benefit framework. Res. Transp. Bus. Manag. 19, 51–64 (2016)
5. Cuadrado, M., Frasquet, M., Cervera, A.: Benchmarking the port services: a customer oriented proposal. Benchmarking Int. J. 11(3), 320–330 (2004)
6. Duran, C., Cordova, F.: Conceptual analysis for the strategic and operational knowledge management of a port community. Informatica Economica 16(2), 35–44 (2012)
7. Fiestras-Janeiro, M.G., García-Jurado, I., Meca, A., Mosquera, M.A.: Cooperative game theory and inventory management. Eur. J. Oper. Res. 210(3), 459–466 (2011)
8. Fook Seng, C.: Trade facilitation single window and port community system. Technical report, CRIMSONLOGIC (2017). http://agpaoc-pmawca.org/admin/app/webroot/upload/files/2017/Single_Windows/Day2/Fook-SW-PCS'17-PMAWCA-Jul'17.pdf
9. Gomez Padilla, A., Gonzalez, R., Alarcon, F., Voß, S.: Contracts for the shipping industry: an option contracts model. In: Proceedings of the IAME 2016 Conference, The Maritime Transport of the Future, The Role of Innovation Uptake, Sustainability and Availability of Shipping Finance (2016)
10. Heilig, L., Lalla-Ruiz, E., Voß, S.: Digital transformation in maritime ports: analysis and a game theoretic framework. NETNOMICS Econ. Res. Electron. Netw. 18(2–3), 227–254 (2017)

11. Heilig, L., Schwarze, S., Voss, S.: An analysis of digital transformation in the history and future of modern ports. In: Proceedings of the 50th Hawaii International Conference on System Sciences, pp. 1341–1350 (2017)
12. Heilig, L., Voß, S.: The intelligent supply chain: from vision to reality. Port Technol. **78**, 80–82 (2018)
13. International Port Community Systems Association (IPCSA) : How to develop a port community system. Report, IPCSA (2015). http://www.ipcsa.international/armoury/resources/ipcsa-guide-english-2015.pdf
14. International Port Community Systems Association (IPCSA): IPCSA Members (2019). https://ipcsa.international/views/about/members/
15. Ishii, M., Lee, P.T.W., Tezuka, K., Chang, Y.T.: A game theoretical analysis of port competition. Transp. Res. Part E Logist. Transp. Rev. **49**(1), 92–106 (2013)
16. Karia, N., Razak, R.C.: Logistics assets that payoff competitive advantage. In: Soliman, K.S. (ed.) Information Management in the Networked Economy: Issues & Solutions, pp. 33–39. IBIMA (2007)
17. Karp, L.S., McCalla, A.F.: Dynamic games and international trade: an application to the world corn market. Am. J. Agric. Econ. **65**(4), 641–650 (1983)
18. Kaselimi, E.N., Notteboom, T.E., De Borger, B.: A game theoretical approach to competition between multi-user terminals: the impact of dedicated terminals. Marit. Policy Manag. **38**(4), 395–414 (2011)
19. Liu, Q., Wilson, W.W., Luo, M.: The impact of Panama Canal expansion on the container-shipping market: a cooperative game theory approach. Marit. Policy Manag. **43**(2), 209–221 (2016)
20. Long, A.: Port community systems. World Cust. J. **3**(1), 63–67 (2009)
21. Moros, A., Amaya, R., Garcia, G., Paternina, C.: Assessing the effect of implementing a port community system platform in the response time of an international terminal: the case of a multi-cargo facility at the Colombian Caribbean coast. In: 2016 International Conference on Industrial Engineering and Operations Management (IEOM), pp. 2896–2905. IEEE Xplore (2017)
22. Moros-Daza, A., Amaya-Mier, R., Paternina-Arboleda, C.: A survey for the development of port community systems in emerging economies (2019, submitted)
23. Moros-Daza, A., Lalla-Ruiz, E., Heilig, L., Amaya-Mier, R., Voß, S.: Feasibility study for the adoption of a port community system: the case of Barranquilla, Colombia (2019, submitted)
24. MSCI Inc.: MSCI Country Classification Standard 2018 (2018). https://www.msci.com
25. MSCI Inc.: MSCI Market Classification Framework (2018). https://www.msci.com
26. PORTEL: Inventory of port single windows and port community systems, November 2009. http://www.eskema.eu/DownloadFile.aspx?tableName=tblSubjectArticles&field=PDF%20Filename&idField=subjectArticleID&id=231
27. Riedl, J., Delenclos, F.X., Rasmussen, A.: To get smart, ports go digital (2018). https://www.bcg.com/publications/2018/to-get-smart-ports-go-digital.aspx
28. Rodríguez, F.F.: Teoría de juegos: análisis matemático de conflictos. Métodos matemáticos en ciencias sociales, economía, finanzas y administración de empresas, p. 27 (2005). https://imarrero.webs.ull.es/sctm05/modulo1lp/5/ffernandez.pdf
29. Saeed, N.: Cooperation among freight forwarders: mode choice and intermodal freight transport. Res. Transp. Econ. **42**(1), 77–86 (2013)
30. Saeed, N., Larsen, O.I.: An application of cooperative game among container terminals of one port. Eur. J. Oper. Res. **203**(2), 393–403 (2010)
31. Saeed, N., Larsen, O.I.: Container terminal concessions: a game theory application to the case of the ports of Pakistan. Marit. Econ. Logist. **12**(3), 237–262 (2010)

32. Schwab, K.: The global competitiveness report 2018. World Econ. Forum (2018). http://www3.weforum.org/docs/GCR2018/05FullReport/TheGlobalCompetitivenessReport2018.pdf
33. Sharkey, W.W.: Network models in economics. In: Handbooks in Operations Research and Management Science, vol. 8, pp. 713–765 (1995)
34. Shi, X., Meersman, H., Voß, S.: The win-win game in slot-chartering agreement among the liner competitors and collaborators. In: Proceedings of the IAME 2008 Conference Sustainability in International Shipping, Port and Logistics Industries and the China Factor. vol. 3, p. D2 (2008)
35. Shi, X., Voß, S.: Iterated cooperation and possible deviations between liner shipping carriers based on noncooperative game theory. Transp. Res. Rec. **2066**(1), 60–70 (2008)
36. Shi, X., Voß, S.: Game theoretical aspects in modeling and analyzing the shipping industry. In: Böse, J.W., Hu, H., Jahn, C., Shi, X., Stahlbock, R., Voß, S. (eds.) ICCL 2011. LNCS, vol. 6971, pp. 302–320. Springer, Heidelberg (2011). https://doi.org/10.1007/978-3-642-24264-9_23
37. Shieh, E.A., et al.: PROTECT: an application of computational game theory for the security of the ports of the United States. In: Proceedings of the Twenty-Sixth AAAI Conference on Artificial Intelligence, pp. 2173–2179 (2012)
38. Shubik, M.: Game Theory in the Social Sciences: Concepts and Solutions. MIT Press Cambridge, Cambridge (1982)
39. Song, D.W., Panayides, P.M.: A conceptual application of cooperative game theory to liner shipping strategic alliances. Marit. Policy Manag. **29**(3), 285–301 (2002)
40. Srour, F.J., van Oosterhout, M., van Baalen, P., Zuidwijk, R.: Port community system implementation: lessons learned from international scan. In: Transportation Research Board 87th Annual Meeting, Washington DC (2008)
41. The International Transport Forum: The impact of alliances in container shipping: Case-specific policy analysis reports. Technical report (2018). https://www.itf-oecd.org/sites/default/files/docs/impact-alliances-container-shipping.pdf
42. The United Nations Economic Commission for Europe (UNECE): Guidelines on establishing a Single Window. United Nations, Geneva (2005)
43. Tsamboulas, D., Ballis, A.: Port community systems: requirements, functionalities and implementation complications. In: Selected Proceedings of the 13th World Conference of Transport Research, Rio de Janeiro, pp. 1–16 (2013)
44. United Nations Centre for Trade Facilitation and Electronic Business (UN/CEFACT): International Trade Procedures Working Group: Case studies on implementing a single window to enhance the efficient exchange of information between trade and government (2005). https://digitallibrary.un.org/record/556563
45. Van Oosterhout, M.P., Veenstra, A., Meijer, M., Popal, N., Van den Berg, J.: Visibility platforms for enhancing supply chain security: a case study in the Port of Rotterdam. In: Proceedings of the International Symposium on Maritime Safety, Security and Environmental Protection, Athens, pp. 20–21 (2007)
46. World Bank: Equal Opportunity for All: Doing Business 2016 (2017). doingbusiness.org/ /media/WBG/DoingBusiness/Documents/Annualeports/English/DB17-Full-Report.pdf
47. World Bank: Doing Business 2018, Comparing Business Regulation for Domestic Firms in 190 Countries: Measuring regulatory quality and efficiency. World bank Group Flagship Report (2018)
48. World Bank: Doing Business 2019 - Training for Reform - Economy profile Colombia (2019). http://www.doingbusiness.org/content/dam/doingBusiness/country/c/colombia/COL.pdf

49. World Economic Forum: Brazil competitiveness rank. Electronic document (2019). https://tradingeconomics.com/brazil/competitiveness-rank
50. Xu, W., Yang, Z., Wang, H.: A Shapley value perspective on profit allocation for RFID technology alliance. In: 2014 11th International Conference on Service Systems and Service Management (ICSSSM), pp. 1–4. IEEE (2014)
51. Zan, Y.: Analysis of container port policy by the reaction of an equilibrium shipping market. Marit. Policy Manag. **26**(4), 369–381 (1999)
52. Zhuang, W., Luo, M., Fu, X.: A game theory analysis of port specialization-implications to the Chinese port industry. Marit. Policy Manag. **41**(3), 268–287 (2014)

Caribbean Ports, Inland Logistics, and the Panama Canal Expansion: A Mode and Port Choice Analysis

Nicolas Gomez-Jacome[1], Guisselle Garcia-Llinas[2](✉) ⓘ,
Carlos D. Paternina-Arboleda[2] ⓘ, and Miguel Jaller-Martelo[3] ⓘ

[1] Department of Industrial Engineering, Corporación Universitaria Americana,
Calle 72 No. 41C-64, Barranquilla 080010, Colombia
ngomez@coruniamericana.edu.co
[2] Department of Industrial Engineering,
Universidad del Norte, Km. 5 Via a Puerto Colombia,
Area Metropolitana de Barranquilla 080010, Colombia
{gagarcia, cpaterni}@uninorte.edu.co
[3] Department of Civil and Environmental Engineering, University of California
Davis (UC Davis), Ghausi Hall, Room 3143, Davis, CA 95616, USA
mjaller@ucdavis.edu

Abstract. This paper develops an international freight transport model for predicting the importers' behavior in the port/path-choice decisions after the Panama Canal expansion. A real-data case study for Colombian ports was developed assuming that the expansion will reduce the time to cross the canal. Two infrastructure investment projects of the Colombian government were evaluated as likely scenarios. Results show the adequacy of the model for its predictive purpose and its usability to carry sensitivity analyses of scenarios. The overriding outcome is that the model could guide governmental decisions on the infrastructures domain by demonstrating their effect in the logistic realm.

Keywords: International freight transport · Port choice · Multimodal network · Discrete choice models · Infrastructure decisions

1 Introduction

According to the World Bank's Logistics Performance Index (LPI), countries with ports in the Caribbean do not rank well in terms of their logistics systems. For instance, a sample of these countries' rankings range between place 40 (Panama, 3.34) and 159 (Haiti, 1.72) out of the top 160 countries, with an average score of 2.6 out of 5. From the six factors evaluated by the LPI, these countries exhibit higher deficiencies associated to their customs, logistics competence, and infrastructure (see Fig. 1). These, among other factors are crucial for the competitiveness of the countries' freight systems and the many supply chains that operate on it. Moreover, improvements are generally needed to "…establish a sustainable transport system that meets society's economic, social and environmental needs…" [1].

© Springer Nature Switzerland AG 2019
C. Paternina-Arboleda and S. Voß (Eds.): ICCL 2019, LNCS 11756, pp. 154–170, 2019.
https://doi.org/10.1007/978-3-030-31140-7_10

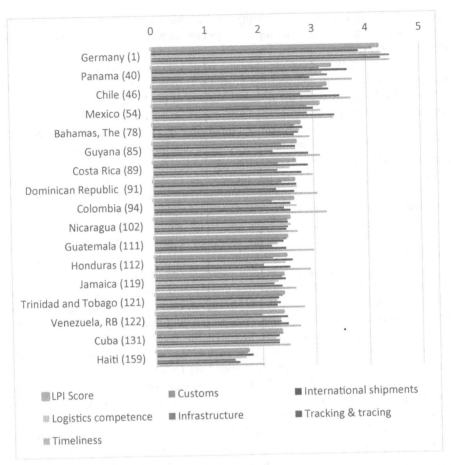

Fig. 1. 2016 LPI Results for countries serving the Caribbean (compared with the top ranked Germany).

While the LPI is just a subjective ranking that could provide a general overview of the logistics systems for imports and exports at international trade gateways, a country's logistics facilities, such as ports, airports, and the availability of intermodal networks also determine the performance of the internal distribution, manufacturing, and consumption economies. Therefore, the supply chains and associated costs may be highly impacted. As a result, the logistics performance of a country is as dependent on its ports and trade gateways as it is on its inland logistics systems. This includes the logistics condition of, without loss of generality: port, transport mode, and path selection. This is extremely important for countries (ports) serving the Caribbean.

On June 26, 2016 the greatly awaited Third Set of Locks Project or Panama Canal Expansion Project began commercial operations. Although global trade factors may have impacted the expected canal's traffic in the short-term; with the expansion, it can now handle 79% of global traffic, compared to the previous 45%, due to vessel size

constraints. There is a great opportunity for Caribbean countries to capitalize on the expansion for both their internal logistics systems as well as their positions on international supply chains.

A number of authors have explored the relationships between port choice, modal transport, path selection, and logistics costs [2–4] in an international context. Others have provided insights about the potential impacts of the Panama Canal expansion on the Caribbean ports [5, 6]. However, these impacts have not been fully studied.

This paper tries to fill this gap by simulating the flow of goods produced and attracted by Caribbean countries and evaluating the impact of the expansion of the Panama Canal. In doing so, we concentrate on the decisions about port selection as well as the impacts of inland transport, i.e., the selection of the point (port) of entry, and the mode and path choices for the internal transport. This is important because some route and mode combinations inside the countries could make a port more competitive than others. On the other hand, these route and mode combinations could generate additional costs and time that remove the efficiency gains from the maritime transport component. In other cases, the cargo destinations inside the countries may be situated at locations that are difficult to access from specific ports of entry.

The authors implement the framework and conduct empirical analyses for the case of Colombia. Colombia is in a strategic position next to Panama, with important ports such as Buenaventura on the Pacific, and fast growing ports in the Caribbean (Cartagena, Barranquilla, and Santa Marta). Moreover, there could even be opportunities for Colombia to play a larger role, with ideas such as the "dry canal," (to link the pacific and the Caribbean through multimodal transport) for example. Using the country's trade and flow data, the authors estimate choice models for port selection and develop country specific approximate cost models. The impacts of the Panama Canal expansion are discussed through a number of scenarios.

In addition, the paper provides a background for the potential impacts that offer the rationale to the improvement scenarios. Section 2 discusses modeling exercises from the literature used to support the framework development. Section 3 describes the proposed methodology. Section 4 explains the case study of inland container transport in Colombia, as well as the data and model calibration. The paper discusses the empirical results from six time-reduction scenarios and two infrastructure investment projects that have been proposed by the Colombian government. Finally, Sect. 5 concludes and discusses the potential directions for further research.

2 Modeling Background

Modeling freight flows and logistics decisions have received special attention during the last decades. Strategic, tactical, and operational planning has been developed in many applications to minimize logistics costs, or to optimize other performance measures.

In general, transport models were initially developed for modeling passenger transport and during the 1970s were initially adapted to freight transport. As a result, many of the models have been proposed considering the customary structure of the four steps model including: production and attraction of freight, trip distribution, mode

choice and trip assignment [7]. Models considered in freight analysis for each of these steps have been extensively reviewed. For instance, econometric models, trend and time series, and system dynamics are common approaches to model freight and freight trip generation. More aggregate models based on gravity theory try to estimate trip distribution between origins and destinations. In some cases these are improved by behavioral and logistics models for destination choice. Similarly, econometric models, usually of the form of multinomial and nested logit models, help to estimate mode and vehicle choice. These models use a cost function for the various modes (or choices) to find the most likely choice [4]. Finally, trip assignment and other optimization models allocate vehicle trips to the transport networks.

During the years, the traditional models have experienced many variations. Other authors [8, 9] present extensions of the four modeling steps in the context of freight transport model systems. Also in [8] the authors investigate the differences between the mode choice variables in short- vs. long-distance intermodal freight transport to overcome some limitations of the traditional modeling approaches.

Discrete choice models have been extensively used to model the cargo behavior considering attributes associated to the freight transport services as is evident in [10–12]. The decisions about port and inland mode selection are commonly taken simultaneously considering costs, time and risk attributes. In [13] the authors indicate that the attributes commonly used to describe the freight transport services are associated to cost, time, punctuality and risk damages. However, nowadays environmental issues are inherent to the transport services, and increase the complexity of models and networks. Externalities are considered to be an important element in transport modeling because they add transportation external cost to the logistic process. In [14] externalities are reviewed in quantitative terms that provide a framework to understand the transportation external costs from the perspective of business.

The integration of networks, policy and freight transport services is a challenge addressed in [15], where the authors propose a freight transport optimization model that incorporates multimodal infrastructure, hub-based service networks structures, and objectives for multiple actors. In the same field, a useful tool that can be used to visualize and analyze the container flow is introduced in [16], where the authors present a case study for the island of Curacao.

Challenges in international freight transport models and approaches to address freight transport studies are presented in [17, 18]. These papers discuss the challenges that need to be undertaken when developing international freight transport models, and identify some of the main issues in freight transportation planning and operations.

An integration of a port supply chain with maritime transport and inland transport is widely studied in [19], where the authors present an approach based on game theory to determine the benefits of cooperation, coordination and such integration.

Usually the logistic services of ports are classified into five modes: rail, road, water (sea or/and river transportation), air and pipeline transportation. The complexity due to the diversity of commodities is a key factor in the port collecting and the distribution network. Considering the complexity associated with the collecting and distributing activities at ports, in [20] the authors propose a Collecting and Distributing System of Container Port Model (CDSCP) for container ports. In their CDSCP, three distribution modes are considered (highway, railway and water transportation). A container

approach is presented in [21], addressing the design of container liner shipping networks taking into consideration container management issues.

Other approaches in port and path selection consider intermodal, multimodal transport and network design as a solution for the commodity flow problem [22, 23]. Due to the complexity of the mathematical problems in these approaches, heuristic algorithms are often used as the solution method.

3   Methodology

As previously discussed, modeling international flows is a complex task. To conduct realistic analyses, under an uncertain scenario, the authors decided to implement an approach traditionally used in transportation problems based on probabilistic decision choice and random utility theory, more specifically, by incorporating decision choice probabilities into a commodity flow problem. The case considered is as follows: imagine two regions X and Y separated by sea and continental mainland, as depicted in Fig. 2. Each region contains several geographic areas (e.g., cities) that generate (produce and attract) international cargo to be imported or exported. The cargoes have to be transported to and from the areas to their international destinations through a combination of inland multimodal transport services (at country of origin and destination) and the maritime transport components between the ports of entry at each region. Without loss of generality, the models consider that these regions have hinterland areas and their cargo can be transported by road or multimodal transport from/to their cities to/from many ports which can be used to serve mainline ships. Currently, the Panama Canal is one of the alternatives to transport the cargo between these regions.

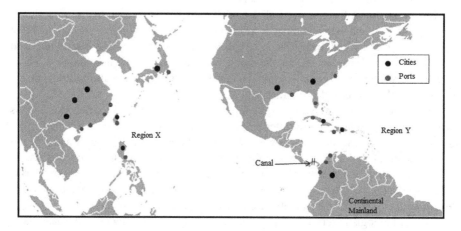

Fig. 2. Problem of transporting import/export cargo between two regions.

The main objective is to find the amount of cargo to be transported from the origins and destinations through the different inland route/mode alternatives at origin and destination, to minimize a generalized cost function. Therefore, it is assumed that each

economic agent will choose the mode and route combination that minimizes costs between origin r and destination s. However, in this approximate model, the decision is simulated based on choice probabilities for three scenarios. The first one is the base scenario which represents the actual behavior of the importers; the second scenario represent the importers' behavior after investment in road infrastructure projects and the third one represent the importers' behavior after the investment in multimodal infrastructure projects.

3.1 Notation

Sets

R: Set of regions (i.e., X and Y previously discussed).

K: Set of origin destination geographic locations. These are assumed to be freight generating cities within the studied regions.

P: Set of maritime ports. These are the maritime ports serving the various markets and geographic locations.

IM: Set of inland combinations of modes and routes connecting specific geographic locations and the maritime ports.

Indexes

r, s: Index for regions. $r, s \in R$.

k, l: Index for geographic locations or cities. $k, l \in K$. Explicitly, K_r is the set of cities within region r.

i, j: Index for maritime ports. $i, j \in P$. Similarly, P_r is the set of ports within region r.

t, f: Index for mode and route combinations. $t, f, z \in M$. IM_r contains the options available within region r.

Parameters

D_{kl}: Total demand in containers to be mobilized from origin city k to destination city l, where $k \in K_r, l \in K_s$.

ω: Length of the containers (feet).

sp_{ij}: Average cost to send a container of length ω from port i to port j.

$Ship\_cost_{ij}$: Average maritime cost per container from port i to port j, where $i, j \in P$.

$Inl\_o\_cost_{ik}^{t}$: Inland transport cost per container between origin city k and port i using the mode/route combination t, where $i \in P_r, k \in K_r, t \in IM_r$.

$Inl\_o\_time_{ki}^{t}$: Average inland transportation time between origin city k and, port i using the mode/route t, where $k \in K_r, i \in P_r, t \in IM_r$.

Mar_time_{ij}: Average maritime transportation time between port i and port j, where $i, j \in P$.

$Inl\_d\_cost_{jl}^{f}$: Inland transport cost per container between port j and destination city l using the mode/route combination f, where $j \in P_s, l \in K_s, f \in IM_s$.

$Inl\_d\_time_{jl}^{t}$: Average inland transportation time between port j and city l using the mode/route f, where $f \in IM_s, j \in P_s, l \in K_s$.

Auxiliary Variables

T_{kijl}^{tf}:   Total transit cost between origin city k, through port i using inland mode/route t, and city l through port j using mode/route f, where $k \in K_r, i \in P_r, t \in IM_r$, $f \in IM_s, j \in P_s, l \in K_s$.

L_{kijl}^{tf}:   Total transit time between origin city k, through port i using inland mode/route t, and city l through port j using mode/route f, where $k \in K_r, i \in P_r, t \in IM_r$, $f \in IM_s, j \in P_s, l \in K_s$.

V_{kijl}^{tf}:   Generalized cost between origin city k, through port i using inland mode/route t, and city l through port j using mode/route f, where $k \in K_r, i \in P_r, t \in IM_r$, $f \in IM_s, j \in P_s, l \in K_s$.

Output Variables

x_{kijl}^{tf}:   Number of containers to send between origin city k, through port i using inland mode/route t, and city i through port j using mode/route f, where $k \in K_r, i \in P_r, t \in IM_r, f \in IM_s, j \in P_s, l \in K_s$.

In the proposed methodology, the probability choice set is an output variable, and the number of containers x_{kijl}^{tf} is defined based on such probabilities and the container traffic demand (see Sect. 3.4 for additional details).

Pr_{kijl}^{tf}:   Probability of choosing the mode and route combination to send maritime containers from origin city k through port i using inland transport option t, to destination city l, through destination port j using the mode/route f, where $k \in K_r, i \in P_r, t \in IM_r, f \in IM_s, j \in P_s, l \in K_s$

3.2   Transportation Costs

In the maritime import/export cargo process there are two major types of costs, the first is associated to the maritime freight rate and the second to the inland multimodal transport. The proposed model considers the monetary costs associated to the various modal/route combinations links available or projected. Inland costs at the origin and destination are modeled following [24]. The methodology estimates generalized costs based on the transportation mode, the average shipment speeds, and the travel times.

Based on these considerations, the total cost to send a container from city k to city l can be expressed as follows:

$$T_{kijl}^{tf} = Inl\_o\_cost_{ik}^{t} + Ship\_cost_{ij} + Inl\_d\_cost_{jl}^{f}, \forall k, l, i, j, t, f \tag{1}$$

3.3   Transportation Time

Similarly, the transportation time is a function of the maritime transit time, and the inland transport time. It is important to recall that one of the key deficiencies for Caribbean countries is the customs time, which is included in the inland transport time.

The total transit time between port i and destination city k was calculated considering the maritime transit time and the transportation time in land, expressed as follows:

$$L_{kijl}^{tf} = Inl\_o\_time_{ki}^{t} + Mar\_time_{ij} + Inl\_d\_time_{ki}^{t}, \forall k, l, i, j, t, f \qquad (2)$$

3.4   Mode and Route Choice Utility and Generalized Cost Function

The model estimates the choice probability considering random utility theory. This requires defining a utility function. In doing so, the authors considered both the transportation cost, and the transportation time derived from the various decisions. It is assumed that inland mode and routes at the origin and destination are independent. Following random utility theory concepts, the utility function has two components, a systematic component which includes the choice factors from the choice set we can explain, and a random component. The random component is included because of uncertainty about the choices, taste variations, measurement errors, and other factors. Mathematically, this can be expressed as:

$$U_{kijl}^{tf} = V_{kijl}^{tf} + \varepsilon_{kijl}^{tf}, \forall k, l, i, j, t, f \qquad (3)$$

Moreover, each combination of port of sail, port of entry and potential inland mode and route options could be considered as an expanded (feasible) choice set M. That is, the combinations of i, j, t and f for each origin k and destination l, result in option m. After defining the transport option M, Eq. 3 reduces to:

$$U_{kl}^{m} = V_{kl}^{m} + \varepsilon_{kl}^{m}, \forall k, l, m \qquad (4)$$

The systematic component as a function of costs and times, assuming M, can be expressed as:

$$V_{kl}^{m} = \beta_0 + \beta_1 T_{kl}^{m} + \beta_2 L_{kl}^{m}, \forall k, l, m \qquad (5)$$

Considering the error term as Gumbel, Weibull, Extreme Value Type I distributed, the authors modeled the choice probabilities as a Multinomial Logit (MNL) model. The MNL model would appropriately illustrate the behavior of importers or exporters (there are different decision making structures in international trade) of general cargo over an international multimodal network. The choices are made taking into account the destination, mode, and routing. The model is capable of estimating probabilities for a discrete choice set. Specifically, the probability to choose destination port j from origin port i, and inland transport mode and route from port j to destination city l, Pr_{kijl}^{tf} or Pr_{kl}^{m} after transforming the choice set, is expressed using the MNL model as:

$$Pr_{kl}^{m} = \frac{\exp\left(V_{kl}^{m}\right)}{\sum_{m=1}^{M} \exp\left(V_{kl}^{m}\right)}, \forall k, l, m \qquad (6)$$

Finally, based on these assumptions, the following equation represents the containers mobilized between any city k and l, using the multimodal path and entry ports combination, m (using M, the decision variable can be simplified to x_{kl}^m):

$$x_{kl}^m = D_{kl} Pr_{kl}^m = D_{kl} \frac{\exp\left(V_{kl}^m\right)}{\sum_{m=1}^M \exp\left(V_{kl}^m\right)}, \forall k, l, m \qquad (7)$$

3.5 Transportation Model

The transport flow model for the containerized traffic could be simplified to a cost minimization of all the movements from the different supply and demand cities that guarantees the demand flows:

$$MinTotal\_Cost_{k,l,m} = \sum_k \sum_l \sum_m V_{kl}^m x_{kl}^m \qquad (8)$$

Subject to:

$$\sum_m x_{kl}^m = D_{kl}, \forall k, l \qquad (9)$$

However, following the proposed approach, the assignment of the flows is determined by the choice probabilities estimated using Eq. 7, thus:

$$Total\_Cost_{k,l,m} = \sum_k \sum_l \sum_m V_{kl}^m x_{kl}^m = \sum_k \sum_l \sum_m V_{kl}^m \left[Pr_{kl}^m D_{kl}\right] \qquad (10)$$

Where the flow constraint Eq. 9 is met because:

$$\sum_m Pr_{kl}^m = 1, \forall k, l \qquad (11)$$

4 Case Study: Colombian Imports

The authors selected Colombia as the case study to analyze the potential impacts of the Panama Canal expansion. Colombia has a strategic location with ports in the Pacific Ocean (Buenaventura Port) and the Caribbean Ocean (Cartagena Port, Barranquilla Port and Santa Marta Port), and some of its maritime imports already use the Panama Canal. Moreover, some of the Colombian ports have access to multimodal transport. For illustration purposes, the authors assumed that the decision reduces to the destination choice from export port i to destination city l, selecting the entry port j and inland transport mode and route f from j to l. This is done because information is only available for containerized cargo from the loading ports and not the origin city. The mathematical implication is that subscript and superscripts k and t are neglected and the

cargo originates at *i*. In addition, the modeling process assumes that 40 feet containers are mobilized fully loaded (FCL), between origins and destinations.

To estimate the models, the authors used the following data and data sources:

- Demand (tons): The data was extracted from the records reported by the Directorate of National Taxes and Customs [25] of Colombia during 2013 and 2014. The demand data has detailed information about the mobilized tons between each city pair (origin-destination). Each origin or destination can be a national city or either international. The data represents the place where the commodity is produced and the place where it will be consumed.
- Maritime freight rate: The monetary cost associated to the sea freight was extracted from SeaRates©. The Panama Canal fee is already included for the origin-destination scenarios in which it is necessary to cross the canal.
- Land transport costs: With respect to transportation cost, it is assumed $170/km-ton for road links, and in the case of the other modes, the results in [26] are considered. Those conclude that when compared to river transportation, the rail transportation is 1.53 times more expensive, while the road transportation is 3.34 times. When travel time is considered, [27] estimate that the parameter associated with travel time cost (based on the average travel time for each route) is $ 360/h-ton. Starting from this value, a growth rate was applied and it was found that this parameter corresponds to $ 417/h-ton at present. With this information, the minimum generalized cost and minimum travel time cost is calculated by ARCGIS 10.2.2 using data about the travel speed of different transportation modes.
- Maritime shipping time: The transit time from port *i* to port *j* was extracted from SeaRates©. The Panama Canal transit time is already included for the origin-destination scenarios in which it is necessary to cross the canal.
- Land transportation time: The transit time between port *i* and city *l* is modeled according to [24]. These times consider the transit time through corridor *j-l* using a specific transportation mode *f*. The time also includes customs related times. The load speed assumed by INVIAS is 40 km/h for flat terrain, 22 km/h for undulating terrain, mountainous terrain is 12 km/h, and finally it is assumed that in double lane, travel speed increases by 8 km/h. Furthermore, river transit time is approximately 17 km/h and many railways operate between 40 and 70 km/h in Colombia.

4.1   Mode Calibration

The data collected for the total import cargo in the period 2013–2014 facilitated the development of the MNL model. The authors used the statistical software Biogeme to estimate the discrete choice model. Table 1 shows the final estimated model. The authors evaluated various structural forms, and considered the log-likelihood results and goodness of fits of the different models. Moreover, the various transport options are mainly dependent on the ports of entry, which are Barranquilla (BAQ), Buenaventura (BUN), Cartagena (CTG), and Santa Marta (SMR).

Table 1. Statistically estimated multimodal path choice model for each corridor separately

Path	Port of entry	Utility function of Multinomial Logit model
1	BAQ	0,237–0,144 * T1 – 0,00367 * L1
2	BUN	0,237–0,144 * T2 – 0,00367 * L2
3	CTG	0,237–0,144 * T3 – 0,00367 * L3
4	SMR	0,237–0,144 * T4 – 0,00367 * L4

T1, T2, T3, T4 $= T_{il}^m, \forall m = Port\ of\ entry\ combination.$
L1, L2, L3, L4 $= L_{il}^m, \forall m = Port\ of\ entry\ combination.$

As mentioned before, we considered cost and time variables for the model. The negative signs in these parameters are logical as they indicate a decrease in utility as cost and time increase. Moreover, the results indicate that there is a preference for cheaper routes over faster routes. The adjusted rho-square of the model was 0.332 which reflects the significance of the model to be used in prediction.

4.2   Impact Scenarios

To evaluate the impact of the Panama Canal expansion on the selection of Colombian ports, the authors considered a number of scenarios. Specifically, the authors evaluated six time-reduction scenarios, for the maritime transit time considering the Canal expansion, and two infrastructure investment projects that have been proposed by the Colombian government. The first investment scenario represents road infrastructure projects (Table 2) and the second investment scenario represents multimodal investment projects (Table 3).

Table 2. Road infrastructure projects

Road infrastructure project	Interventions
Ruta del Sol Project (Sects. 1, 2 and 3)	Build a second lane in the highway between Puerto Salgar-San Roque, and build a road segment between Villetas and Puerto Salgar
Honda-Puerto Salgar Project	Build two bridges in Puerto Salgar and Flandes, and build a double lane highway between Puerto Salgar and Koran
Tunja–Puerto Boyacá Project	Improvements in the road
North Highway Connection	Build a lane between Remedios-Zaragoza and make improvements in the road between Zaragoza and Caucasia.
Mulalo, Loboguerreo, Buga and Buenaventura Project	Build a road segment between Mulalo and Loboguerro
Bogotá-Villavicencio Project	Build a double highway
Pacific Highway Connection (Sects. 1, 2 and 3)	Build a double lane highway between Bolombolo and Primavera, and between La Pintada and Bolombolo. Build a tunnel between Asia and Irra

Table 3. Multimodal infrastructure projects

Rail Infrastructure project	Interventions
Pacific Railway	Build a railway line between La Felisa and Zaragosa and La Tebaida and Zaragosa
Central Railway	Operation, Administration and Maintenance of the line between La Dorada and Chiriguaná, and Bogota and Belencito
Dibulla Port Connection	Build a line between Chiriguaná and Dibulla
Cararé Connection	Build a line between Belencito-La Vizcaya and between Bogota and Santa Sofia
Atlántico Railway	Build a second railway line parallel to the existing one. Build a line between La Loma and Puerto Drummond

The probability port choice is calculated for each port under each investment scenario and considers its time modification effects. This probability contemplates several paths and the transportation modes that can be used to mobilize the freight from a port to a city. In some special cases there are some hinterland areas to which the cargo cannot be mobilized from every port, what implicates that the cargo must be mobilized from the unique port that is accessible. In the other hand, there are some areas that are reachable using all the transport modes and the cargo can be unloaded in any port. In this last case, the selection criterion is the utility perceived by the importer.

The reader is referred to [24] for additional details about the two investment scenarios that consider several road and rail infrastructure projects.

4.3 Empirical Results

The **Base scenario** in Fig. 3 illustrates the actual estimated behavior of the importers in the port choice selection. Nowadays, almost 51% of the importers choose Buenaventura as the port of entry, followed by Cartagena, Barranquilla and Santa Marta. This is consistent with the national trade flows mentioned in Sect. 4. After an analysis of the importers' behavior under several time reduction scenarios due to the Panama Canal expansion, the results show how the reductions in the maritime transit time could move some cargo from the Buenaventura port to the Caribbean ports. However, under the current transportation network, the time reduction necessary to produce a 9% cargo redistribution is 30%. This implies that in the base scenario the cargo will remain almost inflexible to time reduction scenarios. This analysis should be important for the decisions makers since it would allow them to measure the impact of the infrastructure investment projects, proposed by the government, that are analyzed in the Scenarios 1 and 2.

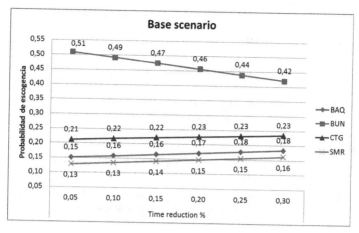

Fig. 3. Probability port choice vs. time reduction in the base scenario

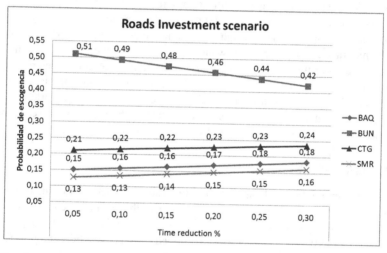

Fig. 4. Probability port choice vs. time reduction in the road investment scenario

The **Scenario 1: Road Investment Scenario**, which implies to invest in the projects presented in Table 2 and also consider 6 time-reduction scenarios, has no significant difference in the cargo distribution under each time-reduction scenario in comparison with the Base scenario, as can be seen in Fig. 4. This implies that the roads investment projects considered by the Colombian government could not produce the expected benefit in freight transport.

Usually the bigger benefits in freight transport are achieved when the transporter integrates multiple transport modes or multimodal transport. In this case, the commodities can be transported in the mode that maximizes the benefits. In contrast with the Base Scenario and the Scenario 1, the **Scenario 2: Multimodal Investment Scenario**, increases the probability to choose Cartagena, Barranquilla and Santa Marta

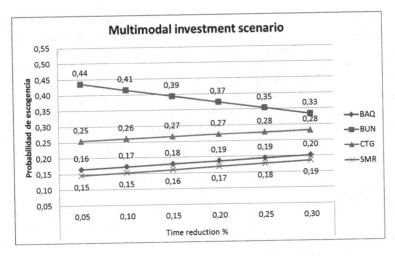

Fig. 5. Probability port choice vs. time reduction in the multimodal investment scenario

ports in a significant quantity, as can be seen in Fig. 5. The multimodal transport modifies not only the port choice probability, but it could also generate more competition for the cargo between port cities, what could produce a reduction in port fees or increases in efficiency (i.e. time reduction).

The results show the impact of the Panama Canal expansion under the considered scenarios. These scenarios correspond to many Colombian government policies related to logistics and transportation. These results and the methodology that we proposed to find these results could be used for the case study, and similar cases, to estimate the impact of governmental policies and to aid in the decision-making process.

Even when the data from the case study only considered general cargo, and knowing that several Colombian ports located in the Caribbean mobilize other types of freight, it is undeniable that after the Panama Canal expansion the flow in the Caribbean will have a substantial redistribution. It is clear that in many instances this redistribution will be a consequence of the maritime time or cost reductions, but in others, the effect will be due to government policies that provide higher benefits to certain ports.

5 Conclusion

The Panama Canal expansion is expected to affect global transportation trade routes. Although it is clear that the impact the Panama Canal expansion will have for shipments from Asia to the U.S. East Coast and the U.S. Intermodal System, the impact on other countries that have access to the Caribbean Sea was not yet clear.

This paper analyzes the relationship between the Panama Canal expansion and the port choice selection in the Caribbean coast, with a case study of the Colombian ports. The case study considers the effects of infrastructure investment in the port choice selection, and the travel time and cost savings due to road projects and multimodal

projects. It also shows an apparent potential improvement on regional (Caribbean coast) competitiveness, motivated by the choice of governmental policies, when a high impact project, like the Panama Canal expansion is executed. We have modeled and proven that, intelligently accompanying the growth in the flow of goods due to the Panama Canal expansion with the investment in new and more efficient multimodal infrastructure might produce a regional advantage based on shifts on inland trade routes. We also showed that investment in new multimodal infrastructure enhances the selection capacity of the frequencies for vessel ship lines on a specific landing port. In addition, we were able to show how the interaction of the Panama Canal expansion with new multimodal infrastructure highly influences the probability choice of a port, now giving priority to those in the Caribbean over landing in the Pacific, which is already happening.

Future work must expand the methodology and models to the capacitated version of the problem, where the capacity of ports, roads, and all transportation infrastructure is limited. This problem is more complex but also closer to the actual conditions in most places. We also plan to focus in the sustainability issues related to the more efficient use of this new multimodal infrastructure and its relationship to the new flow of goods due to the Panama Canal expansion. In addition, we plan to enhance our work as to how this cargo would accommodate along the Caribbean Sea ports. Furthermore, we will explore stronger discrete choice models to explain in a more detailed way other characteristics of transport mode selection, once the port choice has occurred and how regions that had no typical interaction before will now be positively impacted with the generation of new trade routes. We hypothesize that, for those regions located far from seaports and with low efficient transport infrastructure, the likelihood to increase their GDP with transport investment for connectivity is very high.

Acknowledgments. This article is the result of the research project titled "Diamante Caribe y Santanderes, Análisis de los factores clave de competitividad para la construcción de un modelo de territorio Inteligente en la región Caribe y Santanderes, Colombia" (Diamante Caribe y Santanderes, Analysis of the competitiveness key factors for the development of a smart territory in the Caribbean and Santanderes region), supported by the Fund for Science, Technology and Innovation of the General System of Royalties (Departments of the Colombian Caribbean and Santanderes Region).

Project number: 2014000100012. Contract number: 25 of May 30th, 2014.

References

1. CEC: Communication from the Commission: Keep Europe moving – Sustainable Mobility for our Continent. Mid-term Review of the European Commission's 2001 Transport White Paper. Comm. Eur. Communities, Brussels (2006)
2. Tongzon, J.L.: Port choice and freight forwarders. Transp. Res. Part E Logist. Transp. Rev. **45**(1), 186–195 (2009)
3. Tran, N.K.: Studying port selection on liner routes: An approach from logistics perspective. Res. Transp. Econ. **32**(1), 39–53 (2011)

4. Wu, Y., Liu, J., Peng, C.: Analysis of port and inland transport mode selection. In: 11th International Conference on Service Systems and Service Management ICSSSM 2014 – Proceedings (2014)

5. Pagano, A., Wang, G., Sánchez, O., Ungo, R., Tapiero, E.: The impact of the Panama Canal expansion on Panama's maritime cluster. Marit. Policy Manag. **43**, 164–178 (2016)

6. Liu, Q., Wilson, W.W., Luo, M.: The impact of Panama Canal expansion on the container-shipping market: a cooperative game theory approach. Marit. Policy Manag. **43**, 209–221 (2016)

7. Tavasszy, I., Ruijgrok, L.A., Davydenko, K.: Incorporating logistics in freight transport demand models: state-of-the-art and research opportunities. Transp. Rev. **32**, 203–219 (2012)

8. Reis, V.: Analysis of mode choice variables in short-distance intermodal freight transport using an agent-based model. Transp. Res. Part A Policy Pract. **61**, 100–120 (2014)

9. de Jong, G., Gunn, H., Walker, W.: National and international freight transport models: An overview and ideas for future development. Transp. Rev. **24**(1), 103–124 (2004)

10. Serag, M.S., Al-Tony, F.E.: Modeling international freight transport through the ports and lands of Arab countries. Alexandria Eng. J. **52**(3), 433–445 (2013)

11. Arencibia, A.I., Feo-Valero, M., García-Menéndez, L., Román, C.: Modelling mode choice for freight transport using advanced choice experiments. Transp. Res. Part A Policy Pract. **75**, 252–267 (2015)

12. Gelareh, S., Nickel, S., Pisinger, D.: Liner shipping hub network design in a competitive environment. Transp. Res. Part E Logist. Transp. Rev. **46**(6), 991–1004 (2010)

13. Masiero, L., Hensher, D.A.: Freight transport distance and weight as utility conditioning effects on a stated choice experiment. J. Choice Model. **5**(1), 64–76 (2012)

14. Demir, E., Huang, Y., Scholts, S., Van Woensel, T.: A selected review on the negative externalities of the freight transportation: modeling and pricing. Transp. Res. Part E Logist. Transp. Rev. **77**, 95–114 (2015)

15. Zhang, M., Janic, M., Tavasszy, L. A.: A freight transport optimization model for integrated network, service, and policy design. Transp. Res. Part E Logist. Transp. Rev. **77**, 61–76 (2015)

16. Veenstra, A., Mulder, H.M., Alexander Sels, R.: Analysing container flows in the caribbean. J. Transp. Geogr. **13**, 295–305 (2005)

17. Meersman, H., et al.: Challenges and future research needs towards international freight transport modeling. Case Stud. Transp. Policy **4**(1), 3–8 (2016)

18. Crainic, T.G., Laporte, G.: Planning models for freight transportation. Eur. J. Oper. Res. **97**(3), 409–438 (1997)

19. Dong, G.D.G., Wu, Q.W.Q., Zhu, D.Z.D., Li, J.L.J., Li, X.L.X.: Analyzing inland-orientation of port supply chain based on advertising-R&D model. In: 8th World Congress on Intelligent Control and Automation, pp. 3024–3027 (2010)

20. Hui, Z., Rui, L., Yu, Z.: A mode choice method for inland network of collecting and distributing of container port. In: 16th International Conference on Industrial Engineering and Engineering Management (2009)

21. Imai, A., Shintani, K., Papadimitriou, S.: Multi-port vs. hub-and-spoke port calls by containerships. Transp. Res. Part E Logist. Transp. Rev. **45**(5), 740–757 (2009)

22. Crespo Pereira, D., García del Valle A., Rios Prado, R., del Río Vilas, D., Rego-Monteil, N.: Hybrid algorithm for the optimization of multimodal freight transport services. In: 2013 Winter Simulations Conference (WSC) (2013)

23. Chang, T.S.: Best routes selection in international intermodal networks. Comput. Oper. Res. **35**(9), 2877–2891 (2008)

24. Yie, R., Arellana, J., Paternina, C., Saltarin, M. A.: An empirical analysis between freight accessibility and regional productivity: geographical approach. Working paper, Univ. del Norte Barranquilla (2016)
25. DIAN, Official record of import (2014) of the Directorate of National Taxes and Customs of Colombia (2013)
26. Márquez, L.: Optimización De Una Red De Transporte Combinado Para La Exportación Del Carbón Del Interior De Colombia Optimization of a Combined Export Coal Transport Network From the Interior of Colombia. Rev. EIA **16**, 103–113 (2011)
27. Márquez, L., Cantillo, V.: Evaluación de los parámetros de las funciones de costo en la red estratégica de transporte de carga para Colombia. Ing. y Desarollo **29**(2), 286–307 (2011)

Towards a Semantic Intelligence to Support Seaport Governance in Environmental and Ecological Sustainability

Ana X. Halabi-Echeverry[1]([⊠]), Juan Carlos Vergara-Silva[2],
and German A. Ortiz-Basto[3]

[1] Escuela Internacional de Ciencias Económicas y Administrativas,
Universidad de La Sabana, km 7 autopista norte de Bogotá,
D.C. Chía (Cundinamarca), Colombia
ana.halabi@unisabana.edu.co
[2] Facultad de Filosofía y Ciencias Humanas, Universidad de La Sabana,
km 7 autopista norte de Bogotá, D.C. Chía (Cundinamarca), Colombia
juanvs@unisabana.edu.co
[3] Facultad de Ingeniería, Universidad de La Sabana,
km 7 autopista norte de Bogotá, D.C. Chía (Cundinamarca), Colombia
german.ortizl@unisabana.edu.co

Abstract. Taking into account that seaport governance in environmental and ecological sustainability becomes a seaport function of growing public interest, environmental and ecological factors are criteria that will enter into operational choices, capital investments, and cargo routing decisions. Environmental influences as well as factors of institutional governance (regulations and environmental management practices) may lead to common interests among the members. We propose the use of a semantic-based approach to further look at how the knowledge needed to understand the effects observed in seaport decision-making tackle the difficulty of sharing information and finding the best possible representations (model) in order to aid collective decision-making process. The benefits of the semantic use of terminology relevant for the port governance are envisaged in which modeling an ontology will require consensus among domain experts from many and different port areas and the complexity of reaching consensus when multiple conflicting views may exist can threaten consistency across the systems.

Keywords: Seaport governance · Semantic intelligence ·
Environmental and ecological sustainability

1 Introduction

A possible way of sharing the knowledge needed to understand the effects observed in seaport decision-making is providing a cohesive body of knowledge for a more accurate and machine interpretable representation of the port governance. As stated by Karacapilidis (2006) "it is expected that the use of ontologies will result in building more intelligent applications, enabling them to work more accurately at the humans'

© Springer Nature Switzerland AG 2019
C. Paternina-Arboleda and S. Voß (Eds.): ICCL 2019, LNCS 11756, pp. 171–186, 2019.
https://doi.org/10.1007/978-3-030-31140-7_11

conceptual level". An ontology can potentially address the issues of differences in terminology and the different levels of abstraction. As an initial step, we want to capture the semantics behind the data based on the promise of a common decision-making intention and a shared vocabulary provided by an ontology. According to Engers et al. (2008), the first step in the ontological process is to use a rationale of the ontology to learn how it might be structured. Then via an iterative process involving collaboration between experts in the domain, an ontology for that domain will emerge. This paper proposes the use of a common semantic to address this issue. Key value of it is to make the concept of an ontology accessible, hopefully leading to major interest in their application within the domain.

2   Research Approach

How can Semantic Intelligence support port governance to develop capabilities on a common decision-making intention and sharing information with a port partner to envisage sustainable progress?

It is through responsible governance that ports can foresee possibilities for sustainability. The assumption is that port authorities (PAs) will participate in a common decision-making intention and share information with a partner recognising their influences and trade-offs imposed by ecologies and natural borders. To guide port governance in this direction, first is it important to drive knowledge recognising other ports situated in regions/ecosystems in which normative, systemic and procedural dimensions take place. Our approach uses factual or declarative knowledge (i.e. what data is used during decision-making, e.g., verifiable data and regulatory frameworks), needed for decisions concerning seaport environmental and ecological sustainability. We aim to tackle the difficulty of merging information and finding the best possible representations (model) in order to aid the collective decision-making process. The benefits are manifold:

1. Assist in dealing with complex decision-making, especially if uncertainty of inputs/data for decision-making is present.
2. Allow for acquiring knowledge from subjective meaning and intention of the information produced.
3. Provide with a source of quantitative information and periodical results and validation.
4. Allow for handling heterogeneous repositories and merging data with different goals and *foci*.

3   Port Governance

To establish the framework of port governance and its jurisdiction, it is important to recognise the legacy of port governance came from the political reforms in European countries half a century after the Treaty of Rome that formed the basis for the European Union (EU). This fundamental process sought new strategies that led many PAs to

redefine their roles. Verhoeven (2009, 2010) examines the leadership of the EU port authorities to meet contemporary governance challenges differentiating the role of PAs and the EU ports' policy evolution. Thus, an important aspect in port governance lies in legal and regulatory frameworks, namely trade and transport policy, administrative procedures, environment, safety and security regulations.

While it is of relevance to find out the EU port's governance legacy, on the contrary in South America, particularly countries like Colombia, a crisis is being experienced due to the lack of port's policy evolution to carry out general improvements particularly to the seaports. PAs seem to face tensions when referring to ethical and legal principles of governance as important determinants, usually in the sphere of international relations. Disparities, controversies and a number of differences further complicate and obfuscate the possibilities of working out its own emerging dynamics into a functional integrated system. In this respect, negative consequences for the region include (Halabi-Echeverry et al. 2011, p. 1):

- Inability to play an important role in the macro-scale economies
- Lack of rapid growth in international trade
- Reduced capacity to increase wealth and acquire power in the region
- Lower infrastructure investment in the region.

Legal and regulatory frameworks not only relate to ports but also to levels of local and regional government, and in some cases these bodies seek to control areas endorsed and accredited at the international level, also known as Port State Control (PSC). Complementary, Cullinane and Song (2002) point out a starting point of port essential functions: (a) the regulatory function, (b) the landowner function, and (c) the operator function. The regulatory function specifically refers to controlling, surveillance and policing functions of PAs that ensure the safety and security port area. The landlord function refers to the management of areas and activities entrusted to PAs including the implementation of policies. The operator function accounts for the operating factors of the port and their respective profits. Acciaro et al. (2014) distinguish also the community function of PAs that strengthens the management with stakeholders and the community in general, solving conflicts of interest between the parts.

3.1 About Port Jurisdiction

An important aspect raising interest in port governance is the concept of a jurisdiction. A definition provided by Kaye (2015) says that a "jurisdiction is essentially the ability of a State to validly make laws over activities (p. 3)". A jurisdiction is generated in different forms: prescriptive and enforcement. A prescriptive jurisdiction refers to the State's ability to regulate but not necessarily enforce an activity. It may extend to activities taking place within another State. An enforcement jurisdiction refers to the State's ability to actively enforce its laws. This form of jurisdiction cannot be exercised upon another State's sovereignty. Over the sea a complexity arises when referring to an enforcement jurisdiction because more than one State conducts activities at the same time and parts of the ocean. Guy & Lapointe suggest that interregional and inter-jurisdictional perspectives are attracting interest into transportation policies and port planning; however, that poses challenges for "integrating different modal, cargo-based

and regional segments of the transport industry [due to it also] raises governance difficulties because it involves more than one jurisdiction (2011, p. 161)".

An important distinction on territorial boundaries is relevant to a State´s jurisdiction. Maritime zones generate those distinctions creating debates around resources' exploitation, use of ecosystems and territorial protection. Kaye (2015) illustrates in Fig. 1 the Maritime zones available under the United Nations Convention on the Law of the Sea (LOSC). A brief description of each maritime zone follows:

- Internal waters are the innermost zone to the sovereignty of a State. They are treated as equivalent to land.
- Territorial sea (waters) is also under the sovereignty of a State but has been the subject of debate because its breadth lack of consensus among the international community.
- Contiguous zone engages the State in warnings and inspections of infringing vessels, but it has not the right to take enforcement actions against those.
- The Exclusive Economic Zone (EEZ) gives the State an exclusive right of marine resources exploitation for commercial or scientific purposes. It is also a zone of protection and preservation; therefore, enforcement is contained.
- Continental shelf is an extension of the EEZ and gives similar rights to the State for exploitation and exploration of natural resources of the seabed.
- International waters refer to the open seas of the world outside the territorial waters of any State.

The impact of territorial boundaries to the State's jurisdiction has to be considered under three perspectives: (a) A coastal state jurisdiction, (b) A port state jurisdiction, and (c) a flag state jurisdiction. A coastal state jurisdiction is defined by Bautista as a general competence of a State to exercise jurisdiction "over its land territory and internal waters to the territorial sea, and for limited and specified purposes over the other maritime zones (2015, p. 60)". It recognises the State may exercise enforcement whether at the port and harbours or at sea waters. A port state jurisdiction is defined by Rayfuse as "the jurisdiction a State may exercise over vessels visiting its ports…related to the safety and welfare of the State such as health and quarantine requirements as well as immigration and security restrictions (2015, p. 72)". It recognises internal waters as a territorial boundary and that all port states are coastal states. Finally, a flag state jurisdiction is defined by Bateman as an "exclusive jurisdiction over ships flying their flags on the high seas (international waters) (2015, p. 34)". When a vessel leaves the international waters and enters into inner maritime zones, the flag state jurisdiction is no longer exclusive but shared with coastal/port states jurisdictions respectively. Authors such as Keselj (1999) and Molenaar (2007) indicate that ports exercise powers to restrict and control their jurisdictions. These powers have been endorsed and accredited in recent years at the international and local level, including the right of intervention on high seas against vessels having committed violations, such as marine pollution discharges. Molenaar also discusses important exercises of a port jurisdiction which according to his research remain unexplored in the literature. These are (a) departure from state jurisdiction; (b) providing satellite-based vessel monitoring system (VMS) data; (c) penalties for furnishing false information; (d) Lacey Act approaches; and (e) stateless vessels. A brief explanation on these approaches follows (2007, p. 241–244):

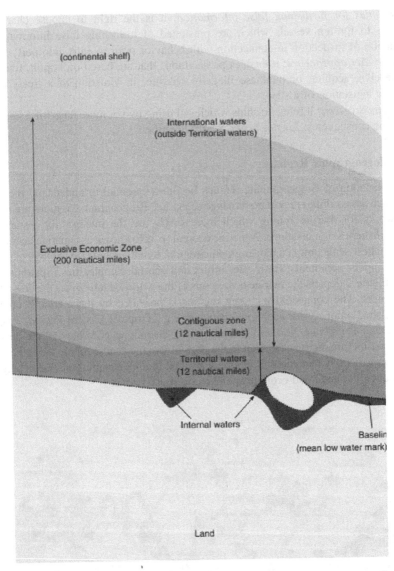

(continental shelf)

International waters
(outside Territorial waters)

Exclusive Economic Zone
(200 nautical miles)

Contiguous zone
(12 nautical miles)

Territorial waters
(12 nautical miles)

Internal waters

Baselin
(mean low water mark)

Land

Fig. 1. Maritime Zones available under LOSC. Taken from Kaye (2015, p. 5)

Departure State Jurisdiction: it is a condition for the departure from the port as a condition for entry. For example, in order to combat vessel-source air pollution, ports could require as a condition for entry into the port that ships off-load all fuels with sulfur contents exceeding limits and to bunker other types of fuel.

Providing VMS data: VMS data as condition for landing a catch, use of port service, or even access to ports to verify if a foreign vessel has engaged in illegal activities.

Penalties for furnishing false information: it is the right to impose charges or penalties to foreign vessels which are presumed of furnishing false information or obstruction of inspection in connection with behavior prior to entry into port.

Lacey Act approaches: it enables prosecutions that are based on import, transport, sell, receive, acquire, or purchase illegally wildlife, in violation of a treaty or the domestic legislation of a state.

Stateless vessels: it refers to ships which sail using two or more flags (of a state) and under the consequence for not being clearly identified.

3.2   Biogeographic Regions

The explanation of Biogeographic regions becomes essential to understand the space into which assess different port governance scenarios. Biogeographic regions are natural frameworks for marine zoning which increasingly are the interest, for instance, of regional fisheries organisations. A marine ecoregion, also known as a marine ecosystem, is the smallest-scale unit in a biogeographic region. It may include (in ecological terms): nutrient inputs, sediments, freshwater influx and coastal complexities (Spalding et al. 2007). Figure 2 depicts the marine ecoregions of the world which covers the coastal and shelf waters. The connectivity among regions is provided by the marine ecoregions distinguished uniquely for ecological patterns (e.g., Tropical Eastern Pacific, Tropical Atlantic and Temperate Northern Atlantic). That distinction goes beyond the designation of coastal jurisdictions (as seen in the port governance subsection). The exercise of coastal jurisdictions becomes an important context in understanding marine ecoregions and its environmental and ecological effects on coastal zones.

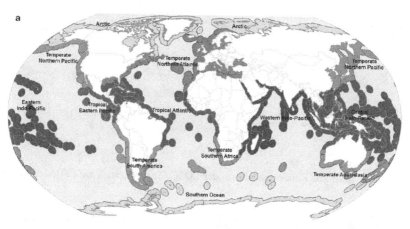

Fig. 2. Marine Ecoregions of the Word. Taken from Spalding et al. (2007, p. 577)

4 Environmental and Ecological Factors as Drivers in Achieving Port Governance

This subsection highlights very recent interesting questions proposed in Hiranandani (2014, p. 127):

1. What are the [factors] in achieving sustainable development in ports?
2. What specific sustainable practices do ports utilise to manage environmental [and ecological] aspects such as air pollution, water quality, ballast water, dredging and disposal of dredge materials, waste disposal, hazardous substances and land/resource use?
3. What policy [and regulatory] frameworks do ports adopt to attain sustainable development?

Ports depend on the environment in which they operate to be sustained. Even if ports compete rather than collaborate in many of their functions, the necessity of ports to collaborate is stronger when the function of the port concentrates on operating under the consideration of climate, water, air, soil and use of biodiversity as resources (Halabi-Echeverry et al. 2012; Verhoeven 2010; Hall et al. 2011). Factors driving environmental and ecological management worldwide enable connections between incentives and international standards that set the parameters in which a seaport can engage in partnership with other seaports. As a result, to integrate environmental management functions among ports such as plans, documents, policy, normativity and performance measurements, is of extreme importance.

Although it is clear a port experiences different environmental questions while in operation, a general awareness exists on three important drivers for its sustainability (APPA 1998; Kruse 2005; Ng and Song 2010; Acciaro et al. 2014; Lam and Notteboom 2014; Hiranandani 2014):

- Climate change (air pollution),
- Water quality (social needs),
- Land use (impacts of growth).

Air emissions becomes an environmental and ecological issue because vessels arriving at ports use their engines to provide heating, cooling and electricity for loading and unloading activities generating in this way emissions (International Council on Clean Transportation -ICCT-, 2007). Other transport and industrial activities within the port area may generate emissions of carbon dioxide (CO_2), nitrogen oxides (NOx), Sulphur oxides (Sox) and particulate matter (PM) (Bailey et al. 2004). Water quality becomes an environmental and ecological issue because there are plenty sources of water pollution at ports, some include: runoff water from storm drains, ship sewage and ballast waters that result in the introduction of non-native or invasive species, port or ship discharges containing detergents, chemicals, etc., and dredging activities consisting on periodic removal of sediments from seabeds to maintain port access channels, the latest affects water quality because increases suspended sediments and releases contaminants (ESPO 2013). Impacts of growth become an environmental and

ecological issue because generally, surrounding communities are increasingly interested in the impacts of port expansion. Congestion, safety, and environmental impacts are derived from port growth.

4.1 Port Environmental Regulatory Frameworks

Specific sustainable practices that ports utilise to manage environmental and ecological factors (namely, air pollution, water quality and land/resource use) also bring into the policy and regulatory frameworks ports adopt to attain sustainable development. According to Acciaro et al. (2014), there is a need to stimulate and facilitate ports in adopting green practices. Pursuing green objectives, the main functions of PAs are likely to influence such practices.

A mandatory or voluntary approach to environmental regulation largely depends on international requirements to enable an organisation to develop and implement green policies and objectives. The International Organisation for Standardisation (ISO), The Environmental Protection Agency (EPA) and The Association of American Port Authorities (AAPA) are some of the advocates on port environmental performance improvement and recognition of the achievements gained. Royson (2008) presents various standards, codes and schemes developed to certify a port's environmental performance. This review also adds voluntary alliances and protocols to accomplish environmental objectives.

ISO 14001:2004 specifies requirements for an environmental management system (EMS) to enable an organisation to develop and implement policy and objectives which take into account legal requirements and other requirements to which the organisation subscribes, and information about significant environmental aspects. It applies to those environmental aspects that the organisation identifies as those it can control and can influence. It does not state specific environmental performance criteria. ISO 14001:2004 is an initiative proposed between the Association of American Port Authorities (AAPA), the Environmental Protection Agency (EPA) and the Global Environment and Technology Foundation (GETF) in the United States.

The International Safety and Environmental Protection Management in Ports (IPSEM) is the minimum requirement needed for a port to operate under environmental standards. The IPSEM is a code of practice and certification scheme provided by Bureau Veritas. Its content is similar to the International Safety Management Code (ISM) for ships.

The Environmental Management System (EMS) Primer for Ports is a formal system for proactively promoting awareness of the environmental footprint in a port. It is promoted by EPA. According to the U.S. Environmental Protection Agency (EPA), an EMS can be seen as a systematic approach to manage environmental issues such as the footprint of a port (EPA 2007).

AAPA/GETF are early initiatives developed by AAPA, EPA and the Global Environment and Technology Foundation (GETF) that brought together nine ports and two federal marine facilities to develop environmental management systems (EMSs).

The World Ports Climate Initiative (WPCI) is a voluntary alliance among ports all over the world signing the World Ports Climate Declaration to work together to reduce the threat of global climate change.

The Climate Registry's General Verification Protocol (GVP) presents the verification requirements for The Registry's voluntary Greenhouse Gases (GHG) emissions reporting program. The Registry developed the general verification protocol (GVP) to provide recognised verification bodies with clear instructions in executing a standard approach to the independent verification of annual GHG emissions. The accreditation process is based on the internationally-recognised ISO 14065 standard.

Other schemes include: The Environmental Improvement Annual Program Winners proposed by AAPA. The European Eco-Management and Audit Scheme (EMAS), specifically with its Code for Port Safety and Health and Environmental Protection Management (PSHEM); Partnerships in Environmental Management for the Seas of East Asia (PEMSEA); The Self Diagnosis Method (SDM) of Eco Ports Foundation (EPF), and the Port Environmental System (PERS) restricted to European Maritime States Members (ESPO). Figure 3 gives an overview of the port policy and regulation described.

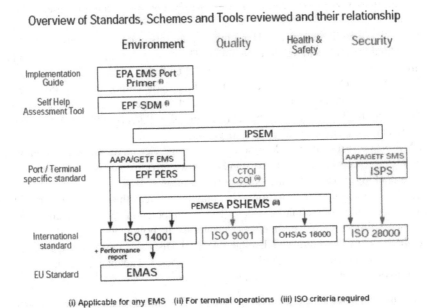

Fig. 3. Overview of Port policy and Regulation. Taken from Green Port Portal (www.greenport.com)

4.2   Environmental Management Practices in Ports

Because an environmental management system links the physical environment and management processes to offer a computerised version of nature, the assumption is that a port would want to participate in a common decision-making intention and share

information with a partner on standards based on desirable practices and requirements for the port's sustainable development. Following Lam and Notteboom (2014) and Verhoeven (2010) it is possible come to the conclusion that ports are able to successfully use an EMS to identify and monitor environmental and ecological challenges, setting standards and accomplishing international initiatives. Cross-cutting seaport EMS programs are the Global Environmental Program (GEP) that encompasses projects such the Land-Ocean Interactions in the Coastal Zones (LOICZ) (Heip and Laane 2011) and The Environmental Information Management System (EIMS) conducted in 2006 by Cambridge Systematics and Venner Consulting Inc. (NCHRP 2007).

The substantial amounts of information that must be gathered, preserved, and used to analyse environmental and ecological factors and drivers, and to account on analyses for decision-making has served as the origin of environmental management practices (EMPs). Some are described as follows:

- EMPs are referred by APPA (1998) after conducting a detailed on-site survey with 30 US ports which still is an obliged source of reference and has a proven track record. This subsection presents the most effective and commonly used EMPs at ports:
- EMP N° 04: Bulk Storage and Handling Liquid: this practice prevents releases from bulk liquid storage and handling facilities. The main potential pollutants are: oil and grease, fuels, organic chemicals, inorganic chemicals.
- EMP N° 05: Chemical Storage and Handling (Non Bulk): this practice prevents releases of pollutants from the handling and storage of chemicals in 55-gallon drums or smaller quantities. The main potential pollutants are: organic chemicals, inorganic chemicals, metals, oils and grease.
- EMP N° 09: Protection of Marine Mammals and Sensitive Aquatic Habitats: this practice protects marine mammals and other sensitive marine life from ship strikes, ship docking procedures, and other port activities.
- EMP N° 014: Ship Air Emissions: this practice reduces the discharge of pollutants to the air from operation of ships. The main potential pollutants are: nitrogen oxides, sulfur dioxides, particulates, hazardous air pollutants.
- EMP N° 016: Dredging and Dredge Material Disposal: this practice reduces the impacts from dredging and dredge materials disposal. The main potential pollutants are: hydrocarbons, heavy metals and pesticides.

It is thus expected for ports working with EMSs to build relationships with business partners to find sustainable solutions to the complex challenges that need to be met. Although, more than a decade ago the issue of EMS's development and adoption was identified as "the need to produce a variety of output products from basic data to high-level value added products aimed at specific user applications (Geerders 1997, p. 60)", there is still debate concerning the value of adopting current EMSs. The contribution of Dinwoodie et al. (2012, p. 114) in "developing environmental awareness in ports and systems-based input-output modelling of port environment management processes" is of major importance. They identify remaining unexplored research on this matter and point out evidence to say that (ibid, p. 123):

- Port authorities have rarely integrated predominantly physical environmental evaluations with business strategy.
- Port authorities need to deploy the business process framework to identify strategic, tactical and operational levels of environmental management processes.

5   A Pathway to Semantic Intelligence

Recently, the IT term *Semantic Intelligence* represents an exclusive concept for the intelligent complex systems way of response to non-predetermined and non-specified ever-changing environment. Semantic Intelligence is autonomous in comprehension and creation of information and discriminating between true and false statements. Through its properties support better understandings and insights in business decision-making (Koleva 2018; Halabi-Echeverry et al. 2018). Understanding port governance can be viewed as a complex system that specifies what kinds of knowledge can be identified in the domain and thus may help when making sense of that domain and acquiring relevant knowledge. In agreement with Bennet and Bennet (2008), a definition of complexity in a decision-making context is given as (p. 5):

> *Complexity is the condition of a system, situation, or organisation that is integrated with some degree of order, but has too many elements and relationships to understand in simple analytic or logical ways...In the extreme, the landscape of a complex situation [system] is one with multiple and diverse connections with dynamic and interdependent relationships, events and processes.*

According to Bennet & Bennet, the decision/problem space $\mathfrak{R}$ for complex systems should include (2008, p. 5):

- Perceived boundaries of the system: underlying structure and dynamic characteristics of the system;
- The semantics of the situation; message and intention;
- Sets of relative data and information: It requires understanding of several sources simultaneously;
- Observable events, history, trends, and patterns of behaviour;
- The identity of the individuals/groups involved.

The framework in Fig. 4 illustrates key subsystems (building blocks) that contribute to the assumption that PAs will participate in a common decision-making intention and share information with a partner recognising their influences and trade-offs imposed by ecologies and natural borders. The key subsystems briefly call the attention on four main matters (CIESIN, 1992): (1) *Economic Systems*: determines how the seaports uses land patterns referring to the spatial patterns of urban expansion affected by physical factors; (2) *Factors of Production and Technology*: determines how the seaports uses this concept to point out port to environmental opportunities, such as the conditions of various pollutants before entering waterways including ballast water onboard vessels typically released nearby the port area; (3) *Political Systems and Institutions*: determines how the seaports uses factors of institutional governance (regulations and environmental management practices) and, (4) *Global Scale*

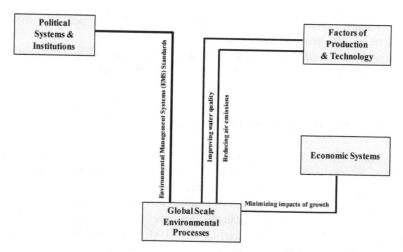

Fig. 4. A possible Semantic Representation (model) ($\Diamond\phi$). (Halabi-Echeverry and Richards 2014)

Environmental Processes: determine how the seaports uses environmental opportunities, such as the reduction of common air pollutants including nitrogen oxides (NOx) and sulfur oxides (SOx).

Outputs from one subsystem act as inputs of another in a dynamic way. Also, factors of one subsystem influence the evolution of factors presented in other systems. Thus, the system is not in equilibrium. Factors driving environmental and ecological management are represented by arrows (linkages) and both: factors and connected subsystems constitute a possible semantic representation (model) ($\Diamond\phi$).

Interesting forms of data are regarded as Graph Data, Text-based Data and Existential Graphs which motivates the analysis for a visual simplification of the decision-making system presented in Fig. 4. Graph Data is defined using the mathematical definition of a graph G corresponding to a collection of vertices or nodes and edges that connect pairs of vertices. Interesting properties adjacent to Graph Theory allows reflecting on Graph data for applications on connectivity information as an important area of research (Smyth et al. 2002). Text-based Data is another important form of data that includes documents as a source. Existential Graphs (EG) introduced by Peirce in 1880 relate to Modal Logic. They also illustrate common syntax and rules of inference. "Unlike the syntax-based approach of most current textbooks, Peirce's method addresses the semantic issues of logic in a way that can be transferred to any notation" (Jimenez 2003).

Thus, the EG in Fig. 5 captures two elements: nodes (K1...K4) and edges. Each edge is an ordered n-tuple or an n-ary relation $< a_1,...a_n > ... < h_1,...h_n >$. The nodes in the EG are indivisible graphs called *atoms*. In the algebraic notation, each atom consists of a single predicate with its associated relations. The n-ary relation introduced can be found in Table 1. For instance, a set of 9-tuples corresponding to $<$ **facilities, inadequacies, CO2, O3comply, O3, SO2, NOx, scientist, type** $>$ connects edges (linkages or semantic relationships) for −***Reducing Air Emissions*** (dash before

"Reducing Air Emissions" is the "line of identity." The identification of linkages or semantic relationships between the nodes in EG describe the predicate as a complete sentence, as shown (this syntax is no simple as we do not discuss it in this paper).

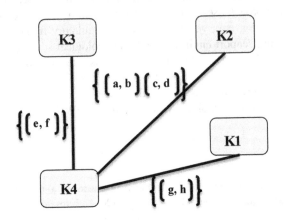

Fig. 5. Existential Graph (EG) for an Exemplary Predicate

Table 1. Linkages or semantic relationships

Reducing air emissions	*facilities, inadequacies, CO2, O3comply, O3, SO2, NOx, scientist,type*
Improving water quality	*needWtTreat, facilities, oils, chemicals, Inadequacies, Scientist, runoff, NMS, dredgeOcean*
Minimizing impacts of growth	*CRP, LandFarms, Scientist, MarketVal, GAPStatus1, GAPStatus2, GAPStatus3, GAPStatus4, CountyArea, LeaseNum, LeaseAcres, LeaseArea*
Port EMS status	*ISO14001, AAPA_GETF, IPSEM, WPCI, Climate Registry*

An Exemplary Predicate: *PAs will share information with a partner if environmental influences as well as factors of institutional governance (regulations and environmental management practices) may lead to common decision-making intentions imposed by ecologies and natural borders.*

6 Conclusions and Future Work

Understanding port governance can be viewed as a complex system that specifies what kinds of knowledge can be identified in the domain and thus may help when making sense of that domain and acquiring relevant knowledge. To guide port governance, first is it important to drive knowledge recognising other ports situated in regions/ecosystems in which normative, systemic and procedural dimensions take place. The assumption is that PAs will participate in a common decision-making intention and

share information with a partner recognising the influences and trade-offs imposed by ecologies and natural borders.

As an initial step, we want to capture the semantics behind the data based on the promise of a common decision-making intention and a shared vocabulary provided by an ontology. Our approach uses factual or declarative knowledge (i.e. what data is used during decision-making, e.g., verifiable data and regulatory frameworks), needed for decisions concerning seaport environmental and ecological sustainability. Finally, we propose Existential Graphs (EG) to conveniently illustrate common syntax and rules of inference.

As part of our future work, we will further develop the syntax-based approach and transfer the notation to an Ontology Web Language (OWL).

References

Acciaro, M., et al.: Environmental sustainability in seaports: a framework for successful innovation. Marit. Policy Manag. **41**(5), 480–500 (2014). https://doi.org/10.1080/03088839.2014.932926

American Association of Port Authorities AAPA. (1998). Environmental Management Handbook

Bailey, D., et al.: Harbouring Pollution: The Dirty Truth About US Ports. National Resources Defence Council, New York (2004)

Bateman, S.: The role of flag states. In: Warner, R., Kaye, S. (eds.), Routledge Handbook of Maritime Regulation and Enforcement, pp. 43–58 (2015). http://mqu.eblib.com.au/patron/FullRecord.aspx?p=2194898

Bautista, L.: The role of coastal states. In: Warner, R., Kaye, S. (eds.), Routledge Handbook of Maritime Regulation and Enforcement, pp. 59–70 (2015). http://mqu.eblib.com.au/patron/FullRecord.aspx?p=2194898

Bennet, A., Bennet, D.: The Decision-making process in a complex situation. In: Handbook on Decision Support Systems 1, pp. 3–20. Springer, Heidelberg (2008). https://doi.org/10.1007/978-3-540-48713-5_1

Consortium for International Earth Science Information Network (CIESIN).—Executive Summary‖. http://www.ess.co.at/GAIA/SPD/spd_image.html

Cullinane, K., Song, D.-W.: Port privatization policy and practice. Transp. Rev. **22**(1), 55–75 (2002)

Dinwoodie, J., Tuck, S., Knowles, H., Benhin, J., Sansom, M.: Sustainable development of maritime operations in ports. Bus. Strategy Environ. **21**(2), 111–126 (2012). https://doi.org/10.1002/bse.718

Engers, T., Boer, A., Breuker, J., Valente, A., Winkels, R.: Ontologies in the legal domain. In: Chen, H., Brandt, L., Gregg, V., Traunmüller, R., Dawes, S., Hovy, E., Macintosh, A., Larson, C. (eds.) Digital Government, 17, pp. 233–261. Springer, US (2008)

Environmental Protection Agency (EPA): An Enviromental Management System Primer for Ports: Advancing Port Sustainability (2007). http://archive.epa.gov/osem/sectors/web/pdf/ems_primer.pdf

ESPO (European Sea Ports Organization): Annex 1: Good practice examples in line with the 5 Es (2013). http://www.ecoports.com/templates/frontend/blue/images/pdf/Annex%201_Good%20Practices_Version%202_July%202013.pdf

Geerders, P.J.F.: Nature's data and data's nature. In: Harmancioglu, N., Alpaslan, M.N., Ozkul, S., Singh, V. (eds.) Integrated Approach to Environmental Data Management Systems, 31, pp. 49–60. Springer, Netherlands (1997). https://doi.org/10.1007/978-94-011-5616-5_5

Guy, E., Lapointe, F.: Building value into transport chains: the challenges of multi-goal policies. In: Comtois, C., Slack, B., Hall, P.V., McCalla, R.J. (eds.) Integrating Seaports and Trade Corridors, pp. 154–164. Ashgate, Farnham, Surrey (2011)

Halabi-Echeverry, A., Richards, D., Bilgin, A.: Proposing a port decision system approach for dynamic integration of South American sea ports. In: Proceedings of IEEE Explorer International Conference on Advances in ICT for Emerging Regions (ICTer). Colombo, Sri Lanka, September 1–2, 2011, pp. 58–64 (2011). https://doi.org/10.1109/icter.2011.6075027

Halabi-Echeverry, A.X., Richards, D., Bilgin, A.: Identifying characteristics of seaports for environmental benchmarks based on meta-learning. In: Richards, D., Kang, B.H. (eds.) PKAW 2012. LNCS (LNAI), vol. 7457, pp. 350–363. Springer, Heidelberg (2012). https://doi.org/10.1007/978-3-642-32541-0_31

Halabi-Echeverry, A.X., Richards, D.: Towards management of the data and knowledge needed for port integration: an initial ontology. In: Kim, Y.S., Kang, B.H., Richards, D. (eds.) PKAW 2014. LNCS (LNAI), vol. 8863, pp. 165–179. Springer, Cham (2014). https://doi.org/10.1007/978-3-319-13332-4_14

Halabi-Echeverry, A.X., Vergara-Silva, J.C., Karray, M.H.: Semantic intelligence in a seaport context. In: IEEE/ACS 15th International Conference on Computer Systems and Applications (AICCSA), Aqaba 2018, pp. 1–2 (2018). https://doi.org/10.1109/aiccsa.2018.8612887. http://ieeexplore.ieee.org/stamp/stamp.jsp?tp=&arnumber=8612887&isnumber=8612765 Echeverry

Heip, C., Laane, R.: Aspects of coastal research in contributions to LOICZ in the Netherlands and Flanders (2002–2010). LOICZ Research & Studies No 38. Helmholtz-Zentrum Geesthacht, p. 184 (2011)

Hiranandani, V.: Sustainable development in seaports: a multi-case study. WMU J. Marit. Aff. **13** (1), 127–172 (2014)

Jimenez, L.: Gráficos Existenciales Gama en Color y Algunos Sistemas de Lógica Modal. Fundación Universitaria Konrad Lorenz, Trabajo de Grado (2003)

Karacapilidis, N.: An overview of future challenges of decision support technologies. In: Intelligent Decision making Support Systems, pp. 385–399. Springer, London (2006). https://doi.org/10.1007/1-84628-231-4_20

Kaye, S.: A zonal approach to maritime regulation and enforcement. In: Warner, R., Kaye, S. (eds.), Routledge Handbook of Maritime Regulation and Enforcement, pp. 3–15 (2015). http://mqu.eblib.com.au/patron/FullRecord.aspx?p=2194898

Keselj, T.: Port state jurisdiction in respect of pollution from ships: the 1982 united nations convention on the law of the sea and the memoranda of understanding. Ocean Dev. Int. Law **30**(2), 127–160 (1999)

Koleva, J.: Semantic intelligence. In: Khorsow-Pour, M. (ed.) Encyclopedia of Information Science and Technology 4-th edn. IGI- Global, Hershey (2017). https://doi.org/10.4018/978-1-5225-2255-3.ch020

Kruse, C.J.: Environmental management systems at ports - a new initiative. In: Proceedings of the 14th Biennial Coastal Zone Conference (2005)

Lam, J.S.L., Notteboom, T.: The greening of ports: a comparison of port management tools used by leading ports in Asia and Europe. Transp. Rev. **34**(2), 169–189 (2014)

Molenaar, E.: Port state jurisdiction: toward comprehensive, mandatory and global coverage. Ocean Dev. Int. Law **38**(1–2), 225–257 (2007). https://doi.org/10.1080/00908320601071520

National Cooperative Highway Research Program (NCHRP): Proto-type software for an environmental information management and decision support system (2007). http://onlinepubs.trb.org/onlinepubs/nchrp/nchrp_rrd_317.pdf

Ng, A.K.Y., Song, S.: The environmental impacts of pollutants generated by routine shipping operations on ports. Ocean Coast. Manag. 53, 301–311 (2010)

Rayfuse, R.: The role of port states. In: Warner, R., Kaye, S. (eds.), Routledge Handbook of Maritime Regulation and Enforcement, pp. 71–85 (2015). http://mqu.eblib.com.au/patron/FullRecord.aspx?p=2194898

Royson, K.: Environmental certification for ports: assessing the options (2008). http://www.greenport.com/news101/Projects-and-Initiatives/environmentalcertification-for-ports-assessing-the-options#sthash.2jWMZuBQ.dpuf

Smyth, P., Pregibon, D., Faloutsos, C.: Data-driven evolution of data mining algorithms. Commun. ACM 45(8), 33–37 (2002)

Spalding, M.D., Fox, H.E., Allen, G.R., Davidson, N., Ferda, X.D., et al.: Marine ecoregions of the world: a bioregionalization of coastal and shelf areas. Bioscience 57(7), 573–583 (2007)

Verhoeven, P.: European ports policy: meeting contemporary governance challenges. Marit. Policy Manag. 36(1), 79–101 (2009). https://doi.org/10.1080/03088830802652320

Verhoeven, P.: A review of port authority functions: towards a renaissance? Marit. Policy Manag. 37(3), 247–270 (2010). https://doi.org/10.1080/03088831003700645

Vehicle Routing Problems

Optimal Solutions for the Vehicle Routing Problem with Split Demands

Hipólito Hernández-Pérez and Juan-José Salazar-González[(⊠)] ⓘ

DMEIO, Facultad de Ciencias, Universidad de La Laguna,
38200 La Laguna, Tenerife, Spain
{hhperez,jjsalaza}@ull.es

Abstract. This paper describes a branch-and-cut algorithm for a generalization of the Capacitated Vehicle Routing Problem where a customer can be visited several times. The algorithm is based on solving a Mixed Integer Programming model for a relaxed problem, and a cutting-plane procedure to eliminate invalid integer solutions. Computational results on benchmark instances show the performance of the algorithm compared with other approaches found in the literature. In particular, the new algorithm is able to solve a testbed SDVRP instance that was unsolved in the literature.

Keywords: Capacitated Vehicle Routing · Split demand ·
Branch-and-cut

1 Introduction

Let us consider a set of customers and a depot, each one in a known location. The travel costs between locations are known. There is a homogeneous fleet of capacitated vehicles originally at the depot. The well-known *Capacitated Vehicle Routing Problem* (CVRP) looks for designing vehicle routes minimizing the total travel cost while satisfying the demand requirement of the customers and the upper limit capacity of the vehicles. A traditional assumption in CVRP is that each customer must be visited by one vehicle with a single visit.

An interesting variant of the CVRP is the *Split Demand Vehicle Routing Problem* (SDVRP), in which a customer can be visited several times serving part of its demand in each visit. It was introduced in [6] and it is NP-hard.

Several exact approaches have been described in the literature. A constraint relaxation branch-and-bound algorithm is designed in [7]. Column generation approaches are described in [17] and [12]. An integer linear programming formulation and a relaxed model for the SDVRP are given in [4]. A dynamic programming approach is described in [13]. A two-stage algorithm is introduced in [11]. A tight formulation with several variables to compute a lower bound for

This work has been partially supported by the research project MTM2015-63680-R (MINECO/FEDER) and ProID2017010132 (Gobierno de Canarias/FEDER).

© Springer Nature Switzerland AG 2019
C. Paternina-Arboleda and S. Voß (Eds.): ICCL 2019, LNCS 11756, pp. 189–203, 2019.
https://doi.org/10.1007/978-3-030-31140-7_12

the SDVRP is provided in [14]. A branch-and-cut-and-price algorithm for two versions of the SDVRP (the limited and unlimited fleet of vehicles) is presented in [1]. Two branch-and-cut algorithms in [2] are based on solving relaxed models for the SDVRP called two-index vehicle flow formulation and single commodity flow formulation. The first relaxation is similar to that proposed by [4]. Finally, [15] propose a new arc formulation for the SDVRP and a relaxation which is similar to the single commodity flow formulation in [2]. They also propose a variant of the SDVRP where an upper limit on the number of visits to a customer is imposed. Heuristic approaches have also been proposed, like in [18] and [5]. See [3] for a survey.

Our manuscript proposes a new exact approach to solve the SDVRP with the assumption that a customer can be visited a maximum number of times. The approach described in our manuscript could easily be adapted to deal with the *preemption* case, where it may be convenient to deliver product in one visit to a customer and collect product from that customer in another visit. Although this case is of practical use in some applications (for example, in bike sharing) the SDVRP in the literature does not allow any preemption, and therefore we present and evaluate the approach in this manuscript for the SDVRP without any preemption. Finally, since benchmark SDVRP instances in the literature are symmetric, we describe the approach assuming symmetric travel costs.

The exact approach is a branch-and-cut algorithm based on solving the two-index vehicle flow formulation in [4] and [2]. The novelty of our algorithm is the procedure for checking whether a relaxed solution is SDVRP feasible. The branch-and-cut algorithm is tested on instances in [2] and indeed it solves to optimality an instance that was previously unsolved.

2 Mathematical Formulations

This section describes Mixed Integer Linear Programming (MILP) models for the SDVRP. Before we need to introduce some notation.

2.1 Problem Definition

Let $I = \{0, 1, \ldots, n\}$ be the set of locations, where the depot is represented by location 0 and customers by locations from 1 to n. For each $i \in I \setminus \{0\}$, let d_i be the demand of customer i. We denote by $d_0 = -\sum_{i \in I \setminus \{0\}} d_i$ the demand of the depot. The capacity of a vehicle is a-priori known, and it is denoted by Q. We denote by m_0 the number of available vehicles, i.e. the fleet size and by m_i the maximum number of visits to customer i for each $i \in I \setminus \{0\}$. The travel cost between location i and location j is c_{ij}.

Let V_i be an ordered set of m_i nodes representing potential visits to location i. Since all nodes in V_i are identical, we intend that the sequence of visits of the vehicles to i is represented by consecutive nodes in V_i, with i_1 representing the first visit. The set $V = \cup_{i \in I} V_i$ is the node set of a directed graph $G = (V, A)$ where A is the arc set connecting nodes associated with different locations.

For a given subset S of nodes, we write $\delta_A^+(S) = \{(v, w) \in A : v \in S, w \notin S\}$ and $\delta_A^-(S) = \{(v, w) \in A : v \notin S, w \in S\}$. Given an arc $a = (v, w)$ we also denote the cost c_a from v to w as the travel cost c_{ij} if $v \in V_i$ and $w \in V_j$.

2.2 Exact Model

Let us consider the following variables. For each arc $a \in A$, a binary variable x_a assumes value 1 if and only if the route includes a, and a continuous variable f_a represents the load of the vehicle when traversing a. For each node $v \in V$, a binary variable y_v assumes value 1 if and only if the route includes v, and a continuous variable g_v determines the number of units delivered when performing the visit v.

We now present a mathematical model for the SDVRP in which the set of routes are described by a single route for one capacitated vehicle passing through the depot at most m_0 times:

$$\min \sum_{a \in A} c_a x_a \tag{1}$$

subject to:

$$y_{i_1} = 1 \qquad \text{for all } i \in I \tag{2}$$

$$\sum_{a \in \delta_A^+(v)} x_a = \sum_{a \in \delta_A^-(v)} x_a = y_v \qquad \text{for all } v \in V \tag{3}$$

$$\sum_{a \in \delta_A^+(S)} x_a \geq y_v + y_w - 1 \qquad \text{for all } S \subseteq V \,,\, v \in S \,,\, w \in V \setminus S \tag{4}$$

$$\sum_{a \in \delta_A^-(v)} f_a - \sum_{a \in \delta_A^+(v)} f_a = g_v \qquad \text{for all } v \in V \tag{5}$$

$$0 \leq f_a \leq Q x_a \qquad \text{for all } a \in A \tag{6}$$

$$\sum_{l=1}^{m_i} g_{i_l} = d_i \qquad \text{for all } i \in I \tag{7}$$

$$g_v \geq 0 \qquad \text{for all } v \in V \setminus V_0 \tag{8}$$

$$g_v \leq 0 \qquad \text{for all } v \in V_0 \tag{9}$$

$$y_v, x_a \in \{0, 1\} \qquad \text{for all } v \in V \,,\, a \in A. \tag{10}$$

Equation (2) ensure that all locations are visited at least once by the vehicle. Equation (3) force the vehicle to enter and leave once each visit v with $y_v = 1$. Inequalities (4) ensure a connected route. Constraints (5)–(7) ensure that the load of the vehicle is able to satisfy the demand decided at each visit. Inequalities (8) and (9) guarantee that the deliveries are only at the customers and the pickups only at the depot.

A related model was presented in [10] for the *One-commodity Pickup-and-Delivery Travelling Salesman Problem* (1PDTSP) and in [16] for the Split-Demand variant (SD1PDTSP) where a customer may be served by several visits.

The SDVRP is a particular case of the SD1PDTSP, thus their model could also be used to solve SDVRP. Still, that model could strongly be simplified to get the model (1)–(10) for the SDVRP, and in addition could be strengthened to remove its major drawback: multiple solutions representing the same routes. This drawback can be reduced with the help of symmetry-breaking constraints like the following ones.

All the nodes V_i can represent a visit to a customer i, and this could create symmetry that can be eliminated with the inequalities:

$$y_{i_l} \geq y_{i_{l+1}} \qquad \text{for all } i \in I, \, l \in \{1, \ldots, m_i - 1\}, \, i_l \neq 0_1. \tag{11}$$

These constraints force a sequence of visits for the vehicle visiting each customer i when starting and ending at 0_1. They guarantee that visit $l+1$ to customer i can only be performed after visit l was previously completed.

Each node in V_0 represents the starting visit of a vehicle route, which is a source of another symmetry in the formulation because all vehicles have the same capacity. Valid inequalities to avoid this situation are:

$$x_{0_k,u} + x_{0_l,v} \leq 1 \quad \text{for all } 1 \leq k < l \leq m_0, \, i > j \,, u \in V_i, v \in V_j \tag{12}$$

$$x_{0_k,u} + x_{v,0_k} \leq 1 \quad \text{for all } 1 \leq k \leq m_0, \, i > j \,, u \in V_i, v \in V_j. \tag{13}$$

Constraints (12) establish a non-decreasing order of the first visited customer in each route, and constraints (13) force that the last visited customer is not smaller than the first visited customer in each route.

2.3   Relaxed Model

We present a relaxed formulation on a smaller graph representation and related to the classical two-index formulation for the CVRP. Similar relaxed formulations have been presented for the SDVRP in [4] and [2]. We contribute with a different procedure to check the feasibility for the SDVRP of a relaxed solution. Since the travel costs c_{ij} are symmetric, we present the relaxed model on a non-oriented graph.

Let $G' = (I, E)$ be the non-oriented graph where each node represents a location (instead of a visit, as in G) and each edge $e \in E$ joins two locations. For a given subset $S \subseteq I$, we write $\delta_E(S) = \{[i, j] \in E : i \in S, j \notin S \text{ or } i \notin S, j \in S\}$. We define the following decision variables. For each edge $e \in E$, let x_e be an integer variable representing the number of times the edge e is traversed by the vehicle. Observe that we are using the notation x for variables in the two models; however, this should not confuse the reader as the subindex identifies the model univocally. For example, x_a is a binary variable in the model (1)–(10), and x_e can be greater than one in a relaxed solution. For each $i \in I$, let z_i be an integer variable representing the number of visits to location i. For each $e = [i, j] \in E$ let c_e be the travel cost c_{ij} from i to j.

Then the relaxed model is the following:

$$\min \sum_{e \in E} c_e x_e \tag{14}$$

subject to:

$$\sum_{b\in\delta_E(i)} x_e = 2z_i \quad \text{for all } i \in I \tag{15}$$

$$1 \leq z_i \leq m_i \quad \text{for all } i \in I \tag{16}$$

$$\sum_{e\in\delta_E(S)} x_e \geq 2 \left\lceil \frac{\sum_{i\in S} d_i}{Q} \right\rceil \quad \text{for all } S \subset I \setminus \{0\} \tag{17}$$

$$x_e \geq 0 \text{ and } x_e \in \mathbb{Z} \quad \text{for all } e \in E. \tag{18}$$

Equation (15) force that the number of edges incident to a location i is two times the number of visits to i. Inequalities (16) limit the number of visits to each location. Inequalities (17) ensure the connectivity of the route and the vehicle capacity. Clearly, all SDVRP solutions satisfy the above formulation, but there may be solutions of (15)–(18) which do not correspond to SDVRP solutions. Indeed this model allows preemption, as defined in the introduction. The next section shows how to check whether a relaxed solution is valid for the SDVRP, and how it can be eliminated from the model with a linear inequality if the relaxed solution is not valid.

As noted in [2], if the travel costs satisfy the triangular inequality, the following inequalities are valid for the relaxed model:

$$\sum_{i\in I\setminus\{0\}} z_i \leq n + m_0 - 1 \tag{19}$$

$$x_e \leq 1 \quad \text{for all } e \in E \setminus \delta_E(\{0\}). \tag{20}$$

3   A Branch-and-Cut Algorithm

The large number of variables in the model (1)–(13) makes it difficult to solve medium-size SDVRP instances to optimality in practice. We propose a different procedure based on solving the relaxed model (14)–(20). The procedure consists of restricting the invalid formulation with a new linear system of linear constraints.

Let (x^*, z^*) be an integer solution of the relaxed model (14)–(20). To check whether it yields a feasible solution of the SDVRP we propose to solve the following linear system. The linear system is taken from the model (1)–(13) but now considering that each customer i is visited z_i^* times. Let G'' be the subgraph of G induced by $V^* := \cup_{i\in I}\{i_1,\ldots,i_{z_i^*}\}$. For each arc a in G'', consider a binary variable x_a assuming value 1 if and only if the route includes a, and a continuous variable f_a representing the vehicle load when traversing a. For each node v in G'', consider a continuous variable g_v representing the $|g_v|$ units of product delivered (if $g_v > 0$) or collected (if $g_v < 0$) when performing the visit v. The linear system contains now the constraints (2)–(13) with V replaced by V^* and the y_i variables fixed to value 1 for all $i \in V^*$. The only variables in the linear system are now x_a, f_a and g_v. The linear system is then

$$\sum_{a\in\delta_A^+(v)} x_a = \sum_{a\in\delta_A^-(v)} x_a = 1 \quad \text{for all } v \in V^* \tag{21}$$

$$\sum_{a\in\delta_A^+(S)} x_a \geq 1 \quad \text{for all } S \subset V^* \tag{22}$$

$$\sum_{a\in\delta_A^-(v)} f_a - \sum_{a\in\delta_A^+(v)} f_a = g_v \quad \text{for all } v \in V^* \tag{23}$$

$$0 \leq f_a \leq Qx_a \quad \text{for all } a \in A \tag{24}$$

$$\sum_{1\leq l\leq z_i^*} g_{i_l} = d_i \quad \text{for all } i \in I \tag{25}$$

$$g_v \geq 0 \quad \text{for all } v \in V^* \setminus V_0^* \tag{26}$$

$$g_v \leq 0 \quad \text{for all } v \in V_0^* \tag{27}$$

$$\sum_{a=(i_k,j_l):[i,j]=e} x_a = x_e^* \quad \text{for all } e \in E \tag{28}$$

$$x_{0_k i_l} + x_{0_{k'} i'_{l'}} \leq 1 \quad \text{for all } 1 \leq k < k' \leq z_0^*, i > i', l = 1\ldots z_i^*, l' = 1\ldots z_{i'}^* \tag{29}$$

$$x_{0_k i_l} + x_{i'_{l'} 0_k} \leq 1 \quad \text{for all } 1 \leq k \leq z_0^*, i > i', l = 1\ldots z_i^*, l' = 1\ldots z_{i'}^* \tag{30}$$

$$x_a \in \{0,1\} \quad \text{for all } a \in A. \tag{31}$$

The explanation of the constraints is the same as the explanation of system (2)–(13). A solution of the new linear system can be seen as a certificate that the solution (x^*, z^*) from (14)–(20) represents a valid route for the SDVRP.

This formulation suggests an algorithm to solve the SDVRP. The algorithm solves the relaxed model (14)–(20) where constraints (17) are heuristically separated. To this end, three heuristic procedures have been implemented.

The first procedure builds a list of candidates S to check constraints (17). Two integer parameters p_1 and p_2 are used for this selection. First, the procedure finds the p_1 subsets of cardinality two $\{i,j\}$ such that $i,j \in I \setminus \{0\}$ and $x_{[i,j]}^*$ are greater as possible. After, each set of cardinality $p_2 \geq k \geq 3$ is generated from a set of cardinality of size $k-1$ in the list, inserting a vertex j verifying that $j \in I \setminus (S \cup \{0\})$, $\sum_{i\in S} x_{[i,j]}^* > 0$ and $S \cup \{j\}$ is not in the list. If there are various vertices j verifying these conditions, the one which maximizes $\sum_{i\in S} x_{[i,j]}^*$ is inserted. Parameters p_1 and p_2 were set to n and $n/2$, respectively.

The second procedure computes max-flow problems to guarantee

$$\sum_{e\in\delta_E(S)} x_e \geq 1.$$

The third procedure computes other max-flow problems to guarantee

$$\sum_{e\in\delta_E(S)} x_e \geq \sum_{i\in S} d_i/Q.$$

Solutions with fractional values are discarded through a classical binary branching procedure. Given an integer solution (x^*, z^*), consider the linear system (21)–(31). If this system is feasible then (x^*, z^*) is feasible for the SDVRP; otherwise, a linear constraint that separates this relaxed solution must be inserted. An example of such constraint was proposed in [2], and is the following:

$$\sum_{e \in E \,:\, x_e^* = 0} x_e \geq 1. \tag{32}$$

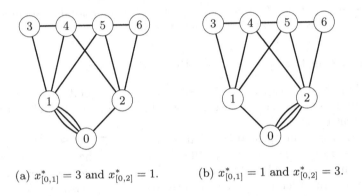

(a) $x_{[0,1]}^* = 3$ and $x_{[0,2]}^* = 1$.   (b) $x_{[0,1]}^* = 1$ and $x_{[0,2]}^* = 3$.

Fig. 1. Two relaxed solutions with the same set of non-zero variables.

However, this constraint can discard more than one relaxed solution. Figure 1 shows two different relaxed solutions where the set of edges $\{e \in E : x_e^* > 0\}$ are the same. The authors in [2] can use the inequality (32) to eliminate the invalid solutions because they enumerate all possible routes using arcs in the set $\{e \in E : x_e^* > 0\}$, and solve a route-based formulation for the SDVRP with those routes. We propose a different alternative procedure based on the fact that the difference in the potential SDVRP is the number of visits to customers, as suggested in Fig. 1. To implement this idea, we modify the relaxed model (14)–(20) adding the following family of constraints (and new variables).

$$z_i = \sum_{k=0}^{\lfloor \log_2 m_i \rfloor} 2^k w_{ik}, \quad \text{for all } i \in I \setminus \{0\}, \tag{33}$$

where w_{ik} is a binary decision variable for all $i \in I \setminus \{0\}$ and $k \in \{0, \ldots, \lfloor \log_2 m_i \rfloor\}$. Note that w_{ik} is a binary representation of the number z_i. Then the inequality to separate an infeasible integer solution (x^*, z^*) of the relaxed model (14)–(20) and (33) is the following:

$$\sum_{\substack{e \in E \setminus \delta(0) \,:\, x_e^* = 0}} x_e + \sum_{\substack{i \in I \setminus \{0\}, \\ k \in \{0, \ldots, \log_2 m_i\}, \\ w_{ik}^* = 0}} w_{ik} + \sum_{\substack{i \in I \setminus \{0\}, \\ k \in \{0, \ldots, \log_2 m_i\}, \\ w_{ik}^* = 1}} (1 - w_{ik}) \geq 1. \tag{34}$$

Moreover, the right hand side of this inequality can be strengthened up to 2. This is because, if the first sum is equal to zero, there must be at least two customers visited a different number of times; whereas if the last two sums are equal to zero, at least two edges have to be different.

4   Strengthening the Branch-and-Cut Algorithm

As it is shown in Sect. 5, solving the linear system (21)–(31) may be quite time consuming on some instances. For that reason, we propose an alternative feasibility-checking procedure based on shrinking a graph and enumerating some routes. This section describes the procedure.

4.1   Shrinking the Graph of the Relaxed Solution

A first simplification consist in shrinking the relaxed solution by joining connected vertices of degree two. A second simplification divides the relaxed solution into different parts.

Figure 2 shows an optimal solution of the relaxed model (14)–(20) for the instance eil30 in our benchmark collection (Sect. 5) when rounded Euclidean distances are considered. All customers are visited once but customer 18 is visited twice. In order to check if it corresponds to a feasible solution of the SDVRP model (21)–(31) we reduce the problem joining connected vertices of degree two to a single vertex. More precisely, sets $S_1 = \{22, 2, 5, 4, 3, 20\}$, $S_2 = \{19, 6, 1, 24, 25, 29, 27, 28, 26\}$, $S_3 = \{15, 16, 13, 7, 17, 9, 14, 8, 12, 11, 10\}$ and $S_4 = \{23, 21\}$ are shrunk to a single vertex each. Figure 3 shows the shrunk solution.

Another observation is that model (21)–(31) can be disaggregated to different parts of the shrunk solution. To be more precise, it can be applied to each connected components of the solution graph removing the depot. For example, if we remove the depot from the graph of Fig. 3, there are two connected components, then the subproblem can be applied to the subgraph induced by vertices 0 and S_1, and by vertices 0, 18, S_2, S_3 and S_4. In addition, when there are not customers visited more than once in a connected component, this component is trivially feasible.

This procedure based on shrinking the relaxed solution yields a stronger inequality than constraint (34) because any solution generating the same shrunk graph would be infeasible, too. Thus, let $S_1, \ldots, S_p$ be the vertices of the shrunk graph; the inequality is:

$$
\sum_{\substack{1 \le k < l \le p \\ i \in S_k \ j \in S_l : x^*(S_k : S_l) = 0}} x_{[i,j]} + \sum_{\substack{i \in I \setminus \{0\}, \\ k \in \{0, \ldots, \log_2 m_i\}, \\ w_{ik}^* = 0}} w_{ik} + \sum_{\substack{i \in I \setminus \{0\}, \\ k \in \{0, \ldots, \log_2 m_i\}, \\ w_{ik}^* = 1}} (1 - w_{ik}) \ge 2.
$$

$$(35)$$

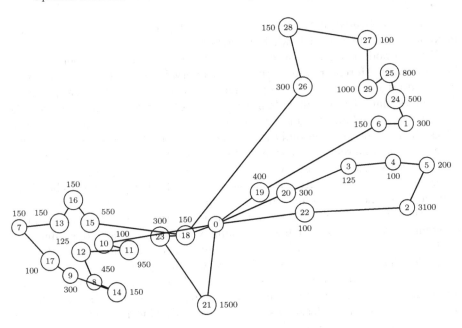

Fig. 2. Solution of the relaxed model for instance ei130.

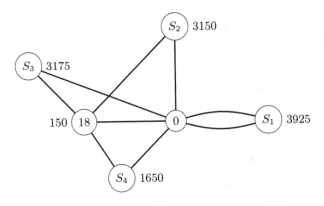

Fig. 3. Shrunk solution of the relaxed model for instance ei130.

4.2   Checking the Feasibility by Enumeration

The procedure explores, in a depth-first way, the sequences of the shrunk supported graph of the relaxed solution and tries to build a sequence of visits to each location $(v_0, \ldots, v_r)$ where each edge $e \in E$ is used x_e^* times by adding nodes iteratively. However, if it detects an infeasibility condition, the exploration is pruned at this point and it continues by other branch. To this aim, the procedure stores at each visit v_p a *demand* interval of the minimum and maximum quantity of

the product delivered at the visit v_p, and a *load* interval of the minimum and maximum load of the vehicle when it leaves the visit v_p.

When a node is added to the sequence, the procedure sets its demand and load intervals. Moreover, the setting of an interval (demand or load interval) may readjust intervals of previous positions in the sequence. If this readjustment determines that a lower bound of an interval is greater than its upper bound, the sequence is pruned and the method continues building the sequence with other edges. If it is not possible build a sequence with all the edges, the procedure concludes that the relaxed solution is infeasible. Otherwise, we cannot conclude that the relaxed solution is feasible for the SDVRP and it takes two steps to determine whether the solution is feasible or infeasible. The first one is a heuristic approach and the second one is an exact approach.

In the first step, while there is a demand interval where the minimum is different to the maximum, it fixes the minimum and maximum to a value and readjust the rest of intervals. If the procedure finds demand and load values, it concludes that the solution is feasible, otherwise it repeats the process setting other values of that interval or of other interval until a parameter which was set to 2000 in this computational results. The second step is executed if the first step does not find a feasible solution. It solves a Linear Programming (LP) problem as it is done in [8] for the SD1PDTSP where preemption is not allowed. This LP problem is called Time Extended Flow Problem because it is built on a time extended network from the sequence of vertices. If the LP problem is infeasible, it means that the sequence is infeasible for the SDVRP, otherwise the sequence is feasible for the SDVRP. Moreover, this last step was not necessary to perform for the benchmark instances of Sect. 5, when a sequence was built first step always determined that it was feasible.

The symmetry-breaking constraints (29) and (30) avoid exploring the whole supporting graph. For example, the edges going out of the depot can be taken in a non-decreasing order.

4.3   Heuristic Algorithm

The effectiveness of a branch-and-cut algorithm does not depend only on the quality of the lower bound, but also on the procedures to find good upper bounds. This paper does not intend to describe a new approach for the SDVRP but to use an existing one to give a reasonable upper bound. For this purpose, we use an algorithm described in [9] for the SD1PDTSP.

The SDVRP is a particular case of the SD1PDTSP where there is only a pickup location (the depot) and a preemption is not allowed. Thus any algorithm to solve the SD1PDTSP can be applied for the SDVRP. [9] describe two algorithms, the first algorithm does not allow any preemption, while the second algorithm refines the solution obtained by the first one allowing preemption. Thus, we only applied the first algorithm to obtain a heuristic approach. Obviously, specific algorithms for the SDVRP are expected to have better performance than those more general for the SD1PDTSP. However, the algorithm described in [9] has a reasonable performance.

The heuristic algorithm transforms an SD1PDTSP instance to a 1PDTSP instance (i.e., the split forbidden version of the SD1PDTSP) and applies a meta-heuristic for the 1PDTSP on this transformed instance. The transformation phase uses a strategy decomposing each customer (and the depot) into a set of nodes with fixed partial demands. The strategies differ in the way of determining these partial demands. We use the strategy labelled as MAX for each customer and the strategy labelled as MIN for the depot.

5   Computational Results

The branch-and-cut algorithm described in the previous sections has been executed on a desktop computer, with a 3.20 GHz Intel Core i7-8700 processor and 16 GB of RAM, running Windows 10 64-bit. The code was written in C++, using IBM ILOG CPLEX 12.8 as MILP solver and run in a single thread of the computer. The time limit was set to 7200 s.

The computational results are compared with the results of [2], where the computer had an Intel Xeon processor W3680, at 3.33 Ghz, with 12 GB of RAM. They used IBM ILOG CPLEX 12.5 with a time limit of 7200 s. They applied two different models with two different configurations in the separation procedures. Thus, they show four computational executions for each instance. They show results taking the Euclidean distances rounded to the nearest integer and unrounded Euclidean distances. For the sake of brevity, we only show results for rounded Euclidean distances and when the gap in [2] is less or equal to the 3%. We compare our results with the best result among their four executions presented for each instance. Based on the single thread rating values https://www.cpubenchmark.net/singleThread.html, our computer is 1.74 times faster than the computer in [2].

Table 1 shows the computational results of the best of the four executions in [2] (columns from 3 to 6) and our two branch-and-cut algorithms. The first branch-and-cut algorithm solves the subproblem (21)–(31) with the CPLEX solver (columns from 7 to 16), whereas the second branch-and-cut algorithm solves the subproblem (21)–(31) by enumeration as it is described in Sect. 4.2 (columns from 17 to 26). The column headings in this table have the following meanings: *Name* is the name of the instance; n is the number of customers (excluding the depot) of the instance; *LB* is the best lower bound obtained by the branch-and-cut algorithm; *UB* is the best upper bound obtained by the algorithm; *Gap* is the percentage between the upper and lower bounds for the branch-and-cut algorithm calculated as $100 \cdot (UB - LB)/LB$ and it is not shown if LB is equal to UB; *Time* is the computational time in seconds of each execution of each algorithm and it is not shown if it is equal to the time limit (7200 s); z_i^* shows the number of customers with split demand 2, 3 and at least 4 visits in the optimal solution; *Fea.* is the number of valid integer relaxed solutions; *Inf.* is the number of invalid integer relaxed solutions (i.e., the number of constraints (35) found); and $T.$ is the time consumed while checking the feasibility of integer relaxed solutions.

Table 1. Results of rounded Euclidean instances with $m_i = m_0$ for all $i \in I$.

Name	n	Best result in [2]				B&C with sublp				z_i^*			Subproblems			B&C with enumeration				z_i^*			Subproblems		
		LB	UB	Gap	Time	LB	UB	Gap	Time	2	3	≥4	Fea.	Inf.	T.	LB	UB	Gap	Time	2	3	≥4	Fea.	Inf.	T.
eil22	21	375.00	375		0.0	375.00	375		0.7	0	0	1	0	0	0.0	375.00	375		0.7	0	0	1	1	0	0.0
eil23	22	569.00	569		0.0	569.00	569		0.0	0	0	1	1	0	0.0	569.00	569		0.0	0	0	1	1	0	0.0
eil30	29	510.00	510		13.5	510.00	510		0.4	1	0	1	1	4	0.1	510.00	510		0.3	1	0	1	1	4	0.0
eil33	32	835.00	835		1.0	835.00	835		0.6	0	0	2	2	1	0.1	835.00	835		0.5	0	0	2	2	1	0.0
eil51	50	521.00	521		28.3	521.00	521		7.9	1	0	2	2	0	0.0	521.00	521		8.1	1	0	2	2	0	0.0
S51D1	50	458.00	458		0.5	458.00	458		1.7	0	0	1	1	0	0.0	458.00	458		1.7	0	0	1	1	0	0.0
S51D2	50	703.00	703		729.8	703.00	703		275.2	6	1	8	8	10	14.3	703.00	703		253.5	6	1	8	8	10	0.0
S51D3	50	925.94	944	1.95		936.12	946	1.06		8	1	5	5	0	44.5	936.15	946	1.05		8	1	5	5	0	0.0
S51D4	50	1547.04	1551	0.26		1512.14	1607	6.27		18	4	0	0	0	7152.2	1551.00	1551		619.3	13	3	5	5	0	0.0
S51D5	50	1306.57	1328	1.64		1278.42	1444	12.95		18	2	0	0	0	7157.8	1318.64	1330	0.86		18	0	11	11	0	0.1
S76D1	75	592.00	592		30.3	592.00	592		57.0	1	0	3	3	8	0.2	592.00	592		56.4	1	0	3	3	8	0.0
S101D1	100	716.00	716		2946.8	716.00	716		3264.7	0	0	6	6	38	2.0	716.00	716		3285.0	0	0	6	6	38	0.0
p01.110	50	458.00	458		0.6	458.00	458		2.1	0	0	1	1	0	0.0	458.00	458		2.1	0	0	1	1	0	0.0
p01.1030	50	753.00	753		2904.6	753.00	753		695.0	9	0	6	6	110	353.5	753.00	753		338.3	9	0	6	6	110	0.0
p01.1050	50	978.63	998	1.98		984.54	1010	2.59		11	2	6	6	0	251.9	984.68	1010	2.57		11	2	6	6	0	0.0
p01.1090	50	1459.58	1481	1.47		1436.50	1594	10.96		21	0	1	0	0	7160.0	1467.15	1481	0.94		8	7	7	7	2	0.6
p01.3070	50	1444.68	1473	1.96		1418.50	1519	7.08		19	3	0	0	1	7137.2	1440.50	1477	2.53		19	2	10	10	1	2.8
p02.110	75	612.00	612		4284.5	612.00	612		3219.9	1	0	4	4	3	0.5	612.00	612		3230.9	1	0	4	4	3	0.0
p03.110	100	740.76	760	2.6		742.09	758	2.14		1	0	2	2	32	3.4	742.13	758	2.14		1	0	2	2	32	0.0
SD1	8	228.00	228		0.0	228.00	228		0.2	4	0	1	1	0	0.0	228.00	228		0.2	4	0	1	1	0	0.0
SD2	16	708.00	708		0.2	708.00	708		1.3	7	2	1	0	0	0.2	708.00	708		1.1	7	2	1	1	0	0.0
SD3	16	432.00	432		0.1	432.00	432		1.1	8	0	1	1	0	0.0	432.00	432		1.1	8	0	1	1	0	0.0
SD4	24	630.00	630		0.2	630.00	630		1.2	12	0	2	2	0	0.1	630.00	630		1.0	12	0	2	2	0	0.0
SD5	32	1392.00	1392		2.9	1392.00	1392		7.7	8	6	1	1	0	0.8	1392.00	1392		6.6	8	6	1	1	0	0.0
SD6	32	832.00	832		1.1	832.00	832		3.8	16	0	1	1	0	0.1	832.00	832		3.7	16	0	1	1	0	0.0
SD7	40	3640.00	3640		2.0	3640.00	3640		20.4	13	4	2	2	0	4.0	3640.00	3640		16.4	13	4	2	2	0	0.0
SD8	48	5068.00	5068		106.8	5068.00	5068		568.5	15	10	1	1	0	515.4	5068.00	5068		53.2	15	10	1	1	0	0.0
SD9	48	2046.00	2046		657.4	2046.00	2046		913.9	28	3	0	0	0	878.9	2046.00	2046		34.3	28	3	0	0	0	0.0
SD10	64	2652.45	2688	1.34		2652.99	2764	4.18		14	12	5	0	0	7171.4	2677.05	2699	0.82		19	12	5	5	0	0.0
SD11	80	13280.00	13280		6370.9	13280.00	13280		6674.6	11	2	1	1	0	75.8	13280.00	13280		6621.2	11	2	11	11	0	22.3

Table 2. Results of B&C with enumeration for different values of m_i.

The first block of result columns corresponds to $m_i = 2$ and the second block to $m_i = 3$.

Name	n	LB	UB	Gap	Time	z_i^* (2)	Fea.	Inf.	T.	LB	UB	Gap	Time	z_i^* (2)	z_i^* (3)	Fea.	Inf.	T.
eil22	21	375.00	375		0.7	0	1	0	0.0	375.00	375		0.7	0	0	1	0	0.0
eil23	22	569.00	569		0.0	1	2	0	0.0	569.00	569		0.0	0	0	1	0	0.0
eil30	29	510.00	510		0.3	1	1	3	0.0	510.00	510		0.3	1	0	1	4	0.0
eil33	32	835.00	835		0.6	0	3	11	0.0	835.00	835		0.5	1	0	1	0	0.0
eil51	50	521.00	521		6.7	1	1	3	0.0	521.00	521		6.1	0	0	2	3	0.0
S51D1	50	458.00	458		1.8	0	1	0	0.0	458.00	458		1.7	0	0	1	0	0.0
S51D2	50	703.00	703		98.9	7	6	32	0.0	703.00	703		134.4	6	1	7	17	0.0
S51D3	50	938.39	942	0.39		9	9	14	0.0	936.17	945	0.94		8	1	12	26	0.6
S51D4	50	1560.00	1560		1127.0	26	8	2	0.9	1552.00	1552		420.6	12	7	8	5	0.6
S51D5	50	1328.00	1328		3762.1	22	7	0	0.0	1315.09	1328	0.98		13	4	8	0	0.3
S76D1	75	592.00	592		44.2	1	3	4	0.0	592.00	592		23.9	2	0	2	0	0.0
S101D1	100	716.00	716		4131.4	2	5	21	0.0	716.00	716		3041.0	2	0	4	10	0.0
p01_110	50	458.00	458		2.1	0	1	0	0.0	458.00	458		2.1	0	1	1	0	0.0
p01_1030	75	753.00	753		519.4	6	14	65	0.0	753.00	753		916.8	8	1	5	122	0.0
p01_1050	50	983.95	1007	2.34		14	6	0	0.0	984.29	1004	2.00		12	1	14	22	0.0
p01_1090	50	1475.38	1484	0.58		24	10	0	1.6	1471.64	1481	0.64		15	5	11	14	29.1
p01_3070	50	1441.19	1483	2.90		23	5	0	0.0	1441.18	1478	2.55		16	3	10	0	1.0
p02_110	50	612.00	612		3342.3	1	4	5	0.0	612.00	612		3358.6	0	0	4	22	0.0
p03_110	100	742.85	753	1.37		0	6	70	0.0	742.14	754	1.60		2	0	4	37	0.0
SD1	8	228.00	228		0.2	4	1	0	0.0	228.00	228		0.2	4	0	1	0	0.0
SD2	16	708.00	708		1.0	10	1	0	0.0	708.00	708		1.1	5	3	1	0	0.0
SD3	16	432.00	432		1.1	8	1	0	0.0	432.00	432		1.1	8	0	1	0	0.0
SD4	24	630.00	630		1.0	12	1	0	0.0	630.00	630		1.1	12	0	2	0	0.0
SD5	32	1392.00	1392		6.6	21	2	0	0.0	1392.00	1392		6.8	11	6	1	0	0.0
SD6	32	832.00	832		3.6	16	3	0	0.0	832.00	832		2.9	16	0	4	0	0.0
SD7	40	3640.00	3640		12.4	28	1	0	0.0	3640.00	3640		16.5	11	9	1	0	0.0
SD8	48	5068.00	5068		31.4	35	1	0	0.0	5068.00	5068		39.9	13	11	1	0	0.0
SD9	48	2046.00	2046		35.7	30	4	0	0.5	2046.00	2046		37.1	19	8	6	0	0.0
SD10	64	2683.19	2699	0.59		43	7	0	4945.9	2677.41	2703	0.96		17	15	12	0	363.1
SD11	80	13280.00	13280		2359.8	59	1	0	0.0	13280.00	13280		3924.9	13	23	1	0	0.3

The branch-and-cut algorithm that solves subproblems by enumeration dominates the branch-and-cut that solves subproblems with the CPLEX solver. Comparing with results in [2], we observe that sometimes the best result in [2] is better than our result and sometimes it is worse. Thus, for example, avoiding instances solved in few seconds, instances S76D1, S101D1, p02_110 and SD11 are solved by an algorithm in [2] faster than our algorithm, and instances S51D2, p01_1030, SD8 and SD9 are solved by our algorithm faster than all algorithms in [2]. In addition instance S51D4 is solved by our algorithm (in 623 s) whereas it was unsolved in [2]. It is worth observing that the optimal SDVRP solutions of some instances (like eil22 and eil23) serve all customers with one visit, so they could be solved with a CVRP code. Still, this observation has only been known after the SDVRP had been solved on those particular instances.

Table 2 shows the computational results of the branch-and-cut algorithm solving subproblems by enumeration for different upper bounds on the number of visits of the customers: $m_i = 2$, $m_i = 3$ and $m_i = m_0$. The meaning of the headings are the same as those in the previous table.

In general, an instance is solved faster when m_i are smaller. There are exceptions to this claim, as for example the instance S101D1, which took longer to be solved when $m_i = 2$ than when $m_i = 3$ and $m_i = m_0$. Another observation is that the problem with small values of m_i is more restricted and therefore the optimal values may be larger when m_i are smaller. Thus, the optimal values of instance S51D4 are 1560, 1552 and 1551 when $m_i = 2$, $m_i = 3$ and $m_i = m_0$, respectively.

6   Conclusions

The SDVRP is a generalization of the CVRP where a customer can be served with more than one visit. In general, this problem is more difficult to solve than the CVRP, and has been investigated by several authors in the recent literature. Our manuscript provides a new algorithm to solve the SDVRP. The algorithm consists of a branch-and-cut procedure applied on a model for a relaxed problem, a checking process to decide whether the relaxed solution is valid for the SDVRP, and a set of valid inequalities to separate invalid solutions. Computational results on benchmark instances show the performance of the algorithm when compared to the best approach in the literature. Our implementation was able to solve to optimality an instance previously unsolved in the literature.

References

1. Archetti, C., Bianchessi, N., Speranza, M.: A column generation approach for the split delivery vehicle routing problem. Networks **58**(4), 241–254 (2011)
2. Archetti, C., Bianchessi, N., Speranza, M.: Branch-and-cut algorithms for the split delivery vehicle routing problem. Eur. J. Oper. Res. **238**(3), 685–698 (2014)
3. Archetti, C., Speranza, M.G.: Vehicle routing problems with split deliveries. Int. Trans. Oper. Res. **19**, 3–22 (2012)

4. Belenguer, J.M., Martinez, M.C., Mota, E.: A lower bound for the split delivery vehicle routing problem. Oper. Res. **48**, 801–810 (2000)
5. Chen, P., Golden, B., Wang, X., Wasil, E.: A novel approach to solve the split delivery vehicle routing problem. Int. Trans. Oper. Res. **24**, 27–41 (2017)
6. Dror, M., Trudeau, P.: Saving by split delivery routing. Transp. Sci. **23**, 141–145 (1989)
7. Dror, M., Laporte, G., Trudeau, P.: Vehicle routing with split deliveries. Discrete Appl. Math. **50**, 239–254 (1994)
8. Erdogan, G., Battarra, M., Wolfler Calvo, R.: An exact algorithm for the static rebalancing problem arising in bicycle sharing systems. Eur. J. Oper. Res. **245**, 667–679 (2015)
9. Hernández-Pérez, H., Salazar-González, J.J., Santos-Hernández, B.: Heuristic algorithm for the split-demand one-commodity pickup-and-delivery travelling salesman problem. Comput. Oper. Res. **97**, 1–17 (2018)
10. Hernández-Pérez, H., Salazar-González, J.-J.: The one-commodity pickup-and-delivery travelling salesman problem. In: Jünger, M., Reinelt, G., Rinaldi, G. (eds.) Combinatorial Optimization — Eureka, You Shrink!. LNCS, vol. 2570, pp. 89–104. Springer, Heidelberg (2003). https://doi.org/10.1007/3-540-36478-1_10
11. Jin, M., Liu, K., Bowden, R.O.: A two-stage algorithm with valid inequalities for the split delivery vehicle routing problem. Int. J. Prod. Econ. **105**(1), 228–242 (2007)
12. Jin, M., Liu, K., Eksioglu, B.: A column generation approach for the split delivery vehicle routing problem. Oper. Res. Lett. **36**(2), 265–270 (2008)
13. Lee, C.G., Epelman, M.A., White, C.C., Bozer, Y.A.: A shortest path approach to the multiple-vehicle routing problem with split pick-ups. Transp. Res. Part B Methodol. **40**(4), 265–284 (2006)
14. Moreno, L., Poggi, M., Uchoa, E.: Improved lower bounds for the split delivery vehicle routing problem. Oper. Res. Lett. **38**, 302–306 (2010)
15. Ozbaygin, G., Karasan, O., Yaman, H.: New exact solution approaches for the split delivery vehicle routing problem. EURO J. Comput. Optim. **6**, 85–115 (2018)
16. Salazar-González, J.J., Santos-Hernández, B.: The split-demand one-commodity pickup-and-delivery travelling salesman problem. Transp. Res. Part B Methodol. **75**, 58–73 (2015)
17. Sierksma, G., Tijssen, G.A.: Routing helicopters for crew exchanges on off-shore locations. Ann. Oper. Res. **76**, 261–286 (1998)
18. Silva, M., Subramanian, A., Ochi, L.: An iterated local search heuristic for the split delivery vehicle routing problem. Comput. Oper. Res. **53**, 234–249 (2015)

A MILP Model and a GRASP Algorithm for the Helicopter Routing Problem with Multi-Trips and Time Windows

André Manhães Machado[✉][iD], Geraldo Regis Mauri[iD],
Maria Claudia Silva Boeres[iD], and Rodrigo de Alvarenga Rosa[iD]

Universidade Federal do Espírito Santo, Vitoria, ES, Brazil
andre.manhaes@gmail.com,
{geraldo.mauri,rodrigo.a.rosa}@ufes.br,boeres@inf.ufes.br

Abstract. This work presents a new mixed-integer linear programming (MILP) model and a GRASP algorithm for the Helicopter Routing Problem (HRP) with multi-trips and time windows. The HRP consists of serving a set of transportation requests, defined as a pair of boarding and landing locations, where helicopters are used as transportation mode. This novel model incorporates the characteristics present in routing helicopters for offshore platforms and, also, a new set of constraints is proposed to handle an issue unnoticed so far for the HRP: offshore platforms can be visited by helicopters one at a time. The main goal of the presented HRP is to minimize the cost of meeting the set of transportation requests. We provide computational experiments for well-known instances available in the literature with up to 80 requests. On average, our GRASP algorithm is capable to solve all instances, whereas CPLEX found solutions only for small instances, lower bounds for small to medium instances and no lower bounds for medium to large instances.

Keywords: Helicopter Routing Problem · GRASP ·
Offshore platforms logistics

1 Introduction

For decades the transportation of employees in the oil industry has been carried out mainly by helicopters to and from offshore drilling platforms. Two reasons for this can be pointed out: it reduces the transit time and increases employee satisfaction [15]. In general, the main objective in this field consists of minimizing the overall cost to meet a set of air transportation requests [6].

The Helicopter Routing Problem (HRP) is defined as the class of problems with the goal of building a set of feasible routes to meet a group of transportation requests, and where each route is served by a helicopter [16]. In line with this definition, the HRP is classified as the General Pickup and Delivery Problem (GPDP) [17]. Several approaches have been proposed to solve the HRP, while mathematical models were developed using distinct formulations [1,6,10,14,15].

© Springer Nature Switzerland AG 2019
C. Paternina-Arboleda and S. Voß (Eds.): ICCL 2019, LNCS 11756, pp. 204–218, 2019.
https://doi.org/10.1007/978-3-030-31140-7_13

Since the HRP is classified as NP-complete and NP-hard [7], heuristics methods are usually devised and applied to instances at medium and large sizes. The transportation of people to offshore platforms in the Rio de Janeiro Basin was formulated as an HRP in [9]. A heuristic method was provided to solve real cases. The problem of constructing feasible routes for helicopters using a Genetic Algorithm (GA) was analyzed in [4, 11, 13] and Ant Colony Optimization (ACO) for the HRP with a single helicopter was presented by [16].

A helicopter routing system to meet transportation requests to offshore platforms including a set of directives for aircraft routing that resulted in safety upgrades was proposed by [12], who presented a Column Generation (CG) technique to solve the HRP. In [3] the HRP was successfully adapted to disaster relief operations using a hybrid between Particle Swarm Optimization (PSO) and a greedy search algorithm. A heuristic of coordination for multiple criteria analysis of helicopter mission planning in disaster relief operations was proposed by [5]. Different approaches to analyze the passenger safety in the HRP for offshore platforms were reported in [15]. A mathematical model to the HRP based on the Dial-a-Ride-Problem (DARP) with several specific constraints and a more realistic objective function was defined in [2]. Furthermore, a Clustering Search (CS) was proposed to solve it. The mathematical model and the CS metaheuristic were tested with real data from a Brazilian oil and gas company.

Given the related works, we aim to extend the one-trip model proposed by [2] to tackle the unique characteristics of offshore platforms. We add a new constraint called platform landing safety time to solve an issue that remains unnoticed until now: offshore platforms can be visited by helicopters one at a time. The published papers about HRP do not impose this rule and they may generate infeasible routes in the real world. Imposing this constraint improves the route safety by preventing simultaneous landings in the same platform and therefore has a high impact in real applications. We also add the possibility to make multiple trips inside each route and enforce time windows to attend passengers. We define a mixed-integer linear programming (MILP) model to represent this new problem. So far, no GRASP heuristic was proposed to tackle the HRP. For this reason, we implement a GRASP algorithm to solve it. Besides, a set of adapted instances representing the helicopter routing problem in the main oil exploration basin in Brazil was used to validate the new model.

The main contributions of this study are: the new platform landing safety time constraint, to the best of our knowledge, has not yet been researched for helicopter routing to offshore platforms; unlike [2], a helicopter can perform several trips within the route and there are time windows for each request; a GRASP heuristic to solve medium and large size instances, whereas [2] proposed a linear CS algorithm.

The paper is structured as follows. We provide a description of the problem in Sect. 2 and define the proposed MILP model in Sect. 3. Our GRASP algorithm is presented in Sect. 4. Section 5 reports the performance of the proposed model and algorithm using a set of instances available in the literature, followed by our conclusions in Sect. 6.

2   Problem Description

The Helicopter Routing Problem (HRP) can be described as follows. Every day, a set of transportation requests must be served by a heterogeneous fleet of helicopters respecting a set of operational constraints. A transportation request, or request for short, represents a passenger demand which is associated with boarding and landing locations, i.e., an airport or a platform, for example. Besides, each location has a time window linked to the passenger who is being attended.

Each helicopter makes a route per day and a route is formed by a pre-determined number of trips. A trip is a sequence of visits beginning and ending at the airport. After each trip, the helicopter must be landed at the airport for a minimum time called airport service time. If the helicopter boards or lands at a platform, it has a fixed service time that is independent of the number of passengers that will board or land at the platform. Moreover, the helicopters have the following properties: (h.1) a maximum weight limit; (h.2) a maximum fuel load; (h.3) a crew; (h.4) a maximum number of seats available to transport passengers and (h.5) a maximum number of trips per day.

There are two types of operational constraints in the HRP: (c.1) technical constraints and (c.2) safety constraints. The technical constraints of the problem are: (t.1) the helicopters cannot exceed their maximum weight limit during the route, and the total weight must be calculated considering the weights of the crew, the helicopter, the fuel and the requests to be met; (t.2) the number of passengers in the helicopter cannot exceed the limit of available seats during the route and (t.3) the helicopters can take off from the airport strictly between pre-established departure and arrival times.

The safety constraints are: (s.1) each helicopter must return to the airport with a minimum amount of fuel for a safety additional flight time; (s.2) each helicopter can fly up to a bounded hour per trip; (s.3) the duration of a trip cannot exceed a pre-established limit and (s.4) the new constraint proposed by this work to tackle the issue of simultaneous visits on the same platform. This is a requirement not yet implemented in any model for helicopter routing. Besides, the new feature also has the objective to prevent the collision between distinct helicopters when they visit the same platform by ensuring an interval time between consecutive landings on any platform.

3   Mathematical Model

The proposed model is based on the formulation proposed by [2] and it has the objective to meet a set R of flight (transportation) requests using a fleet S of helicopters. The helicopters in S make per day up to t trips and a trip is defined as a sequence of visits beginning and ending at the airport. A flight request $r(i, j) \in R$ specifies a boarding node $i \in N_P$ and a landing node $j = i+n, j \in N_D$ and $n = |R|$. The airport is designated as node 0 and node $2n+1$, referring to the initial departing and final arrival locations, respectively. Therefore, the proposed HRP can be represented by a complete graph $G(N, A)$ where $N = N_P \cup N_D \cup N_G$,

Table 1. List of parameters for each node $i \in N$ and arc $(i,j) \in A$.

Par	Description	Par	Description
q_i	Number of passengers boarding or landing at node i	w_i	Weight of the passenger and his luggage at node i
$c_{i,j}$	Distance between nodes $i,j \in N$	$d_{i,j}$	Service time in node $j \in N$ when the helicopter is departing from node $i \in N$
L_i	Lower bound of pick-up or drop-off time	U_i	Upper bound of pick-up or drop-off time

with $N_P = \{1, 2, \ldots, n\}$, $N_D = \{n+1, n+2, \ldots, 2n\}$, $N_G = \{0, 2n+1\}$ and $A = N \times N$. Table 1 shows the list of parameters for node $i \in N$, where the column Par designates the name of the parameter.

Since all helicopters have the same number of trips in their route, we create a new set K of helicopters to represent the trips. By doing so, there is no need to add a new index to the variables in the MILP model to control the trip of each helicopter. In this way, we maintain the proposal of [2] as formulating a 3-index model. This is done as follows. Let $V_s = \{k_{s_1}, k_{s_2}, \ldots, k_{s_t}\}$, $|V_s| = t$, be the set of trips of s, where k_{s_1} is the first trip of s, k_{s_2} is the second trip of s and so on. The new set K of helicopters is defined as the union of every V_s. In other words, $K = \cup_{s \in S} V_s$ and $|K| = t \times |S|$. Then, we handle each trip in V_s as a helicopter with the same properties of s. Therefore, the trips from V_s must be grouped in a sequence which will define the route of $s \in S$, i.e., the route of helicopter $s \in S$ is defined as the sequence of trips executed by each one of the "new" helicopters $k_{s_1}, k_{s_2}, \ldots, k_{s_t}$. Besides that, let $K_1 \subset K$ be the set containing the first trip k_{s_1} of each helicopter $s \in S$.

Given that new modeling proposition, it is still necessary to ensure that each trip of the helicopter $s \in S$ is sequenced as required. To this effect, without loss of generality, we define the parameter $y_{k,l}$ to control the correct order of the helicopters $k, l \in K$ within a route of $s \in S$ in the following manner. The parameter $y_{k,l}$ is equal to 1 if the helicopters $k, l \in V_s$, for some $s \in S$, and k must be a flight trip preceding the trip flew by l. Otherwise, it is equal to 0. Besides that, a route of the helicopter $s \in S$ is considered as valid when the trips are used in sequence, i.e., the trip $k_{s_{i+1}}$ can only meet any request when the trip k_{s_i} meets at least one transportation request.

Table 2 shows the list of parameters defined for each helicopter $k \in K$, where the column Par designates the name of the parameter. In the following, we only consider the helicopter in the sets K and K_1. The parameter platform landing safety time (σ_P) is the minimum time between two different helicopters landing on the same platform. The parameter airport service time (σ_G) is the minimum time between the end of trip k and start of trip l when $y_{k,l} = 1$, $k, l \in K$. The parameter safety flight time (σ_S^k) is used to ensure that the helicopter has sufficient fuel to complete the trip.

Table 2. List of parameters for each helicopter $k \in K$.

Par	Description	Par	Description
a^k	Weight of the crew	g^k	Maximum fuel capacity
f^k	Maximum flight time	s^k	Maximum seating capacity for passengers
p^k	Weight of the helicopter	P^k	Weight limit of the helicopter during the flight
z^k	Average hourly fuel consumption	v^k	Average flight speed
H_{sr}^k	Earliest time to take-off	H_{sd}^k	Latest time to return to the airport
α^k	Cost to use the helicopter in a route	β^k	Cost per travelled distance
θ^k	Conversion of fuel volume to fuel weight	$y_{k,l}$	Define if the trip k must be travelled before l
σ_S^k	Safety flight time	σ_G	Airport service time
σ_P	Platform landing safety time		

In a graph $G(N, A)$, each passenger is represented by two nodes: one of them represents the passenger at the pickup platform (or airport) and the other represents the same passenger at the delivery platform (or airport). If two different passengers are located at the same platform (or airport), the distance between their respective nodes representation in the graph should be equal to zero, $c_{i,j} = 0$. If they are located on different platforms, the distance between them in the network or graph is greater than zero, $c_{i,j} > 0$. The parameter $d_{i,j}$ represents the service time to serve node $j \in N$ when the helicopter departs from node $i \in N$ and it is independent of passenger numbers. In addition, the service time $d_{i,j}$ should be greater than zero only if $c_{i,j} > 0$. In other words, the service time $d_{i,j}$ will be considered only if the helicopter flies from one platform to another. For all nodes $i \in N_P \cup N_D$ the number of boarding passengers $q_i = -q_{n+1}$ and the weight of the passenger and his luggage $w_i = -w_{n+1}$.

There are three distinct transport requests in the proposed HRP model: (i) passengers start their trip at the airport and go to some platform; (ii) passengers start their trip at a platform with destination to the airport; and (iii) passengers start their trip at a platform and go to another platform. These requests generate the following assumptions: (a) for all requests starting at the airport (node 0), the passenger is represented by node $i \in N_P$, the distance between node 0 and node i is $c_{0,i} = 0$ and the service time is also equal to zero, $d_{0,i} = 0$; (b) for all requests finishing at the airport (node $2n + 1$), the passenger is represented by node $j \in N_D$ and the distance between node j and node $2n + 1$ is equal to zero, $c_{j,2n+1} = 0$, and the service time is also equal to zero, $d_{j,2n+1} = 0$.

The decision variables used in the proposed model are described below:

- $x_{i,j}^k$: binary variable equal to 1 if helicopter $k \in K$ flies from node $i \in N$ to node $j \in N$, $j \neq i$. Otherwise, it is equal to 0;

- u^k: binary variable equal to 1 if helicopter $k \in K$ is used to meet any request. Otherwise, it is equal to 0;
- B_i^k: indicates the time when helicopter $k \in K$ starts serving node $i \in N$;
- Q_i^k: indicates the number of occupied seats after helicopter $k \in K$ visits node $i \in N$;
- W_i^k: indicates the total flight weight of helicopter $k \in K$ after visiting node $i \in N$ (the total weight is calculated as the sum of the weight of the passengers (including luggage) plus the weight of the helicopter crew plus the weight of the helicopter itself);
- F_i^k: indicates the amount of fuel of helicopter $k \in K$ when visiting node $i \in N$.

Based on the previous definitions, the proposed mathematical model for the HRP is defined as follows. The objective function (1) must be minimized and it is composed of two terms. The first one is associated with the use of the helicopter $k \in K_1$. As the helicopter $s \in S$ only flies when the first trip represented by the helicopter k_{s_1} is used, this controls the cost of using a new helicopter in the solution. The second term expresses the traveled distance by the helicopter $k \in K$.

$$\Gamma = \sum_{k \in K_1} \alpha^k u^k + \sum_{k \in K} \sum_{i \in N} \sum_{j \in N, j \neq i} \beta^k c_{i,j} x_{i,j}^k \tag{1}$$

The objective function is subject to the following constraints. Constraints (2) keep an account if any helicopter $s \in S$ is used to serve some transportation request. This is done by checking if the first trip of s is used, i.e., the helicopter $k_{s_1} \in V_s$ meets some request.

$$u^k = \sum_{j \in N_P} x_{0,j}^k, \quad \forall k \in K_1 \tag{2}$$

Constraints (3) make sure that if helicopter $l \in K$ meets some request then the helicopter k must meet at least one request, too.

$$\sum_{j \in N_P} x_{0,j}^k >= \sum_{j \in N_P} x_{0,j}^l, \quad \forall k, l \in K \mid y_{k,l} = 1 \tag{3}$$

Constraints (4) guarantee that all requests will be met, i.e., each node is visited exactly once. Constraints (5) ensure that the same helicopter visiting the pickup node $i \in N_P$ will attend the respective delivery node $i + n \in N_D$. Constraints (6) and (7) certify that the trip of each helicopter starts and ends at the airport, and constraints (8) ensure the flow conservation.

$$\sum_{k \in K} \sum_{j \in N, j \neq i} x_{i,j}^k = 1, \quad \forall i \in N_P \tag{4}$$

$$\sum_{j \in N, j \neq i} x_{i,j}^k - \sum_{j \in N, j \neq i} x_{n+i,j}^k = 0, \quad \forall i \in N_P, k \in K \tag{5}$$

$$\sum_{j \in N_P} x_{0,j}^k \leq 1, \quad \forall k \in K \tag{6}$$

$$\sum_{i \in N_D} x_{i,2n+1}^k \leq 1, \quad \forall k \in K \tag{7}$$

$$\sum_{j \in N, j \neq i} x_{j,i}^k - \sum_{j \in N, j \neq i} x_{i,j}^k = 0, \quad \forall i \in N_P \cup N_D, k \in K \tag{8}$$

Constraints (9) forbid the generation of sub-routes, where ψ is a sub-graph of $G(N, A)$ and $|\psi|$ is the number of vertices in ψ.

$$\sum_{i \in \psi} \sum_{j \in \psi, j \neq i} x_{i,j}^k \leq |\psi| - 1, \quad \psi \subset N, 2 \leq |\psi| \leq |N| - 1, \forall k \in K \tag{9}$$

Constraints (10) keep record of occupied seats in the helicopter $k \in K$. Constraints (11) guarantee that the number of occupied seats in helicopter does not exceed the number of seats limit s^k of each helicopter $k \in K$. Constraints (12) and (13) make sure that the helicopter leaves and arrives at the airport (nodes $N_G = \{0, 2n+1\}$) with no passengers. As mentioned before, the nodes of requests departing or arriving at the airport are represented as nodes in N_P or N_D, respectively, with the distance from the airport equal to zero.

$$Q_j^k \geq \left(Q_i^k + q_j\right) x_{i,j}^k, \quad \forall k \in K, i, j \in N, j \neq i \tag{10}$$

$$Q_i^k \leq s^k, \quad \forall k \in K, i \in N \tag{11}$$

$$Q_0^k = 0, \quad \forall k \in K \tag{12}$$

$$Q_{2n+1}^k = 0, \quad \forall k \in K \tag{13}$$

Constraints (14) guarantee the weight of passengers leaving or boarding is reflected in the weight of the helicopter after visiting each node. Constraints (15) ensure that the weight limit of a helicopter does not exceed the maximum weight P^k after visiting each node $i \in N$. Constraints (16) set the initial weight of the helicopter $k \in K$ as the weight of the helicopter plus the weight of the crew.

$$W_j^k \geq \left(W_i^k + w_j\right) x_{i,j}^k, \quad \forall k \in K, i, j \in N, j \neq i \tag{14}$$

$$W_i^k + \theta^k F_i^k \leq P^k, \quad \forall k \in K, i \in N \tag{15}$$

$$W_i^k \geq p^k + a^k, \quad \forall k \in K, i \in N \tag{16}$$

Constraints (17) model the fuel consumption of the helicopter $k \in K$ after visiting each node $i \in N$ considering the travelled distance and the service time at the node i. The term $z^k \frac{c_{i,j}}{v^k}$ represents the amount of fuel consumed and it is based on the time that a helicopter needs to travel distance $c_{i,j}$ at an average speed of v^k. The term $-z^k d_{i,j}$ identifies the amount of fuel consumed while helicopter k is landed at platform $j \in K$ when departing to another platform i.

Constraints (18) define that the amount of fuel at each node must be equal or lower than the maximum fuel capacity g^k.

$$F_j^k \leq \left(F_i^k - z^k \frac{c_{i,j}}{v^k} - z^k d_{i,j} \right) x_{i,j}^k, \quad \forall k \in K, i, j \in N, j \neq i \tag{17}$$

$$F_i^k \leq g^k, \quad \forall k \in K, i \in N \tag{18}$$

Constraints (19) set that the time required for a helicopter $k \in K$ to travel the trip at an average speed v^k must be equal or lower than the maximum flight time f^k. Constraints (20) assure that there is sufficient fuel F_i^k in each node $i \in N$ visited by the helicopter $k \in K$ to execute the flight with safety flight time σ_S^k.

$$\sum_{i \in N} \sum_{j \in N, j \neq i} \left(\frac{c_{i,j}}{v^k} + d_{i,j} \right) x_{i,j}^k \leq f^k, \quad \forall k \in K \tag{19}$$

$$F_i^k \geq z^k \sigma_S^k, \quad \forall k \in K, i \in N \tag{20}$$

Constraints (21) compute the arrival time in the node $j \in N$ when the helicopter $k \in K$ travels from node $i \in N$ to the node $j \in N$. The arrival time is represented by the travelled time ($\frac{c_{i,j}}{v^k}$) plus the service time ($d_{i,j}$). Constraints (22) ensure that the time to visit the landing node $i + n \in N_D$ must be equal or greater than the time to visit the boarding node $i \in N_P$. Constraints (23) guarantee that the helicopter $k \in K$ leaves the airport after the earliest time to take-off H_{sr}^k. Constrains (24) make sure that the helicopter returns back to the airport before latest time of returning H_{sd}^k.

$$B_j^k = \left(B_i^k + d_{i,j} + \frac{c_{i,j}}{v^k} \right) x_{i,j}^k, \quad \forall k \in K, i, j \in N, j \neq i \tag{21}$$

$$B_{i+n}^k \geq B_i^k, \quad \forall k \in K, i \in N_P \tag{22}$$

$$B_0^k \geq H_{sr}^k \, x_{0,j}^k, \quad \forall k \in K, j \in N_P \tag{23}$$

$$B_{2n+1}^k \, x_{i,2n+1}^k \leq H_{sd}^k, \quad \forall k \in K, i \in N_D \tag{24}$$

Constraints (25) ensure that if helicopter $l \in K$ starting the next trip of helicopter $k \in K$ ($y_{k,l} = 1$) then the initial time B_0^l of helicopter l must be equal or greater than the arrived time B_{2n+1}^k of the helicopter k in the airport plus an airport service time σ_G. Constraints (26) determine that distinct helicopters $k, l \in K$ visiting the same platform designated by the nodes $i \neq j$, $i, j \in N_D \cup N_P$ must have a difference between their visiting times greater than σ_P. The condition of representing the same platform by different nodes $i \neq j$ is achieved when $d_{i,j} = 0$. Constraints (27) set the time windows of each request.

$$B_0^l \geq B_{2n+1}^k + \left(1 - x_{l,2n+1}^k \right) \sigma_G, \quad \forall k, l \in K \mid y_{k,l} = 1 \tag{25}$$

$$\mid B_i^k - B_j^l \mid \geq \sigma_P, \quad \forall k \neq l \in K, i < j \in N_P \cup N_D \mid c_{i,j} = 0 \tag{26}$$

$$L_i \leq B_i^k \leq U_i, \quad \forall k \in K, i \in N_P \cup N_D \tag{27}$$

Constraints (28)–(32) define the domain of the variables.

$$x_{i,j}^k \in \{0,1\}, \quad \forall k \in K, i,j \in N \tag{28}$$

$$u^k \in \{0,1\}, \quad \forall k \in K \tag{29}$$

$$Q_i^k \in \mathbb{R}^+, \quad \forall k \in K, i \in N \tag{30}$$

$$W_i^k \in \mathbb{R}^+, \quad \forall k \in K, i \in N \tag{31}$$

$$F_i^k \in \mathbb{R}^+, \quad \forall k \in K, i \in N \tag{32}$$

Despite constraints (10), (17), (19), (21) and (24) are nonlinear, they can be linearized by using a large constant value M, as it is commonly used in other routing problems such as [2]. Constrains (26) are also nonlinear, but they need to use the large constant M as well as new decision variables. Let $m_{i,j}^{k,l}$ be a binary variable relating the nodes $i < j$, $i,j \in N$ and the helicopters $k \neq l$, $k,l \in K$. Then, constraints (26) are rewriting as follows:

$$B_i^k - B_j^l \geq \sigma_P - M\left(m_{i,j}^{k,l}\right), \quad \forall k \neq l \in K, i < j \in N_P \cup N_D \mid c_{i,j} = 0 \tag{33}$$

$$-B_i^k + B_j^l \geq \sigma_P - M\left(1 - m_{i,j}^{k,l}\right), \quad \forall k \neq l \in K, i < j \in N_P \cup N_D \mid c_{i,j} = 0 \tag{34}$$

$$m_{i,j}^{k,l} \in \{0,1\} \quad \forall k,l \in K, i,j \in N_P \cup N_D \tag{35}$$

4   The Proposed GRASP Algorithm for the HRP

The proposed algorithm to solve the HRP is inspired by the GRASP (Greedy Randomized Adaptive Search Procedure) metaheuristic initially proposed by [8]. GRASP is an iterative process composed of two phases: construction and local search. In the construction phase, the objective is to build a feasible solution using a procedure that is random by definition. In the local search, the objective is to explore the neighborhood of a solution until a local optimal solution is found. These two phases are repeated by a predetermined number of iterations and the best solution found is returned as the final solution. Next, we detail the proposed GRASP algorithm for the HRP.

A solution is represented as a set of routes traveled by the helicopters. The routes consist of trips, where each trip begins and ends at the airport. A helicopter that is not attending any requisition has a route formed by trips with only the start and end nodes. In both cases, they represent the airport. Figure 1 shows an example of a route compound of two trips.

The search space is restricted to feasible solutions. We do not allow any violations of the helicopter or passenger constraints as defined in the MILP model explained in Sect. 3. The reason that we always maintain the feasibility of solutions is that it is quite difficult and time-consuming to perform feasibility repairing after the solutions are unfeasible in the search.

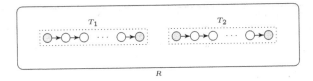

Fig. 1. Example of a route R with two trips T_1 and T_2. The grey circle represents the airport node.

The initial solution of GRASP is built executing the insertion of transportation requests not served into the current solution of the problem as represented in Algorithm 1. Given the set R of requests, while there is some request $r \in R$ to be served (lines 2–10), the procedure $GreedyInsert$ (line 5) evaluates how to add r in the current solution Λ (lines 4–6). For each trip t in Λ, this involves examining every possible way to insert the pickup node of r and delivery node of r in the trip t. The procedure $GreedyInsert$ returns a list of valid movements for each request r not yet met and based on that it is built a list of candidates (line 6). After each insertion of $r \in R$ in Λ is evaluated, the Restricted Candidate List (RCL) is generated using the value ω, $0 \leq \omega \leq 1$, in the function $MakeRCL$ (line 7). In this case, the returned RCL contains every movement $m \in CL$ in such way that $\omega f(m) < f^*$, where $f(\cdot)$ and f^* are the function that return the increasing cost when m is inserted in solution Λ and f^* is the best cost of insertion in CL, respectively. Finally, a random movement δ_r which represents the insertion of $r \in R$ in Λ, is chosen from RCL, δ_r is applied in the current solution Λ and the request r is removed from the set R (lines 8–10).

Algorithm 1. ConstructionPhase(ω,HRP)

Input: ω, HRP
1 $\Lambda \leftarrow \emptyset$, $R \leftarrow HRP.Requests$
2 **while** $R \neq \emptyset$ **do**
3      $CL \leftarrow \emptyset$
4   **for** $r \in R$ **do**
5          $insertions_r \leftarrow GreedyInsert(r, \Lambda)$
6          $CL \leftarrow CL \cup insertions_r$;
7      $RCL \leftarrow MakeRCL(\omega)$
8      $\delta_r \leftarrow SelectRandomn(RCL)$
9      $\Lambda \leftarrow ApplyMove(\delta, \Lambda)$
10     $R \leftarrow R \setminus \{r\}$

11 **return** Λ

The GRASP local search procedure is described in Algorithm 2. Two types of neighborhoods are considered: (n.1) request reinsertion neighborhood and (n.2) trip swap neighborhood. The request reinsertion neighborhood, which is generated by the function $RequestReinsertion$, analyzes for each request r in Λ the removal from its original trip t_r and its insertion in a different trip t_d. The trip swap neighborhood, which is built by the function $TripSwap$, exchanges two distinct trips t_i and t_j in the solution Λ since that t_i and t_j does not belong to the same route.

Algorithm 2. LocalSearchPhase(Λ)

Input: a feasible solution Λ
1 $\Theta \leftarrow RequestReinsert(\Lambda) \cup TripSwap(\Lambda)$
2 **while** $\Theta \neq \emptyset$ **do**
3      $\delta \leftarrow SelectMov(\Theta)$
4      $\Lambda \leftarrow ApplyMove(\delta, \Lambda)$
5      $\Theta \leftarrow RequestReinsert(\Lambda) \cup TripSwap(\Lambda)$
6 **return** Λ

The Algorithm 3 shows the general GRASP algorithm using the construction and local search phases as previously described. It receives the maximum number of iterations $maxit$, the parameter ω used in the construction phase to build Restricted Candidate List (RCL) and the data that represents the problem.

Algorithm 3. GRASP($maxit$, ω, HRP)

Input: $maxit$, ω, HRP
Output: a feasible solution
1 $\Lambda \leftarrow \emptyset$, $Best \leftarrow \emptyset$
2 **for** $i \leftarrow 1$ **to** $maxit$ **do**
3      $\Lambda \leftarrow ConstructionPhase(\omega, R)$
4      $\Lambda \leftarrow LocalSearchPhase(\Lambda)$
5   **if** $cost(\Lambda) < cost(Best)$ **then**
6          $Best \leftarrow \Lambda$
7 **return** $Best$

In Algorithm 3 it is possible to see that the GRASP can be easily parallelized because it is composed of a loop of independent building and improvements procedures. So, we take advantage of that and define a parallel GRASP algorithm. Given the numbers of threads $numthreads$ and the parameters ω and $maxit$, we divide the iterations among the threads and run an independent version of the GRASP into each one. After the completion of the threads, the best solution between them is returned as the final solution. Algorithm 4 presents the parallel GRASP algorithm proposed for the HRP.

Algorithm 4. Parallel GRASP Algorithm

Input: $numthread$, $maxit$, ω, HRP
1 **for** $i \leftarrow 1$ **to** t **do**
2      $T_i \leftarrow Thread.new(GRASP(maxit, \omega, HRP))$
3      $T_i.run()$; /* non blocking method */
4 Wait each T_i to end
5 $\Lambda \leftarrow$ Get best solution return by the threads T_i, $1 \leq i \leq t$
6 **return** Λ

---

5 Computational Experiments

We considered a subset of the real-based instances proposed by [2]. There are three cases of time windows in the data set: (t.1) morning time window with interval from 7:30 AM to 11:00 AM; (t.2) afternoon time window with interval from 2:00 PM to 5:00 PM and (t.3) day time window with interval from 9:00 AM to 5:00 PM. The morning and afternoon time windows are related to transport from or to the airport and it depends on the platform being visited. The

day time window is only applied to transport requests between platforms. The maximum number of trips allowed for each helicopter in the fleet is fixed and it is set as 3. Table 3 shows the list of used instances. The column I represents the instance identification. Columns $|R|$ and $|S|$ indicate the number of requests and the number of helicopters, respectively. Columns W_1, W_2, W_3 named the number of the requests with morning, afternoon and day time window, respectively. The cost to use a helicopter for executing a route is defined as 750.0 in each instance. For each transportation request, two time windows are specified: one time window for the client pickup and another time window for the client delivery. The pickup and delivery events must take place inside of their respective time windows.

Table 3. Set of instances.

I	$|R|$	$|S|$	W_1	W_2	W_3	I	$|R|$	$|S|$	W_1	W_2	W_3
1	5	10	2	3	0	8	40	10	10	18	12
2	10	10	4	5	1	9	45	10	2	20	13
3	15	10	5	7	3	10	50	10	12	24	14
4	20	10	5	12	3	11	55	10	12	27	16
5	25	10	6	13	6	12	60	10	13	30	17
6	30	10	8	14	8	13	70	10	16	32	22
7	35	10	9	15	11	14	80	10	16	37	27

The proposed GRASP algorithm was coded in C++11 and the MILP model was solved by CPLEX 12.8. The computational tests were performed on an Intel Core AMD FX-8320E Eight-Core Processor of 1.4 GHz with 14 GB RAM. The GRASP was run 10 times with the following parameters: (g.1) $\omega = 0.7$; (g.2) maximum number of iterations (*maxit*) equal to 50 and (g.3) 8 threads. The CPLEX was parameterized in the following manner: (c.1) the maximum execution time of 2 h (7200 s); (c.2) tree memory limit of 80 GB; (c.3) work memory of 10 GB; (c.4) memory emphasis set as true and (c.5) the number of threads equal to 8. The reason for those parameters for CPLEX is to handle the out of memory situation normally found in larger models. Since CPLEX, by default, uses all threads of the computer, we run our GRASP (Algorithm 4) also using the same number of threads, i.e. both CPLEX and GRASP take advantage of the same computer's processing capabilities. The values of GRASP parameters ω and *maxit* were determined by tuning the algorithm using Instances 1 and 14. The tuning was executed using the combination of values $\omega \in \{0.1, 0.2, \ldots, 0.9\}$ and *maxit* $\in \{10, 25, 50, 100\}$. The pair $(\omega, maxit)$ with the best value in the tuning was chosen for use in the computational experiments.

5.1   Results

Table 4 presents the comparison between the solutions obtained by the mathematical model and the GRASP algorithm proposed. The column I indicates the instance. The columns UB, LB, GAP and H represent the objective function (upper bound), lower bound, gap and number of helicopters used by the model, respectively. The columns $Best$, Avg, $Worst$, H_{min} and H_{max} indicate the best solution, the average solution, the worst solution, the minimum number of used helicopters and the maximum number of used helicopters returned by the heuristic. The column Dev denotes the deviation and it is defined as $Dev = 100(worst - best)/best$. The deviation is equal to zero when all the obtained solutions have the same value. The columns T and T_{avg} designates the total and the average execution time (in seconds) of the CPLEX and GRASP algorithm, respectively.

Table 4. Comparison of the results from the MILP and the GRASP algorithm.

I	CPLEX					GRASP						
	UB	LB	T	GAP (%)	H	$Best$	Avg	$Worst$	Dev	T_{avg}	H_{min}	H_{max}
1	2,732	2,732	17	0.00	1	2,732	2,732	2,732	0.00	<1	2	2
2	3,095	961	7,200	68.94	2	2,806	2,813	2,829	0.81	3	2	2
3	-	280	7,200	-	-	3,086	3,182	3264	5.77	12	2	2
4	-	98	7,200	-	-	3,174	3,373	3553	11.94	45	2	2
5	-	-	7,200	-	-	3,749	4,277	4,773	27.31	88	2	3
6	-	-	7,200	-	-	4,488	5,344	5,631	25.46	140	2	3
7	-	-	7,200	-	-	6,060	6,401	6918	14.15	211	3	4
8	-	-	7,200	-	-	6,271	6,881	7509	19.74	334	3	4
9	-	-	7,200	-	-	7,264	7,655	7,957	9.54	478	4	5
10	-	-	7,200	-	-	7,459	8,170	8,656	16.05	699	4	5
11	-	-	7,200	-	-	8,803	9,113	9,491	7.81	938	5	5
12	-	-	7,200	-	-	9,102	9,844	10,274	12.87	1,262	5	5
13	-	-	7,200	-	-	11,742	12,169	12,587	7.19	1,845	6	7
14	-	-	7,200	-	-	13,185	13,924	15,095	14.49	2,484	7	9

Since the proposed HRP mathematical model represents an NP-hard problem, it was expected that an exact approach could not solve medium to large instances. This indeed happened and it can be seen in Table 4, where CPLEX found the optimal solution only for Instance 1 (the smallest one with 10 requests) and a non-optimal solution for Instance 2 (the second smallest one with 15 requests). Even in small instances, Instances 3 and 4, varying from 15 to 20 requests, CPLEX could not find any feasible solution after running for 2 hours. For medium to large instances, Instance 5 to 14, varying from 25 to 80, CPLEX did not find even a lower bound after the time limit. Besides, CPLEX was only able to find upper bound values for Instance 1 to 2 and lower bound values

for Instance 1 to 4. This shows that the proposed model is hard to be resolved exactly.

As can be observed in Table 4, the GRASP algorithm found solutions for all instances, including the optimal solution for Instance 1. For Instance 2, the solution found by GRASP was between the LB and the UP found by CPLEX. In the remaining instances, there is no available comparison to the MILP model using the lower bound, because the CPLEX either shows an LB than 750 or no value at all. For instances of medium or large size, our GRASP shows a deviation from 11.94% to 27.31%. The main reason for that is the high price related to add or not a new helicopter into the solution. The parallel GRASP heuristic gives good results at an affordable time when compared with the 2 hours of CPLEX execution. Besides, when GRASP runs with only one thread for the larger instance (Instance 14), the average time is 679% greater than the time spent by parallel GRASP with eight threads and with similar values to best, average and worst solution as represented in Table 4.

The proposed model can be used in any offshore oil and gas company and its approach represents a significant advance for preventing the generation of planned routes where is possible the collision of helicopters landing on the same platform.

6 Conclusion

This paper proposed a mixed-integer linear programming (MILP) model and a GRASP algorithm to address the Helicopter Routing Problem (HRP) with the addition of new constraints related to transportation requests for offshore platforms. In particular, a not yet modelled constraint to prevent the simultaneous use of platforms by helicopters was proposed and the possibility to perform multi-trips with time window was defined. The computational experiments showed that CPLEX could solve only small instances for the problem and that our GRASP algorithm showed the effectiveness and efficiency of the proposed model.

The main results of this study are: a new constraint to prevent different helicopters landing simultaneously on the same platform and a model allowing several trips within the same route with time windows for each request. Besides, a parallel GRASP algorithm to solve medium and large size instances. The proposed model can be used in any offshore oil and gas company and its approach represents a significant advance for preventing the generation of planned routes where it is possible the collision of helicopters attending the same platform.

We propose for future works: (w.1) to apply the proposed model to instances based on data of humanitarian aid routing; (w.2) to model the weight and fuel constraints as non linear constraints and therefore more accurate to reality and (w.3) to include client satisfaction related to time windows as usually found in DARP models.

Acknowledgments. We want to express our thanks to the National Council for Scientific and Technological Development – CNPq (processes 301725/2016-0 and 307439/2016-0) and FAPES (process 75528452/2016) for financial support.

References

1. Abbasi-Pooya, A., Kashan, A.H.: New mathematical models and a hybrid grouping evolution strategy algorithm for optimal helicopter routing and crew pickup and delivery. Comput. Ind. Eng. **112**, 35–56 (2017)
2. de Alvarenga Rosa, R., Machado, A.M., Ribeiro, G.M., Mauri, G.R.: A mathematical model and a clustering search metaheuristic for planning the helicopter transportation of employees to the production platforms of oil and gas. Comput. Ind. Eng. **101**, 303–312 (2016)
3. Andreeva-Mori, A., Kobayashi, K., Shindo, M.: Particle swarm optimization/greedy-search algorithm for helicopter mission assignment in disaster relief. J. Aerosp. Inf. Syst. **12**(10), 646–660 (2015)
4. Armstrong-Crews, N., Mock, K.: Helicopter routing for maintaining remote sites in Alaska using a genetic algorithm. In: Proceedings of the National Conference on Artificial Intelligence, vol. 20, p. 1586. AAAI Press, Menlo Park; MIT Press, Cambridge, 1999 (2005)
5. Barbarosoğlu, G., Özdamar, L., Cevik, A.: An interactive approach for hierarchical analysis of helicopter logistics in disaster relief operations. Eur. J. Oper. Res. **140**(1), 118–133 (2002)
6. Brachner, M., Hvattum, L.M.: Combined emergency preparedness and operations for safe personnel transport to offshore locations. Omega **67**, 31–41 (2017)
7. Díaz-Parra, O., et al.: Oil platform transport problem (OPTP) is NP-hard. Int. J. Comb. Optim. Probl. Inform. **8**(3), 2–19 (2017)
8. Feo, T.A., Resende, M.G.C.: A probabilistic heuristic for a computationally difficult set covering problem. Oper. Res. Lett. **8**(2), 67–71 (1989)
9. Galvão, R.D., Guimarães, J.: The control of helicopter operations in the Brazilian oil industry: issues in the design and implementation of a computerized system. Eur. J. Oper. Res. **49**(2), 266–270 (1990)
10. Gribkovskaia, I., Halskau, O., Kovalyov, M.Y.: Minimizing takeoff and landing risk in helicopter pickup and delivery operations. Omega **55**, 73–80 (2015)
11. Hernadvolgyi, I.T.: Automatically generated lower bounds for search. Ph.D. thesis, University of Ottawa, Ottawa, ON, Canada, aAINR01708 (2004)
12. Menezes, F., et al.: Optimizing helicopter transport of oil rig crews at Petrobras. Interfaces **40**(5), 408–416 (2010)
13. Motta, A., Vieira, R., Soletti, J.: Optimal routing offshore helicopter using genetic algorithm. In: 2011 6th IEEE Joint International Information Technology and Artificial Intelligence Conference (ITAIC), vol. 2, pp. 6–9. IEEE (2011)
14. Ozdamar, L.: Planning helicopter logistics in disaster relief. OR Spectr. **33**(3), 655–672 (2011)
15. Qian, F., Gribkovskaia, I., Halskau Sr., Ø.: Helicopter routing in the Norwegian oil industry: including safety concerns for passenger transport. Int. J. Phys. Distrib. Logist. Manag. **41**(4), 401–415 (2011)
16. Rosero, V., Torres, F.: Ant colony based on a heuristic insertion for a family of helicopter routing problems. In: Conference Proceedings from the Third International Conference on Production Research-Americas Region, pp. 1–12 (2006)
17. Savelsbergh, M.W., Sol, M.: The general pickup and delivery problem. Transp. Sci. **29**(1), 17–29 (1995)

Metaheuristics for the Generalised Periodic Electric Vehicle Routing Problem

Tayeb Oulad Kouider[✉], Wahiba Ramdane Cherif-Khettaf, and Ammar Oulamara

Université de Lorraine, Lorraine Research Laboratory in Computer Science and Its Applications - LORIA (UMR 7503), Campus Scientifique, 615 Rue du Jardin botanique, 54506 Vandœuvre-les-Nancy, France
{tayeb.ouladkouider,ramdanec,oulamara}@loria.fr

Abstract. This paper presents the Generalized Periodic Electric Vehicle Routing Problem, named GPEVRP, which is an extension of the Periodic Electric Vehicle Routing Problem (PEVRP) proposed in previous research. The GPEVRP is defined by a planning horizon of several periods typically "days", with the constraints that each customer must be visited a required number of times, which must be related to a pattern, that is, a combination of days chosen among a set of allowed combinations of days, must be satisfied every time by its required demand, and that a limited fleet of electric vehicles (EVs) available each day at the depot must be respected. The EVs could be charged during their trips at the depot and in the available external charging stations. The objective of the GPEVRP is to minimize the total cost of routing and charging over the time horizon. We propose three metaheuristics to solve the problem. The first two methods are global approaches, allow solving simultaneously the planning level (pattern assignment) and the operational level (routing and charging). These methods are Large Neighbourhood Search (LNS) and Adaptive Large Neighbourhood Search (ALNS) for which we have proposed a new metric for the selection of patterns. The third method is a two-phase approach, called Multi-start Guided LNS, for which a perturbation phase changes the pattern and manages the planning level, and an intensification phase based on LNS optimizes the operational level and guides the perturbation through a memory. The results show that a simple global approach such as LNS leads to better results on a very constraining problem such as the problem of our study.

Keywords: Periodic vehicle routing · Electric vehicle · Charging station · Large Neighborhood Search · Adaptive Large Neighbourhood Search

1 Introduction

Electric vehicles have a positive impact on the economy and the environment since they use alternative energy sources. They reduce dependence on oil,

C. Paternina-Arboleda and S. Voß (Eds.): ICCL 2019, LNCS 11756, pp. 219–232, 2019.
https://doi.org/10.1007/978-3-030-31140-7_14

decrease gas emissions and improve air quality. Although the acquisition cost is still high, this cost is nevertheless offset by the use of these vehicles, where the operating cost is very low compared to combustion engine vehicles, mainly due to the price of electrical energy and reduced maintenance costs.

In order to accelerate the reduction of the acquisition costs of electric vehicles, the automotive industry must shift to massive production of electric vehicle models with an appropriate volume and capacity load, with a priority target of enterprise fleets. Although the available electric vehicles meet the needs of professionals, mainly for planned activities, three weaknesses remain in their use, namely, limited range, charging time and the availability of charging infrastructures. These are part of the research and development challenges of all car manufacturers and public research centers. Despite these shortcomings, electric vehicles have emerged as credible alternative solution in last mile logistics as 80% of daily covered distances do not exceed 150 km [20]. Furthermore, the development of low-emission zones (LETs), where the most polluting vehicles are regulated, forces logistics professionals to switch to electric vehicles.

In this paper, we focus our study on specific purposes in which electric vehicles are most appropriate, such as parcel or mail delivery. More precisely, we consider the periodic electric vehicle routing problem, in which client locations, together with their requested products and frequency of visits are known. Customers must be assigned to a feasible visit option. Our objective is minimizing the total cost including charging cost and routing over the planning period. The objective of this study is to evaluate the performance of three metaheuristics. The question we are addressing here is, knowing that our problem is very constraining, does the Large Neighbourhood Search (LNS) proposed in a previous study [17] remain competitive compared to the two metaheuristics we propose here which are Adaptive Large Neighborhood Search (ALNS) and a hybrid two-phase method based on LNS, called Multi-start Guided LNS.

The rest of the paper is organized as follows. Section 2 provides a selective review on the studied problem. Section 3 gives more details on constraints and characteristics of our problem. Section 4 proposes solution approaches based on LNS. Section 5 presents experimental results. Section 6 concludes this study with a short summary and some perspectives.

2   Related Work

The Periodic Vehicle Routing Problem (PVRP) has been introduced in [6]. The objective of the PVRP is to find a set of routes over a time horizon of h periods of days that minimizes total travel time while satisfying vehicle capacity, predetermined visit frequency for each client, and spacing constraints. More and more variants of PVRPs have been proposed in the literature to address real issues such as the routing of healthcare nurses [11], and the transportation of elderly or disabled persons [7]. Since the PVRP is an NP-hard problem, most of the methods proposed in the literature are based on heuristic and metaheuristic approaches [1,2,8,19]. A survey on the PVRP can be found in [12].

Electric vehicle routing problems (EVRP) have attracted close attention from researchers and business organizations in recent years. Several variants of EVRPs have been studied in the literature. The EVRP with time windows and recharging stations was introduced in [28]. [13] studied the EVRP with time windows and mixed fleet of electric and conventional vehicles. The EVRP with a heterogeneous fleet of vehicles that differ in their capacity, battery size and acquisition cost was addressed in [15]. In [10], the authors present an electric vehicle routing variant in which different charging technologies are considered and partial EV charging is allowed. In [23, 24] a real application of EVRP is studied. The authors propose a rich variant in which several real constraints are included like mixed fleet of conventional and heterogeneous electric vehicles, different charging technologies, partial EV charging, compatibility between vehicles, and different charging costs. The tourist trip problem for the EVRP with time windows and range limitations is proposed in [29]. In [16] a variant of a two-echelon electric vehicle routing problem is proposed. [25, 26] present a location-routing approach that considers simultaneous decisions on routing vehicles and locating charging stations for strategic network design of electric logistics fleets. The EVRP with pickup and delivery and time windows is considered in [14]. The most recent survey on the EVRP is presented by [9] and [21].

In summary, the current EVRP literature is limited to daily or strategic planning. Although EVs routing problems have attracted close attention from researchers and business organizations in recent years, the periodic extension of the EVRP has received very little attention in the literature, and was studied only in [17, 18]. The authors propose to deal with the tactical and operational decision levels for electric vehicle routing and charging in which the frequency visit was fixed to one. Constructive heuristics were proposed to solve the problem in [18]. The result was improved with LNS in [17]. Another study addressing the multi-periodic aspect for electric vehicles could be found in [30], but in this study the routing and the charging over the period are not considered.

In our study presented below, we propose a new variant named Generalised Periodic Electric Vehicle Routing Problem (GPEVRP), which extends the PEVRP by considering a visit frequency greater than or equal to one, so each customer has a set of patterns. The resolution of the GPEVRP consists of assigning customers to the patterns (planning level), and solving the routing and EV charging problem (operational level). We compare two global approaches that simultaneously optimize the planning and operational level with a two-phase approach that optimizes the two levels separately. The two global approaches consist of Large Neighbourhood Search (LNS) proposed in [17] which has been generalized to the case $f \geq 1$ using the new metric proposed in Sect. 4, and an ALNS approach for which we propose three destroy and three repair operators. The two-phase approach is a two-phase hybrid heuristic that is based on LNS with guided perturbation, named Multi-start GLNS (multi-start Guided LNS) in which routing with charging and planning are performed separately. The first phase of the Multi-start GLNS allows to improve the routing by LNS, and to save a memory. This memory is used to calculate adaptability indicators and guide

the second phase, which consists of modifying the patterns of some customers of the best solution obtained by LNS and thus restarting LNS.

3 Problem Definition

The Generalized Periodic Electric Vehicle Routing Problem is an extension of the Periodic Electric Vehicle Routing Problem (PEVRP) introduced in [18] and [17]. It is defined on a complete directed graph $G = (V, A)$. $V = C \cup B \cup \{0\}$, where the set C of n vertices represents the customers, the set B of ns vertices denotes the external charging stations, which can be visited during each day of the planning horizon, and the vertex 0 represents the depot, which contains charging stations allowing charging at night and during the day. A is the set of arcs, where each arc (i, j) has a travel cost c_{ij}, a travel distance d_{ij} and a travel time t_{ij}. When an arc (i, j) is travelled by an electric vehicle (EV), it consumes an amount of energy $e_{i,j} = r \times d_{i,j}$, where r denotes a constant energy consumption rate. This common simplification of energy consumption is used in the most studies of the literature on the EVRP. For more details see [24].

We consider a time horizon H of np periods typically "days", in which each customer i has a frequency $f(i)$ of visits. This means that customer i must be serviced $f(i)$ times over the time horizon, where $1 \leq f(i) \leq np$, but at most once per day. Then the total number of visits nv to be performed over H is $nv = \sum_{\forall i \in C} f(i)$. Each customer i has a set $comb(i)$ of visit patterns; each pattern has $f(i)$ days of visits, for example, $(1, 3)$ means that the customer is visited the first and the third day ("Monday" and "Wednesday"). At each visit, a customer i requires a quantity q_i of goods and a service time s_i. A fixed charging cost Cc is considered, that neither depends on the amount of energy delivered by charging stations nor on the time needed to charge the vehicle [23, 24]. The power of energy delivered to each vehicle k during a day h is a decision variable $P_{h,0,k}$ that setting the initial state of charge of vehicle k before starting routes of day $h + 1$, $h \in \{1...np\}$.

The GPEVRP consists in assigning to each customer a pattern of its allowed combination set $comb(i)$ that minimize the total cost of routing and charging over H. A feasible solution of the GPEVRP must satisfy the following set of constraints: (i) each route must start and end at the depot, (ii) each customer i should be visited $f(i)$ times during the planning horizon according to one pattern of $comb(i)$, (iii) the customer demand q_i must be completely fulfilled during each visit, (iv) no more than m electric vehicles are used, (v) the total duration of each route, calculated as the sum of travel duration required to visit customers, time required to charge the vehicle during the day, and the service time of each customer, could not exceed T; (vi) the overall amount of goods delivered along the route, given by the sum of demands q_i of visited customers, must not exceed the vehicle capacity Q. The objective function to be minimized is $f(x) = \alpha \times f_1(x) + Cc \times nbs(x)$ where, $f_1(x)$ is the total distance of the solution x over the planning period H, and $nbs(x)$ is the number of visits to charging stations in solution x over the planning period, and α is a given weight representing the cost of one unit of distance.

4   Solution Approaches

In this section we develop an extension of the LNS heuristic called Multi-Start Guided LNS (Multi-start GLNS) and an Adaptive Large Neighbourhood Search to solve our generalized electric vehicle routing problem. ALNS and LNS are efficient methods for solving several variants of vehicle routing problems [5,22], and several LNS hybridizations have been investigated and have proved to be very successful in solving vehicle routing problems [3,23,27].

The Multi-start GLNS is a two-phase hybrid heuristic, in which the planning level (pattern assignment) and the operational level (routing and charging) are performed separately, while the ALNS method allows both planning and routing levels to be processed simultaneously.

4.1   Multi-Start Guided Large Neighbourhood Search

The Multi-Start Guided LNS method is a two-phase hybrid metaheuristic. The goal of the first phase is to improve the routing and the EV charging without changing the assignment of customers in days (fixed patterns) and the second phase aims to modify the planning level by changing the patterns for some customers while relaxing routing and energy constraints. The objective is to alternate between a routing and charging improvement phase and a guided perturbation planning phase.

Algorithm 1 provides the pseudo-code of the proposed approach. The details of the algorithm's steps is presented in the next subsections. The Multi-start GLNS begins by calling the construction heuristic $Hconst$ (detailed below), which generates an initial solution. The main loop is executed until a stop criterion is met (maximum time limit or a maximum number of iterations). Each iteration of the while loop repeats two phases whose objective has been given above. The first phase uses a local search based on LNS, $LNS(S0, Adaptability, S1)$, applied to intensively improve routing and charging without changing the planning level. During this LNS phase, we use a memory that allows us to save certain values to compute an indicator that measures the adaptability of the patterns used in LNS named $adaptability(i, h)$. The second phase is a perturbation phase allowing to change the patterns of some customers and to modify the planning level of the best solution found by LNS. The idea is to use the results of adaptability to decide which customers should change their patterns. The modified solution will be used as a new solution to restart LNS.

Construction Heuristic '$Hconst$': The heuristic proceeds in two steps: a first step whose objective is to smooth the load of customers on the horizon by assigning customers to the days of the planning horizon (pattern choice), and a second phase builds the routes of each day. Let Sol be a sequence of ns services that represents a complete solution of the GPEVRP, partitioned into np successive sub-lists Sol_h (one per day h). Each customer i for which the insertion is successful will appear $f(i)$ times in total, i.e., in $f(i)$ days corresponding to one allowed pattern contained in $comb(i)$. If customer i has failed to be inserted,

---

Algorithm 1. Multi-Start Guided LNS

1: **Input:** A GPEVRP instance
2: **Output:** The best solution found S^*
3: Generate an initial solution
4: **while** Stopping criteria is not met **do**
5:     Execute LNS(S_0, $Adaptability$, S_1)
6:     Update the best solution S^*
7:     Perturbation(S_1, $Adaptability$, S_2)
8:     $S_0 := S_2$
9: **end while**

---

it will be inserted into a list NS of non-inserted customers. At the beginning $Sol := \emptyset$, $NS := \emptyset$. The details of the two steps are as follows:

Step 1. The objective of this step is to build the sub-lists Sol_h of Sol by relaxing energy and routing constraints while respecting the total available capacity per day ($m \times Q$). The heuristic starts by sorting customers in a decreasing order according to two criteria (frequency and demand) in a list L (initially $L := C$), then the heuristic repeats the following process until L is empty: (a) select a customer i from the head of L, if for each pattern $cb \in comb(i)$, client i does not verify the capacity constraint of each day as in formula (1), it is added to a list NS of non-inserted clients, and the algorithm returns to (a). Otherwise, the insertion cost is calculated for each $cb \in comb(i)$ satisfying constraint (1), the customer i is then inserted in each day of the combination cb^* selected according to formula (2), delete i from L and return to (a).

$$q_{i,h} + \sum_{j \in Sol_d} q_{j,h} \leq m \times Q \qquad \forall h \in cb \tag{1}$$

$$cb^* = \arg \min_{cb \in comb(i)} \left\{ \sum_{h \in cb} \sum_{j \in Sol_h} d_{i,j} \right\} \tag{2}$$

Step 2. The patterns being fixed by *Step1*, we need to construct routes for each day, while satisfying the following constraints: (1) energy constraint, (2) limit on the number of vehicles, and (3) capacity and time limit for each vehicle. For that purpose, we have adapted the Best Insertion Heuristic (BIH) proposed for the PEVRP [18]. For this, we propose a new metric to calculate the cost of inserting a customer i in the $f(i)$ days of a given pattern cb, noted $InsertCost(i, cb)$ (see formula (3)), where $CostDay(i, h)$ represents the insertion cost of the client i in a day h for a given pattern cb. This metric integrates the frequency variability between customers without automatically benefiting customers with a high frequency.

$$InsertCost(i, cb) = \sum_{h \in cb} \frac{CostDay(i, h)}{f(i) * \sqrt{f(i)}} \tag{3}$$

Local Search Phase - LNS: Our local search is based on LNS. The LNS starts from a given initial solution and improves it using the Destroy-Repair process. Indeed, LNS removes a relatively large number of customers from the current solution and tries to reinsert them into different positions. This leads to a completely different solution, that helps the heuristic to escape local optima. In the proposed approach, we use LNS as a local search so the number of ejected customers must remain relatively small [4,24]. The LNS must not change customer patterns. We have chosen to apply LNS over the entire horizon instead of applying it on a daily basis to have a more global insight into the impact related to the insertion of a client according to a given pattern. The LNS receives as input the solution S_0 which can be a partial solution (case where $NS \neq \emptyset$). The NS list must then be integrated into the list of customers to be ejected. Then we use the Random Removal operator to eject the necessary number of customers, and Regret insertion as repair operator. This choice is guided by our previous study [17] which analyzed several destroy and repair operators on the PEVRP and confirmed the efficiency of the random removal and the regret insertion. These operators use specific procedures "AdjustIncreaseCharging" and "AdjustDecreaseCharging" to manage the energy constraint (see Sect. 4.2). A new solution found during the Destroy-Repair process is accepted if and only if it is better than the previous one. In such case the Destroy-Repair process continues with this new solution.

At the end of each LNS iteration it, we measure the adaptation of each customer i in each day h of its pattern adapt(i, h) (see formula 4) by computing the distance to his close neighborhood (predecessor and successor of i in the route of day h). We then compute adapt(i,h) at each iteration of LNS, and we calculate an average value of adapt(i,h) over all LNS iterations, named adaptability(i,h). This last value measures the proximity of the customer in his routes, a high value could mean that the customer generates a very high cost in his current routes and that it would be more appropriate to move it to other routes or other days. We can assume that the least adapted customers in the current solution are those with a high value of adaptability(i,h). Hence the usefulness of changing the pattern for these customers in the perturbation phase to better guide LNS in the next restart.

$$adapt(i, h) = d_{k,i} + d_{i,l} - d_{k,l} \tag{4}$$

where k (respectively, l) is the successor (respectively, the predecessor) of i in the route that is including i in the day h.

Perturbation Phase: This phase applies the Perturbation(S_1,Adaptability(.)) procedure that receives as an input the solution S_1 obtained by LNS and the adaptability matrix described above. First it calls procedure SelectLessAdapted(Adaptability, L',nc) and then the procedure ChangeAssignment(S_1,L). SelectLessAdapted(Adaptability, L',nc) returns a list L of nc worst pairs ($i \in C$, $h \in H$) that should be excluded in the current solution (see Algorithm 1), and procedure ChangeAffectation(S_1,L) allows to disrupt the solution S_1 by excluding

all fixed assignments in L (output of procedure SelectLessAdapted(Adaptability, L',nc)) and replace them by one of their pattern.

Algorithm 2. SelectLessAdapted(Adaptability, L',nc)

1: **Input:** The matrix Adaptability, the list L',nc
2: **Output:** The list L
3: $L := \emptyset$, $Nbselect := 0$, $SelectedClient := \emptyset$
4: **while** $Nbselect < nc$ **do**
5:    Select $l = (i, h)$ from the head of L',   ▷ i: customer, h: a day
6:    Delete l from L'
7:    **if** $(|comb(i)| > 1)$ & $(i \notin SelectedClient)$ **then**
8:       $Nbselect + +$
9:       Add l to L, add i to $SelectedClient$
10:    **end if**
11: **end while**
12: Return L

Algorithm 3. ChangeAffectation(S_1,L)

1: **Input:** solution S_1', list L ▷ $|L| = nc$
2: **Output:** solution S_2
3: $S_2 := S_1$
4: **for all** $(i, h) \in L$ **do**   ▷ i: a customer, h: a day
5:    $List_{cb} := \emptyset$ ▷ $List_{cb}$: a list of potential patterns
6:    **for all** $cb \in comb(i)$ **do**
7:       **if** $h \notin cb$ **then**
8:          $List_{cb} = List_{cb} \cup cb$
9:       **end if**
10:    **end for**
11:    **if** $List_{cb} \neq \emptyset$ **then**
12:       Choose the new pattern $cb^*(i)$ randomly from $list_{cb}$
13:       Delete the customer i from S_2
14:    **end if**
15: **end for**
16: Reinsert the deleted customers according to the pattern cb^* in S_2 using BIH   ▷
    See Construction heuristic
17: Return S_2

4.2 ALNS Method

In this section we use the ALNS metaheuristic to solve the GPEVRP. We follow the main framework used in [22].

ALNS manages different Destroy-Repair operators. At each iteration, a pair of destroy and repair operators is randomly chosen from a collection of operators. An adaptive mechanism is used to promote more efficient pairs of operators.

Destroy Operators. At each iteration, we choose a number γ of customers to remove.

We implemented three destroy operators: random removal, worst removal, and cluster removal. In the following, we give a description of these operators. Let $S = \{S_1, .., S_{np}\}$ the initial feasible solution, and S_h a set of routes in the day h, $S_h = \{T_{1h}, .., T_{kh}..T_{mh}\}$. After removing γ customers, a new partial solution S' is generated. $S' = \{T'_{1h}, .., T'_{kh}, .., T'_{mh}\}$, such that $T'_{kh} = T_{kh}$ if no customer has been deleted from T_{kh}, and $T'_{kh} = T^-_{kh}$ if one or more customers have been deleted from T_{kh}. Each destroy operator uses a repair procedure, named AdjustDecreaseCharging(Tr), to adjust the energy of each route $T^-_{kp} \in S'$. For more details on this repair procedure, we refer the reader to [17].

Random Removal. This operator randomly chooses γ customers and removes them from the solution; then a procedure AdjustDecreaseCharging(Tr) that adjusts the energy of the vehicle is applied to each route $T^-_{kp} \in S'$. The procedure estimates the unused energy in T^-_{kp}, and decides in which station we will reduce their energy retrieved by the vehicle, or which stations can be removed when station removal is possible.

Cluster Removal. This operator starts by choosing a customer i randomly, then selects $\gamma - 1$ customers nearest to i (in terms of distance). The procedure AdjustDecreaseCharging(Tr) is applied to each route $T^-_{kp} \in S'$.

Worst Removal. For each customer $i \in S$, the operator simulates the removal of i in each route where i is visited and applying AdjustDecreaseCharging(Tr) to each one of them. The customer i^* whose removal produces the largest value of metric (3) is chosen. The algorithm stops when γ customers are deleted.

Repair Operators. We have proposed three repair operators, namely, First Improvement, Best Improvement, Regret Insertion. After each customer insertion, a repair procedure named $AdjustIncreaseCharging(Tr)$ is applied to adjust the energy of each modified tour $Tr \in S$. For more details on this repair procedure, we refer the reader to [17]. If the insertion of the customer i fails in the existing routes, a new route containing the depot and the customer i can be built if the maximum number of vehicles is not reached. When all customers have been reinserted back into the solution, the new solution is compared with the original solution. If it is impossible to insert all ejected clients, the solution may become unfeasible. In order to insert customers removed by destruction operators, we define two strategies to evaluate the insertion of a customer i in a route Tr. The first one denoted insert_All, simulates the insertion of i in all the positions of Tr and returns the best insertion in terms of the routing and charging costs. The second strategy is denoted insert_2, where we select from nodes of Tr the two nearest nodes (customers or charging stations) and simulate the insertion before and after these two nodes. For each insertion operator described below we use the two strategies described above. Let L be the list of

customers not present in the current solution. For each customer i in L we define $P(i) = \{h | h \in cb, \forall cb \in comb(i)\}$.

First Improvement. This operator selects randomly a customer from L, simulates its insertion in every day k of $P(i)$ and every route of k, calculates the metric value of each pattern $cb \in comb(i)$ and selects the pattern that generates the minimal cost. A customer will be inserted in the best route of each day of the selected pattern.

Best Improvement. This operator repeats the following steps until $L = \emptyset$. (i) compute the minimum metric insertion cost of each customer $i \in L$, (ii) the customer i^* with the minimum cost is inserted in its best position on each day of its best pattern, (iii) update L and S.

Regret Insertion. This operator repeats the following steps until $L = \emptyset$: (i) compute the gap between the two best insertion costs of each customer $i \in L$, denoted δ_i, (ii) customer i^* with the maximum value of δ_{i*} is inserted in its best position in the days of its best pattern, (iii) update L and S.

5 Data Sets and Computational Experiments

Our methods are implemented using C++. All computations are carried out on an Intel Core (TM) i7- 5600U CPU, 2.60 GHz processor, with 8GB RAM memory. To test our methods we extend instances proposed for a PEVRP in [10] with 100 customers and 9 charging stations in each instance. We generate frequencies of visits in the set $\{1 \ldots 5\}$ and we fix patterns of each customer with respect to the proportion provided in Table 1. Each pattern is then generated randomly with respect to the customer frequency.

After different tests, we fixed the number of vehicles to 5. The other parameter settings of our instances are similar to those of [18]. We generated 52 instances as explained before. In order to evaluate methods in the same conditions, we considered a time limit of two hours as stopping criterion.

First, we set the local search phase parameter in the Multi-start GLNS. This parameter γ being the number of customers to remove-reinsert in each iteration.

Table 1. Instance settings

Frequency	% of customers	Number of patterns
1	30	2
2	25	3
3	20	3
4	15	2
5	10	1

We proposed and tested three values of γ, namely, 5, 7 and 10. For each value, we derive the following metrics: (i) **Best Solution Occurrence**: reports the number of times the method provides a better solution, (ii) **Average Gap**: returns the average gap among all instances compared to the best solution obtained on all tests, (iii) **Average Gap Without Best**: returns the average gap among the result of instances where the best is not attained, and (iv) **Average Number of Restarts**: the average number of restarts of the method in the time limit condition. We can see in Table 2 that $\gamma = 7$ provides better results either in the number of best solutions found or in the average gap.

Table 2. Multi-start GLNS results

	$\gamma = 5$	$\gamma = 7$	$\gamma = 10$
Best Solution Occurrence	21	**31**	0
Average Gap	0,82%	**0,63%**	11,91%
Average Gap Without Best	**1,38%**	1,55%	11,91%
Average Number of Restarts	48	35	27

In the second step of our experiments, we compare results of ALNS with an extension of the LNS that we have proposed for the PEVRP [17]. This LNS uses as a pair of operators Random Removal-Regret Insertion; this choice is motivated by our previous work [17]. The LNS has been generalized to the case $f \geq 1$, by using the metric (see formula (3)) in the repair operator. The ALNS as described in Sect. 4.2 manages three destroy operators and three repair operators. In Table 3 we report four metrics, the first three being the same as in the experiment of Multi-Start GLNS and the fourth is the average number of iterations where the best solution is reached. We can see that the LNS gives the best solution for 80% of the instances, and it is at most 1.96% of the best solution found by ALNS on the other cases. This result shows that a simple approach like LNS succeeds in being more competitive than ALNS, which uses an adaptive mechanism to choose its operators. This can be explained by the fact that the problem is very difficult due to the constraints of energy, routing and planning, and that the number of unfeasible solutions can be very large. This can limit the search space and has not allowed ALNS to exceed the LNS results. We can conclude that ALNS needs more diversification to cover a larger area (Table 3).

Finally, we compare the five different methods. The results in Table 4 show that the LNS method performs well to find the best solution among the other methods ($\simeq 60\%$) and also to approach the best average gap when the other methods found a better solution. As mentioned before, a strong intensification is more adapted and useful for this problem. Furthermore, the three constraint classes being strongly dependant, managing separately each of them, as in the Multi-start GLNS, limit the solutions exploring and does not give good results. To sum up, the change of patterns in the first method, and the change of operators in the ALNS gives less time to explore the neighbourhood.

Table 3. ALNS/LNS comparison

	ALNS	LNS
Best Solution Occurrence	11	**41**
Average Gap	1,91%	**0,41%**
Average Gap Without Best	2,42%	**1,96%**
Average Best Iteration	194	**166**

Table 4. Comparison of all methods

	GLNS5	GLNS7	GLNS10	ALNS	LNS
Best Solution Occurrence	5	7	0	9	**31**
Average Gap	2,71%	2,51%	13,99%	2,36%	**0,86%**
Average Gap Without Best	3,00%	2,90%	13,99%	2,85%	**2,13%**

6 Conclusion

In this paper we considered the Generalized Periodic Electric Vehicle Routing Problem. It consists in managing the scheduling of multiple visits to different customers and optimizing the routing and charging of a limited fleet of electric vehicles. We proposed to compare and evaluate three metaheuristics. The first method is LNS which has been proposed in [17]; it is a global approach that allows managing the planning level and the operational level simultaneously. We have generalized LNS to our problem by using a new metric in the construction operator proposed in this paper. The second metaheuristic is an ALNS approach, in which three construction and destruction operators were used. This method is also a global approach like LNS. The third approach is a hybrid method, in which the operational level and the planning level are handled separately. This method, named Multi-Start Guided LNS (Multi-Start GLNS) uses a perturbation phase to modify the planning level and an intensification phase based on LNS to optimize the routing and charging phase. The results revealed that LNS remains competitive compared to ALNS and Multi-start GLNS. This can be explained by the fact that the search space is quite limited because the problem is very complex. This result is quite interesting and shows that a simple approach like LNS can be very competitive. In our future work, we want to integrate several diversification techniques into ALNS and Multi-Start GLNS, and determine if the results of these two methods can be improved in comparison to LNS.

References

1. Archetti, C., Fernández, E., Huerta-Muñoz, D.L.: A two-phase solution algorithm for the flexible periodic vehicle routing problem. Comput. Oper. Res. **99**, 27–37 (2018)

2. Baldacci, R., Bartolini, E., Mingozzi, A., Valletta, A.: An exact algorithm for the period routing problem. Oper. Res. **59**(1), 228–241 (2011)
3. Chentli, H., Ouafi, R., Ramdane-Chérif-Khettaf, W.: Behaviour of a hybrid ILS heuristic on the capacitated profitable tour problem. In: Proceedings of the 7th International Conference on Operations Research and Enterprise Systems, ICORES 2018, Funchal, Madeira, Portugal, January 2018, pp. 115–123 (2018)
4. Chentli, H., Ouafi, R., Ramdane Cherif-Khettaf, W.: Impact of iterated local search heuristic hybridization on vehicle routing problems: application to the capacitated profitable tour problem. In: Parlier, G.H., Liberatore, F., Demange, M. (eds.) ICORES 2018. CCIS, vol. 966, pp. 80–101. Springer, Cham (2019). https://doi.org/10.1007/978-3-030-16035-7_5
5. Chentli, H., Ouafi, R., Ramdane Cherif-Khettaf, W.: A selective adaptive large neighborhood search heuristic for the profitable tour problem with simultaneous pickup and delivery services. RAIRO-Oper. Res. **52**(4–5), 1295–1328 (2018)
6. Christofides, C., Beasley, J.: The period routing problem. Networks **14**, 237–256 (1984)
7. Cissé, M., Yalçındağ, S., Kergosien, Y., Şahin, E., Lenté, C., Matta, A.: OR problems related to home health care: a review of relevant routing and scheduling problems. Oper. Res. Health Care **13–14**, 1–22 (2017)
8. Dayarian, I., Crainic, T., Gendreau, M., Rei, W.: An adaptive large-neighborhood search heuristic for a multi-period vehicle routing problem. Transp. Res. Part E **95**, 95–123 (2016)
9. Erdelić, T., Carić, T.: A survey on the electric vehicle routing problem: variants and solution approaches. J. Adv. Transp. **2019**, 48 (2019). https://doi.org/10.1155/2019/5075671. Article ID 5075671
10. Felipe, M., Ortuno, T., Righini, G., Tirado, G.: A heuristic approach for the green vehicle routing problem with multiple technologies and partial recharges. Transp. Res. Part E Logist. Transp. Rev. **71**, 111–128 (2014)
11. Fikar, C., Hirsch, P.: Home health care routing and scheduling: a review. Comput. Oper. Res. **77**, 86–95 (2017)
12. Francis, P.M., Smilowitz, K.R., Tzur, M.: The period vehicle routing problem and its extensions. In: Golden, B., Raghavan, S., Wasil, E. (eds.) The Vehicle Routing Problem: Latest Advances and New Challenges, vol. 43. Springer, Boston (2008). https://doi.org/10.1007/978-0-387-77778-8_4
13. Goeke, D., Schneider, M.: Routing a mixed fleet of electric and conventional vehicles. Eur. J. Oper. Res. **245**(1), 81–99 (2015)
14. Goeke, D.: Granular tabu search for the pickup and delivery problem with time windows and electric vehicles. Eur. J. Oper. Res. **278**, 821–836 (2019)
15. Hiermann, G., Puchinger, J., Ropke, S., Hartl, R.: The electric fleet size and mix vehicle routing problem with time windows and recharging stations. Eur. J. Oper. Res. **252**(3), 995–1018 (2016)
16. Jie, W., Yang, J., Zhang, M., Huang, Y.: The two-echelon capacitated electric vehicle routing problem with battery swapping stations: formulation and efficient methodology. Eur. J. Oper. Res. **272**(3), 879–904 (2019)
17. Kouider, T.O., Ramdane Cherif-Khettaf, W., Oulamara, A.: Large neighborhood search for periodic electric vehicle routing problem. In: Proceedings of the 8th International Conference on Operations Research and Enterprise Systems - Volume 1: ICORES, pp. 169–178. INSTICC, SciTePress (2019)

18. Kouider, T.O., Ramdane-Cherif-Khettaf, W., Oulamara, A.: Constructive heuristics for periodic electric vehicle routing problem. In: Proceedings of the 7th International Conference on Operations Research and Enterprise Systems - Volume 1: ICORES, pp. 264–271. INSTICC, SciTePress (2018)
19. Mancini, S.: A real-life multi depot multi period vehicle routing problem with a heterogeneous fleet: formulation and adaptive large neighborhood search based matheuristic. Transp. Res. Part C **70**, 100–112 (2016)
20. Optimisation de la logistique urbaine dans les villes de tailles moyenne à petite, direction regional de l'Environnement, de l'Amenagement. Misistère des transport, France, October 2018
21. Pelletier, S., Jabali, O., Laporte, G.: 50th anniversary invited article-goods distribution with electric vehicles: Review and research perspectives. Transp. Sci. **50**(1), 3–22 (2016)
22. Ropke, S., Pisinger, D.: An adaptive large neighborhood search heuristic for the pickup and delivery problem with time windows. Transp. Sci. **40**(4), 455–472 (2006)
23. Sassi, O., Ramdane Cherif-Khettaf, W., Oulamara, A.: Iterated tabu search for the mix fleet vehicle routing problem with heterogenous electric vehicles. In: Le Thi, H.A., Pham Dinh, T., Nguyen, N.T. (eds.) Modelling, Computation and Optimization in Information Systems and Management Sciences. AISC, vol. 359, pp. 57–68. Springer, Cham (2015). https://doi.org/10.1007/978-3-319-18161-5_6
24. Sassi, O., Ramdane Cherif-Khettaf, W., Oulamara, A.: Multi-start Iterated local search for the mixed fleet vehicle routing problem with heterogenous electric vehicles. In: Ochoa, G., Chicano, F. (eds.) EvoCOP 2015. LNCS, vol. 9026, pp. 138–149. Springer, Cham (2015). https://doi.org/10.1007/978-3-319-16468-7_12
25. Schiffer, M., Walther, G.: The electric location routing problem with time windows and partial recharging. Eur. J. Oper. Res. **260**, 995–1013 (2017)
26. Schiffer, M., Walther, G.: Strategic planning of electric logistics fleet networks: a robust location-routing approach. Omega **80**, 31–42 (2018)
27. Schmid, V., Doerner, K.F., Hartl, R.F., Salazar-González, J.J.: Hybridization of very large neighborhood search for ready-mixed concrete delivery problems. Comput. Oper. Res. **37**(3), 559–574 (2010)
28. Schneider, M., Stenger, A., Goeke, D.: The electric vehicle routing problem with time windows and recharging stations. Transp. Sci. **75**, 500–520 (2014)
29. Wang, Y.W., Lin, C.C., Lee, T.J.: Electric vehicle tour planning. Transp. Res. Part D Transp. Environ. **63**, 121–136 (2018)
30. Zhang, A., Kang, J.J.E., Kwon, C.C.: Incorporating demand dynamics in multi-period capacitated fast-charging location planning for electric vehicles. Transp. Res. Part B **103**, 5–29 (2017)

Electric Bus Scheduling and Optimal Charging

Bilal Messaoudi[1,2](✉) and Ammar Oulamara[1]

[1] Universiy of Lorraine - LORIA UMR 7503 Laboratory, Campus Scientifique,
615 Rue du Jardin-Botanique, 54506 Vandoeuvre-les-Nancy, France
bilal.messaoudi@univ-lorraine.fr
[2] Antsway SA, ARTEM - 92, Rue du Sergent Blandan, 54042 Nancy Cedex, France

Abstract. In this paper, we study an operational problem of optimizing the deployment of electric buses in an urban public transport service. We propose a modeling aimed at optimizing the assignment of buses to line services of a public transport network, then the allocation of parking places to buses taking into account the typology of depots and finally the optimization of bus charging while complying with technical constraints. The model is based and tested on a concrete case study of a public transport company. The results show that optimization models can address the weaknesses of electric buses, namely limited range and long charging times, and thus provide public transport organizations with bus use strategies and bus charging policies.

Keywords: Electric buses · Charging · Scheduling · MILP · Heuristics

1 Introduction

Road transport is responsible for a large part of greenhouse gas emissions. These emissions have a negative impact on the health of citizens and cause major problems for the world's governments in defining public health and environmental policies. Several cities have already adopted drastic policies by introducing low-emission zones where the most polluting vehicles are regulated. In such zones, vehicles with higher emissions either cannot enter the area or have to pay expensive fees if they enter the low emission zone. On the other hand, incentives are provided for vehicles with alternative fuels, such as electric cars and hydrogen cars. For example, the bonus for the acquisition of an electric vehicle in France and Germany, free parking for electric cars in Norway, etc. In addition, the maturity of electric vehicle technology allows to overcome the barriers concerning their adoption by users, particularly regarding their range, where currently reasonable distances close to those of diesel vehicles can be achieved.

Several professionals have already switched to the use of electric vehicles or offer services with electric vehicles (car sharing, etc.) as well as urban transport operators are testing electric buses in demonstration phases or pilot programs

© Springer Nature Switzerland AG 2019
C. Paternina-Arboleda and S. Voß (Eds.): ICCL 2019, LNCS 11756, pp. 233–247, 2019.
https://doi.org/10.1007/978-3-030-31140-7_15

around the world [1–3]. Indeed, electric buses have been successfully tested in several projects and experiments. In addition, with the reduction in manufacturing costs of battery systems, electric buses have become increasingly competitive with diesel buses. However, the reduction in the operational performance of electric buses, particularly the limited range, remains an obstacle to their massive deployment, nevertheless this obstacle can be overcome by intelligent management of the use of the fleet between transport service and bus charging operations. The objective of this work is to provide tools for optimizing and managing a fleet of electric buses in order to schedule transport and charging operations, based on a concrete case study of an urban transport operator in France.

The paper addresses operational electric bus planning by focusing on the electric bus charging and scheduling in which we will consider problems of allocation of electric buses on transport services, allocation of parking spaces to electric buses in the depot, and optimizing charging of electric buses considering technical constraints of the electric grid.

The paper is structured as follows. Section 2 provides an informative review on the related works. Section 3 gives a general description of the problem and introduces different components of the urban transport system. Section 4 provides the notation used throughout the paper and present our mathematical formulation. In Sect. 5, we propose our solution method using a decomposition approach. In Sect. 6, we provide the settings and present the results of our computational study on real instances. Section 7 summarizes the work and provides a conclusion.

2   Related Works

Although electric buses have many benefits including the absence of gas emissions and noise reduction, one of the main concerns of urban transport operators is the anxiety associated with autonomy. In general, the maximum range of electric buses currently manufactured can reach 280 km [4], close to that of diesel buses. However, under real driving conditions, this autonomy decreases due to the consumption of accessories (heating and air conditioning), which makes it difficult to use them continuously without recharging [5]. In order to guarantee a high quality of operational service, buses must be constantly recharged using fast chargers [5]. Furthermore, as public transport networks have fixed routes and timetables, several research in the literature are focused on tactical or operational optimization issues of electric buses, mainly, analysis of electric bus route design [6,7], charging infrastructure [8,9] and scheduling (including charging scheduling) [10,11].

For the electric bus route design, Lin et al. [7] develop a vehicle-scheduling model for electric transit buses with fast charging at a battery station. Considering the maximum route distance constraint, authors showed the NP-hardness of the problem and develop a column-generation-based algorithm to solve the scheduling problems. For electric bus route design, authors in [6] develop a

genetic algorithm to solve the transit network design for electric buses by determining the optimal bus route in an urban context. Others studies on electric bus route design are conducted by works [12,13].

For the charging station allocation problems, several studies investigated the siting of charging infrastructure for the electric buses. Kunith et al. [8] developed an optimization model to identify the appropriate trade-off between an efficient layout of the charging infrastructure and an adequate sizing of battery capacity in order to minimize the total cost of ownership and to enable an energetically feasible bus operation. In [9], authors developed an optimization charging scheduling framework for electric buses in an urban transit network, where a mixed integer programming optimization model is developed with the objective of minimizing total costs of operating an electric bus recharging system and determines the planning decisions (i.e., location and capacity of the charging stations) and operational decisions (i.e., recharging schedules). In [14], authors present an optimization model to identify the location of fast-charging stations for fueling the bus network in an urban area with application to the city of Stockholm. Authors showed in their model that total costs of a partially electrified bus system in both optimization cases considering cost and energy differ only marginally from the costs for a 100% diesel system. These three studies only considered fast-charging stations along the route and did not include the overnight in-depot charging stations at bus garages. Another relevant studies on electric bus charging infrastructure can be found in [15–17].

Another thread is electric bus scheduling problem in which given a timetable with trips with fixed travel times and start and end locations, we must find an assignment of trips to vehicles with the objective of minimizing the total costs while respecting the following constraints: Each trip is covered by one vehicle, each vehicle performs a feasible sequence of trips, the maximum driving range of each vehicle task is not exceeded, every vehicle task must start and end at the depot and the maximum number of vehicles at the depot should not be exceeded. In [10], authors consider the bus scheduling problem that includes electric buses and diesel buses. The electric buses can be charged at designated terminals with the objective of maximizing the number of electric buses used. A k-greedy algorithm was developed, which makes use of the k-nearest neighbour technique to find the best departure point. In [11], authors consider the single depot electric bus scheduling problem with homogeneous fleet. They developed a column generation approach to minimize the total cost. Other relevant studies on bus scheduling problem are considered in [6,12,13]. All of these studies consider bus charging during service. However, charging on routes requires a comprehensive infrastructure planning as bus lines have to be equipped with a sufficient number of charging points. This requires a significant investment in the deployment and maintenance of the charging infrastructure on the public area. Moreover, this infrastructure is not protected from acts of vandalism, which leads to an additional cost for their replacement. Therefore, it is important to investigate the electric bus solution with a overnight charging in the depots including sufficient battery capacity to provide the services without the need for recharging.

3 Problem Description and Notations

In this section, we lay the basis of our case study of public transportation service with electric buses. We start by defining the different components of the system then we present technical considerations related to the use of electric buses. All information presented here was collected in close collaboration with an urban public transport operator.

The public transportation operator plans to replace all conventional diesel buses with electric buses. To ensure a smooth transition, the existing diesel buses will be replaced exactly with the same number of electric bus, and electric buses will first operate on the already established services (trips) and timetables. Thus, in this study, we are dealing with the operational context of managing the use of electric buses rather than strategic context of re-sizing the fleet of electric buses. So, we are focusing on problems of allocating buses to services, locating buses in the parking depot and optimizing the charging of those buses at the depot. Before developing optimization problems we are considering, we describe below the transportation system and its components.

The transportation system is composed of several resources, namely, electric buses, parking spaces, charging infrastructure, and services (trips) of bus line and their timetables. Details of these components is presented below:

- *Urban transportation network.* The urban transportation network is structured into several bus lines. Each line is characterized by a departure station (the beginning of the line) and an arrival station (the end of the line) and a set of stops to be served by a bus along its route between departure and arrival stations. A bus assigned to a line makes round trips between the departure station and the arrival station. A bus service is the time interval between the beginning of the first trip starting from the depot and the end of the last trip returning to the depot. For a given line, a set of services (trips) are defined, and for each service j the timetable specifies a start time d_j and an end time f_j, and a required energy e_i. Thus, services are assigned to buses such that each service is covered exactly once and each bus performs a feasible sequence of services during the day.
- *Electric buses.* A set of homogeneous electric buses is available providing the transport service. All electric buses have the same driving range of up to 250 km, which allows them to ensure a transportation service without recharging along the routes. In addition, buses are equipped with battery with capacity equal to B (kWh) and buses are only charged in the depot when they return to the parking.
- *Parking at the depot.* The depot contains several parking places organized so that a bus can easily park on each place, and there is no unused space in the depot. The topology and organization of the depot requires an efficient allocation of parking places to buses so that a bus can easily leave the depot as soon as it starts working. In fact, depending on the layout of the depot, we distinguish four types of parking places: (i) *free place* which is independent from the other places, where a bus can park on this place at any time without

any constraint (ii) *entry-blocked place* is an inaccessible place for bus parking if the places upstream of this place are occupied by buses, (iii) *exit-blocked place* is a place where the bus parked there cannot leave the depot if the places downstream of the blocked place are occupied by buses. (iv) *entry-exit-blocked place* is a place of type *entry-blocked* and *exit-blocked place*. Figure 1 provides an example of depot and type of places.

– *Charging infrastructure.* The charging infrastructure is composed of fast chargers. Each parking place is equipped with a charging spot that can delivers up to 90 kW of power. In addition, the power delivered by all charging stations must not exceed the capacity of the depot's electrical grid.

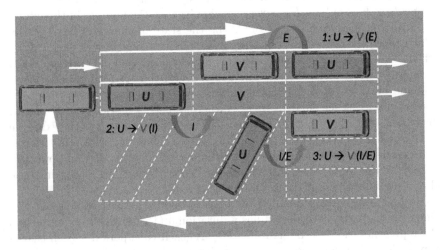

Fig. 1. Layout example of parking places. (1) Exit-blocking constraint, the bus in place V cannot exit when place U is occupied (2) Entry-blocking constraint, the bus in place U blocks entrance of other buses to place V (3) Entry-Exit-blocking constraint, combination of Exit-blocking and Entry-blocking constraints

The purpose in this problem is to provide a feasible assignment of services to buses, allocation of parking places to buses as well as charging schedule of buses at the depot in order to ensures services, while avoiding blocking constraints and respecting energy power limits with the objective of minimizing charging costs. Objective function is divided in two parts. Minimize number of used buses in priority, and then minimize charging costs.

4 Mathematical Formulation

In this section, we propose a mathematical modeling of the optimization problem described in Sect. 3. The problem data used in the mathematical model are as follows. Let R be the transportation network consisting of a set of numbered

lines. Each line is characterized by a departure station and an arrival station and a set of stops to be served by a bus along its route between departure and arrival stations. Each line has a set of services (trips) that should be operated by electric buses. Let S denotes the set of all services of all lines. Each service j, $j \in S$, has a start time d_j, an end time f_j, and an energy requirement e_j depending on total distance that the bus will travel during the service. We assume that services are indexed in non-decreasing order of their start times. Let $N(S_j)$ be a set of services which overlap with service j, i.e. $N(S_j) = \{S_{j'} / [d_j, f_j] \cap [d_{j'}, f_{j'}] \neq \emptyset\}$.

Services are carried out by a set V of homogeneous electric buses. Electric buses have a capacity battery of B, a minimal and a maximal state of charge SoC_m and SoC_M, respectively. SoC_m (SoC_M) means that at any time the amount of energy available in the battery of a bus cannot go less (higher) than $SoC_m \times B$ ($SoC_M \times B$). SoC_m and SoC_M take their values in the interval $[0,1]$, with $SoC_m < SoC_M$. Let $SoC_{i,t}$ be the state of charge of bus i at time t, which specifies the amount of energy available in the battery at time t. Buses start and end their services at the depot, furthermore, they are charged during the parking period at the depot. The depot contains a set P of parking places. We define by $G = (P, A, X)$ the blocking graph of parking places, where P is the set of places, A set of edges (u, v, x) where a place u blocks a place v for the event $x \in X$ and $X = \{I, E, IE\}$, where I is the entry event, E is the exit event and IE the entry-exit event. Each parking place is equipped with a fast charger which can deliver up to p_M kW of power and a minimal p_m kW of power. We denote by g_t the total power available at the depot at time t for recharging buses and c_t represents the cost per kWh during period t.

The planning horizon denoted by $[0, T]$ is divided into H equidistant time periods $t = 1, \ldots, H$, each period has a duration δ (15 min in our experiments), and t represents the time interval $[t - 1, t]$. At each time period t, each charger can provide a charging power within the interval $[p_m, p_M]$. These values allow the modulation of the bus recharging. Thus, if a bus i is charged during the time interval t with a power p_M, it recovers an amount of energy equal to $\delta \times p_M$ (kWh).

Let us consider the following decision variables:

- x_{ikt}: equal to 1 if bus i is assigned to place k during the time period t, 0 otherwise.
- y_{ij}: equal to 1 if bus i is assigned to service j, 0 otherwise
- w_i: equal to 1 if bus i performs at least one service, 0 otherwise.
- z_{it}: equal to 1 if bus i is charged during the time period t, 0 otherwise.
- p_{it}: charging power of bus i during time period t.

The problem is formulated as a Mixed Integer Linear Program (MILP). Its mathematical formulation is as follows:

$$\min \quad \alpha \cdot \sum_{i \in V} w_i + \beta \cdot \sum_{i \in V} \sum_{t \in H} (p_{it} \cdot c_t)$$

$$\sum_{i \in V} y_{ij} = 1 \qquad\qquad \forall j \in S \quad (1)$$

$$y_{ij} + y_{ik} \leq 1 \qquad\qquad \forall i \in V, \forall j \in S, \forall k \in N(S_j) \quad (2)$$

$$\sum_{j \in S} y_{ij} \leq S_{max} \cdot w_i \qquad\qquad \forall i \in V \quad (3)$$

$$(f_j - d_j)y_{ij} + \sum_{k \in P} \sum_{t=d_j+1}^{f_j} x_{ikt} \leq (f_j - d_j) \qquad\qquad \forall i \in V, \forall j \in S \quad (4)$$

$$\sum_{k \in P} x_{ikt} \leq 1 \qquad\qquad \forall i \in V, \forall t \in H \quad (5)$$

$$\sum_{i \in V} x_{ikt} \leq 1 \qquad\qquad \forall k \in P, \forall t \in H \quad (6)$$

$$\sum_{k \in P} x_{ikt} + \sum_{j \in S/t \in [d_j,f_j]} y_{ij} \geq 1 \qquad\qquad \forall i \in V, \forall t \in H \quad (7)$$

$$x_{ikt} + \sum_{k' \neq k, k' \in P} x_{ik'(t+1)} \leq 1 \qquad\qquad \forall i \in V, \forall k \in P, \forall t \in H \quad (8)$$

$$x_{iv(t-1)} + 2 - \sum_{i' \neq i, i' \in S} \left(x_{i'u(t-1)} + x_{i'ut} \right) \geq x_{ivt}$$
$$\forall i \in V, \forall t \in H, \forall(u,v,I) \in G \quad (9)$$

$$x_{iv(t-1)} - 2 + \sum_{i' \neq i, i' \in S} \left(x_{i'u(t-1)} + x_{i'ut} \right) \leq x_{ivt}$$
$$\forall i \in V, \forall t \in H, \forall(u,v,E) \in G \quad (10)$$

$$z_{it} \leq \sum_{k \in P} x_{ikt} \qquad\qquad \forall i \in V, \forall t \in H \quad (11)$$

$$p_m \cdot z_{it} \leq p_{it} \qquad\qquad \forall i \in V, \forall t \in H \quad (12)$$

$$p_M \cdot z_{it} \geq p_{it} \qquad\qquad \forall i \in V, \forall t \in H \quad (13)$$

$$\sum_{t=1}^{d_j} \delta \cdot p_{it} - \sum_{l \in S/f_l \leq f_j} e_l \cdot y_{il} \geq B(SoC_m - SoC_{i,0}) - (1 - y_{ij}) \cdot M$$
$$\forall i \in V, \forall j \in S \quad (14)$$

$$\sum_{t=1}^{d_j} \delta \cdot p_{it} - \sum_{l \in S/f_l \leq d_j} e_l \cdot y_{il} \leq B(SoC_M - SoC_{i,0}) + (1 - y_{ij}) \cdot M$$
$$\forall i \in V, \forall j \in S \quad (15)$$

$$\sum_{i \in V} p_{it} \leq g_t \qquad\qquad \forall t \in H \quad (16)$$

Constraints (1) ensure that each service is assigned to exactly one electric bus. Constraints (2) prevent overlapping services from being served by the same bus. Constraints (3) set a maximum number of S_{max} services that can be served by each bus ($S_{max} = 3$ in our case study). Constraints (4) ensure that a bus cannot be in the parking during its service time. Constraints (5) guarantee that, at each time period, at most one parking place is allowed to each bus. Constraints (6) ensure that, at each time period, at most one bus is assigned to each parking place. Constraints (7) set a parking place to each bus when no service is done at each time period. Constraints (8) allow the same parking place for a bus during the whole period of time between two consecutive services. Constraints (9) guarantee that if (u, v, I) is an edge of the blocking graph then place v cannot be allocated to a bus at time period t if u is occupied in periods $t - 1$ and t. Note that if u is occupied in period $t - 1$ but becomes free at t, then v can be assigned to a bus by simultaneously managing entry to v and exit from u. Constraints (10) ensure that if (u, v, E) is an edge of the blocking graph then v cannot be allocated to a bus at time period t if u is occupied in periods $t - 1$ and t. We note that for an entry-exit event both constraints (9) and (10) are activated. Constraints (11) ensure that a bus cannot be charged if it is not in the parking during time period t. Constraints (12) ensure that a bus recharges with at least the minimum charging power p_m. Constraints (13) guarantee that a bus recharges with at most the maximum charging power p_M. Constraints (14) ensure that at any time the state of charge of bus i is at least SoC_m. Constraints (15) guarantee that at any time the state of charge of the bus i does not exceed SoC_M. Constraints (16) ensure that, at each time period t, the total power used to charge electric buses does not exceed the grid capacity.

5   Solving Method

The mathematical model presented in Sect. 4 allows only very small instances to be solved. In this section, we propose an approximate solution to the problem using a decomposition method. The decomposition method consists of three steps. In the first step, we provide an assignment of services to buses with the objective of minimizing the number of buses and ensuring a sufficient charging time of buses between services. The second step assigns buses to parking places taking into account the services already assigned to buses in the first step. The third step optimizes bus charging based on the assignment of buses to services (step 1) and the assignment of buses to parking places (step 2).

5.1   Assigning Services to Buses

The service selection problem is modeled here as a clique cover problem in a non-oriented graph $R = (S, A)$ where S is the set of services and A is the set of edges. Two vertices (services) u and v are connected by an edge if services u and v are disjoint, i.e., $[d_u, f_u] \cap [d_v, f_v] = \emptyset$. The service selection problem is then equivalent to the problem of partitioning the graph R into a minimal

number of cliques with additional constraints that ensures bus charging. Each clique of partition correspond to a set of services that are disjoints and assigned to each bus. Then, the number of used buses corresponds to the number of cliques in the partition. Although the clique partitioning problem of an interval graph is a polynomial [18], adding feasibility charging constraints related to the existing of charging schedule changes the status of the problem to NP-hard. Our heuristic constructs cliques iteratively, where at each step the algorithm constructs a maximum clique, removes the vertices of the clique from the initial graph and searches again for the new maximum clique, until the initial graph becomes empty.

In the following, we detail our local search based heuristic to find a maximum clique in the graph R. Our clique construction heuristic is composed of two steps. In the first step, a clique is built using a greedy algorithm. In the second step, the size of clique C is improved by a process of removing and adding one or more vertices. The process is repeated until all conditions are met. The clique with the best size is updated at each iteration. The details of the two steps of the algorithm are as follows:

1. **Initial Solution.** Initially, a clique C is empty. The algorithm randomly selects a vertex i and adds it to C. A list of candidate vertices $L = \{s \in S | (s$ is connected to all vertices of $C) \wedge (s \notin C)\}$ is created. The algorithm selects a vertex s from list L with the highest degree that satisfies feasibility conditions related to the of charging schedule and adds it to C. The list L is updated and the procedure is repeated until L becomes empty or all vertices of L do not satisfy feasibility conditions.

2. **Local search procedure.** For each $i \in C$, we determine a list of nodes $N_i = \{s \notin C | s$ is connected to all vertices of $C - \{i\}\}$, then we remove from C vertex i with $|N_i| \geq |N_j|, \forall j \in C$. A new list L of candidate vertices is determined and the best vertex of L is added to C as described in step 1. The process of adding/removing vertices is repeated until a time limit is reached or if the size of the clique found is equal to the size of the graph or to the maximum number of services (3 services by bus). When the procedure stops, the best clique is updated, the current clique is deleted and the algorithm restarts with the first step until the stop condition is reached. Finally, the algorithm generates the best clique and removes the vertices of the clique from R.

Feasibility Conditions. The charging feasibility conditions ensure that a bus can provide all the services assigned to it without interruption due to insufficient energy while having enough time for recharging between consecutive services. Let Q_j be the amount of energy available in the battery of a bus before the beginning of the jth service of the current clique C, then $Q_j = \sum_{i=0}^{j-1}(r_i - e_i)$ where e_j is the energy required by the jth service, and r_j is the maximum energy retrieved from the grid by a bus between the $(j-1)^{th}$ and the jth service, where $r_j = \max\{B, (d_j - f_{j-1}) \cdot p_M\}$. Thus, a clique is feasible if, for each $j \in C$, $Q_j \geq 0$.

5.2   Allocation of Parking Places to Buses

At the end of the services assignment algorithm, we obtain for each bus a list of services to be performed, and each bus returns to the parking depot at the end of each service. Let I be the set of all parking time intervals of all buses. As the list of services assigned to each bus is known, then the start and the end time of each interval $I_i \in I$ are known and let $[h_i, g_i]$ be the time interval corresponding to I_i. In the following, we present the MILP model for the assignment of parking places to buses. Let us consider the following decision variable:

x_{ik}: equal to 1 if the parking interval I_i is assigned to place k, and 0 otherwise

$$\sum_{k \in P} x_{ik} = 1 \qquad\qquad \forall I_i \in I \quad (17)$$

$$x_{ik} + x_{jk} \leq 1 \qquad \forall k \in P, \forall I_i, I_j \in I : [h_i, g_i] \cap [h_j, g_j] \neq \emptyset \ (18)$$

$$x_{iu} + x_{jv} \leq 1 \qquad \forall I_i, I_j \in I : h_j > h_i, g_i > h_j, \forall (u, v, E) \in G \quad (19)$$

$$x_{iu} + x_{jv} \leq 1 \qquad \forall I_i, I_j \in I : g_j < g_i, h_i < g_j, \forall (u, v, S) \in G \quad (20)$$

Constraints (17) provide one parking place for each parking interval i. Constraints (18) prevent overlapping parking intervals to be assigned to the same parking place k. Constraints (19) prevent assignment of intervals I_i and I_j to places u and v respectively, when u blocks v for the entry and I_j overlaps and comes after I_i. Similarly, constraints (20) prevent assignment of intervals I_i and I_j to places u and v respectively, when u blocks v for the exit and I_j overlaps and comes before I_i.

5.3   Buses Charging

Now that we have buses assigned to parking places, we proceed in optimizing buses charging.

Having a set of services assigned for each bus, the charging algorithm determines an optimal charging schedule for all buses. Precisely, let L_j be the set of services assigned to bus j, with $j \in \{1, \ldots, m\}$, charging optimization problem is modeled as a minimum-cost flow problem in a network $R = (X, A)$ defined as follows:

- A set of vertices X is composed of: (1) a source s, (2) nodes h_t representing time periods $[t-1, t]$, (3) nodes $v_k^j \in L_j$ representing services in non-decreasing order, with $k = \{1, \ldots, |L_j|\}$, $j \in \{1, \ldots, m\}$ and f_k^j (resp. d_k^j) denotes end time (resp. start time) of the kth service of bus j, and (4) a sink t.
- A set of directed edges A with limited capacity is composed of: (1) edges (s, h_t) having a capacity of g_t and a cost c_t corresponding to energy cost of time period t, (2) edges (h_t, v_k^j) when $h_t \leq d_k^j$ and $h_t \geq f_{k-1}^j$ with a capacity of $p_M \times \delta$ and zero cost, (3) edges (v_{k-1}^j, v_k^j), $k = \{2, \ldots, |L_j|\}$, with a capacity and a lower bound equal to $B_j - E_{k-1}$ and zero cost, and (4) edges (v_k^j, t) with a capacity of E_k and zero cost.

A generalization of the Ford-Fulkerson algorithm [19] is applied to this network to solve the minimum cost problem. Let $f(i,j)$ be the flow passing through arc $(i,j) \in A$ in an optimal solution of the problem with a total cost of $\sum_{t=1}^{H} c_t \times f(s, h_t)$. Then, we define an optimal charging schedule as follows: For each time period h, apply on bus j charging power of $p_{j,t} = \frac{f(h_t, v_k^j)}{\delta}$, where $(h_t, v_k^j) \in A$.

6 Computational Experiments

In this section, we provide experiments results of our considered approach. Algorithms were developed using Java 8 on a mobile workstation with 3.10 GHz Intel Core i7-7920HQ processor and 32 GB RAM running on Windows 10. MILP models are solved with CPLEX 12.8, and all tests was run in a single-threaded mode. In the following, we detail the instances used in our experiments.

6.1 Instance Description

For confidentiality reasons, we use random instances instead of real instances received from the company. However, those random instances are generated from the real ones with different sizes.

In order to validate efficiency of our approach, we consider 2 types of instances belonging to two different service areas. The first type, called 'Hard', is characterized by a high number of blocking constraints, while the second type, called 'Easy', has a reasonable number of blocking constraints. Each type is subdivided into 3 different sizes depending on the number of services, the number of available buses and the number of parking places.

Each instance represents 24 h of service starting from 12am (midnight), and contains both night and daytime services. We keep the same ratio between them as in the original instance. The characteristics of the instances are summarized in Table 1.

We note that the services used in our instances are randomly selected from the actual instances. In addition, we note that the number of parking places is higher than the number of buses since we also take into account the maintenance places which are also used for bus parking in case there is no maintenance scheduled. Also, we keep the same electricity power parameters in the two depots of the two types of instances. Electricity costs are represented by a piecewise function following the peak/off-peak hours ranges of the french electric grid operator [20]. Battery capacities and charging parameters are reported in Table 2. Finally, we assume that buses are available from 6pm, with $SoC_0 = SoC_m$, so they have time to recharge for night services. Note that in real application, availability of buses is retrieved from previous day results.

Table 1. Instance characteristics

Instance type	Hard			Easy		
Instance name	Small	Medium	Large	Small	Medium	Large
# Services	10	60	130	10	60	130
# Buses	12	70	150	12	70	150
# Parking places	20	100	200	20	100	200
Night services ratio	20%	10%	10%	20%	10%	10%
# Entry-blocking edges	69	510	835	9	58	184
# Exit-blocking edges	6	183	315	8	45	136
# Blocking-free places	3	17	33	5	34	60

Table 2. Electricity parameters

Site parameters	Value	Bus parameters	Value
p_m	0 kW	SoC_m	10%
p_M	90 kW	SoC_M	90%
g_t	10 MW	B	285 kWh

6.2   Planning Results

For each instance, we perform 10 execution tests with different random seeds. We present in Table 3 the running times of our approach with the number of buses used in the solution. It can be observed that 'Hard' instances take longer to solve than 'Easy' instances, and also require more buses to perform all services.

We are now analyzing the computing time required by our approach. Figure 2 shows that most of the time is spent in step 2, particularly on model construction. For example, the construction of a model of one large instance takes 30 s, while the solving phase takes less than one second. This is due to the large number of blocking constraints in the allocation model. This explains the difference in running time between the 'Easy' instances and the 'Hard' instances noted in Table 3.

In order to examine the quality of the solutions found by our approach, we tried to solve the instances using the global MILP model of Sect. 4 with CPLEX. Unfortunately, the instances could not be solved optimally even for the smallest ones. This is why we have relaxed blocking constraints allowing us to find lower bounds to our problem. In this way, we obtain optimal solution for the relaxed problem. Even if these bounds do not tell us exactly how far the solutions of instances are from being optimal, but in some cases, we can obtain optimality proofs on some instances when our approach's results are equal with the lower bounds.

Lower bounds are calculated using CPLEX with a time-limit of 1 h. Results are reported in Table 4 and show a comparison with our approach for each

Table 3. Planning results

Instance family	Hard			Easy		
Instance size	Small	Medium	Large	Small	Medium	Large
Avg. computing time (s)	<0.10	2.50	26	<0.10	2.07	21.11
# Used buses	[7–8]	[48–52]	[96–105]	[6–9]	[40–46]	[90–96]

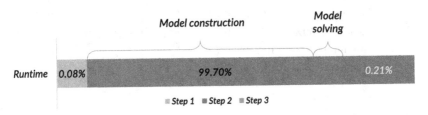

Fig. 2. Analysis of running time over solving steps (logarithmic scales)

objective. Concerning objective 1 (minimization of the number of buses), we present the optimality ratio (percentage of optimal solutions obtained with our approach) and the maximum gap compared to the lower bounds for all the instances. While for objective 2 (minimization of recharging costs), after setting the number of buses used in the relaxed problem, we report only the results on the instances whose optimal solutions on objective 1 are achieved, and we compare the solutions of these instances with the lower bounds for objective 2 using gaps which are computed as follows:

$$\text{Gap} = \frac{\text{Decomposition method objective value} - \text{MILP lower bound}}{\text{MILP lower bound}} \cdot 100$$

We present optimality ratio, average gap, and maximum gap for each instance family. Lower bounds gaps obtained by CPLEX are also reported.

Table 4 shows that our approach allows us to obtain optimal solutions for certain small instances. We cannot confirm whether this is true for medium and large instances because the MILP relaxed problem model cannot be solved optimally. The large values of the maximum deviations observed for Objective 1 in small instances are due to the small number of buses used. For example, a difference of a bus with a lower limit of 5 buses leads to a difference of 20%. We also observe that we obtain relatively good results for the first objective since the maximum deviations do not exceed 20% of the lower bounds, even if these bounds are not close to the optimal one.

Table 4. Analysis of solution quality

Instance		Objective 1			Objective 2			
Family	Size	Opt. ratio	Max gap	Bound gap	Opt. ratio	Avg. gap	Max gap	Bound gap
Hard	Small	80%	16.67%	0%	37.5%	0.47%	1.03%	0%
	Medium	20%	4.35%	0%	/	/	/	>100%
	Large	10%	7.61%	0.72%	/	/	/	>100%
Easy	Small	50%	20%	0%	20%	0.80%	2.77%	0%
	Medium	30%	10.25%	0%	/	/	/	>100%
	Large	0%	18.18%	2.78%	/	/	/	/

7 Conclusion

In this paper, an optimization model that tackles operational problems of buses allocation to services, depot places assignment to buses and charging scheduling for electric buses in an urban public transit network was proposed. A mixed integer programming optimization model which considers assignment of buses to services, assignment of buses to parking places and recharging of buses was developed. The proposed optimization model minimized the total costs of operating an electric bus recharging system. The proposed optimization model is hard to solve even for small instances. A decomposition approach is proposed to tackle these problems separately. Results of our approach are compared with relaxed problem on small, medium and large instances based on instances of real-world transit network based in France. For future study, the MILP model will be revised so as to determine the optimal solutions to the global problem. Second, the decomposition approach will be improved by merging the second and the third steps in one solving method.

References

1. DesignLine: Eco-smart i. https://cptdb.ca/wiki/index.php/DesignLine_Eco-Smart_I. Accessed 19 Apr 2019
2. ZeEUS: Zero emission urban bus system. https://zeeus.eu. Accessed 18 Apr 2019
3. ELIPTIC: Electrification of public transport in cities. https://www.eliptic-project.eu. Accessed 19 Apr 2019
4. BlueBus: 12-meter bus. https://www.bluebus.fr/bluebus-12-metres. Accessed 18 Apr 2019
5. Mahmoud, M., Garnett, R., Ferguson, M., Kanaroglou, P.: Electric buses: a review of alternative powertrains. Renew. Sustain. Energy Rev. **62**, 673–684 (2016)
6. Pternea, M., Kepaptsoglou, K., Karlaftis, M.G.: Sustainable urban transit network design. Transp. Res. Part A: Policy Pract. **77**, 276–291 (2015)
7. Li, J.Q.: Transit bus scheduling with limited energy. Transp. Sci. **48**(4), 521–539 (2014)

8. Kunith, A., Mendelevitch, R., Goehlich, D.: Electrification of a city bus network: an optimization model for cost-effective placing of charging infrastructure and battery sizing of fast charging electric bus systems. SSRN Electron. J. **11**, 707–720 (2016)

9. Wang, Y., Huang, Y., Xu, J., Barclay, N.: Optimal recharging scheduling for urban electric buses: a case study in davis. Transp. Res. Part E Logist. Transp. Rev. **100**, 115–132 (2017)

10. Paul, T., Yamada, H.: Operation and charging scheduling of electric buses in a city bus route network. In: 17th International IEEE Conference on Intelligent Transportation Systems (ITSC). IEEE, October 2014

11. van Kooten Niekerk, M.E., van den Akker, J.M., Hoogeveen, J.A.: Scheduling electric vehicles. Public Transp. **9**(1–2), 155–176 (2017)

12. Fusco, G., Alessandrini, A., Colombaroni, C., Valentini, M.P.: A model for transit design with choice of electric charging system. Procedia Soc. Behav. Sci. **87**, 234–249 (2013)

13. Perrotta, D., et al.: Route planning for electric buses: a case study in Oporto. Procedia Soc. Behav. Sci. **111**, 1004–1014 (2014)

14. Xylia, M., Leduc, S., Patrizio, P., Kraxner, F., Silveira, S.: Locating charging infrastructure for electric buses in Stockholm. Transp. Res. Part C Emerg. Technol. **78**, 183–200 (2017)

15. Rogge, M., Wollny, S., Sauer, D.: Fast charging battery buses for the electrification of urban public transport–a feasibility study focusing on charging infrastructure and energy storage requirements. Energies **8**(5), 4587–4606 (2015)

16. Miles, J., Potter, S.: Developing a viable electric bus service: The Milton Keynes demonstration project. Res. Transp. Econ. **48**, 357–363 (2014)

17. Mohamed, M., Farag, H., El-Taweel, N., Ferguson, M.: Simulation of electric buses on a full transit network: operational feasibility and grid impact analysis. Electr. Power Syst. Res. **142**, 163–175 (2017)

18. Balas, E., Yu, C.S.: On graphs with polynomially solvable maximum-weight clique problem. Networks **19**(2), 247–253 (1989)

19. Edmonds, J., Karp, R.M.: Theoretical improvements in algorithmic efficiency for network flow problems. J. ACM **19**(2), 248–264 (1972)

20. Enedis: Peak/off-peak hours ranges. https://www.enedis.fr/heures-creuses. Accessed 02 May 2019

A Two-Pheromone Trail Ant Colony System Approach for the Heterogeneous Vehicle Routing Problem with Time Windows, Multiple Products and Product Incompatibility

Andres Palma-Blanco⬤, Esneyder Rafael González$^{(\boxtimes)}$⬤,
and Carlos D. Paternina-Arboleda⬤

Universidad del Norte, Km 5 vía Puerto Colombia, Barranquilla, Atlántico, Colombia
eponzon@uninorte.edu.co
https://www.uninorte.edu.co

Abstract. We consider a variant of the vehicle routing problem subject to time windows for every customer, multiple products and incompatibility between them, using a heterogeneous fleet of vehicles. Since the problem is NP-hard, the approach to achieve feasible solutions is an ant colony system with two-pheromone to minimize routing cost and vehicle fleet size. Experiments conducted using instances from the literature show that the proposed approach provides competitive solutions in relatively short computational times.

Keywords: Vehicle routing · Time windows · Multiple products · Compatibility constraints · Ant colony system

1 Introduction

Flow of goods (FoGs) outside and within the supply chain is one of the most important activities in business logistics management since companies are not able to operate without the movement of raw material or finished products. In particular, freight movement represents one- to two-thirds of total logistic costs [4], making FoG the most representative action for a large number of organizations. Thus, companies are in need of conducting detailed studies of goods transportation in order to design a methodology that minimizes the costs produced by the transport and significantly decrease the total supply chain costs.

The Vehicle Routing Problem (VRP) [14] seeks to minimize transport costs in the supply chain logistics by determining the optimal set of routes performed by a fleet of vehicles with determined (fixed) capacity. Each route starts from a single depot, visits a determined number of nodes and finishes at the starting depot. There are some variations of the problem that consider time windows,

© Springer Nature Switzerland AG 2019
C. Paternina-Arboleda and S. Voß (Eds.): ICCL 2019, LNCS 11756, pp. 248–264, 2019.
https://doi.org/10.1007/978-3-030-31140-7_16

the maximum number of drivers working, multiple depots, among others. These variants seek to bring the VRP closer to reality, increasing its complexity.

The first VRP mathematical algorithm was proposed in [14], who applied this approach to deliver gas to service stations. Five years later, Clarke and Wright [11] proposed a heuristic to achieve solutions to the VRP based on savings.

Supporting these two influential contributions and due to its potential applications in many real cases [13,39], a large amount of procedures have been proposed to approach VRP optimal solutions in different versions, applying optimization methods such as linear programming and branch and bound. However, these approaches have mostly been used in small-medium size problems.

The Ant Colony System (ACS) metaheuristic is one of the most used approaches to solve the VRP. Multiple successful applications demonstrate its effectiveness [3,17,21,28,40]. In this paper a two-pheromone trail ACS approach for a practical variant of the VRP is developed in which a fleet of vehicles with limited transport capacity delivers multiple products to customers with respect to time windows and incompatibility of products. Also we used a modified version of the 2-opt local search procedure to refine the ACS solution. The problem is defined as the Heterogeneous Vehicle Routing Problem with Time Windows, Multiple Products and Incompatibility Constraints (HVRPTWMPIC).

The paper is organized as follows. Section 2 describes how the VRP and its variants have been addressed by other authors. Section 3 describes the problem under study. Section 4 presents the sequential two-pheromone trail ACS approach to solve the HVRPTWMPIC. Experiments are presented in Sect. 5. Finally, conclusions and future research guidelines are presented in Sect. 6.

2   Previous Work

The Vehicle Routing Problem, known to be an NP-hard problem [20,29] has been studied by researchers for decades due to its potential applications [39].

Different approaches have been explored, from exact methods such as linear programming and branch and bound procedures [2,26,30] applied in small-to-medium-sized problem instances with simple constraints, to the use of heuristics and metaheuristics that provide agile and competitive results applied to medium-and-large-sized instances with complex constraints [24].

The most used heuristic/metaheuristic approaches to solve the VRP are those based on Simulated Annealing (SA), Tabu Search (TS), Genetic algorithms (GA) and ACS. The SA metaheuristic has been applied to solve the VRP with time windows [10,37] and the capacitated hybrid VRP [42].

GA are also used frequently to provide solutions to the VRP. Thangiah [36] uses a Cluster first - Route second approach to solve the VRP via GA, Braysi and Gendreau [9] also used GA to solve a VRP with time windows. Jeon, Leep and Shim [23] used a GA to solve a multiple depot VRP with heterogeneous fleet. Similar problems have been solved via GA like a multiple depot VRP with multiple products and heterogeneous fleet [41]. TS is frequently used to solve routing problems with time windows [19,32] and multiple depots [16].

Finally, one of the most studied metaheuristics for this kind of problems is the ACS. It is an agent based method inspired by the observation of real ants, particularly, the way they find the shortest path between food sources and their nest trough a communication system (pheromone trail). Generally, ACS procedures place a predefined number of agents on a set of demand points. Each agent builds a feasible solution to the problem by iteratively applying a decision rule taking into account short and long term convenience. To build a solution, each agent updates a pheromone trail leading other agents to build their other solutions, resulting in finding satisfactory solutions in relatively short time.

Typical ACS applications have been proposed to the VRP in [6,35]. Other authors considered expanding the number of restrictions, giving solutions to the VRP with time windows [18], and some others even included multiple products in their scope [1,13].

Some of the more recent advances in the VRP with time windows are presented in [5] which shows the benefits of multi-start strategies for enhancing the capacities of three evolutionary heuristic methods, [7] where split deliveries are considered and [25] where multiple depots are taken into consideration and also make use of ant algorithms.

3   Problem Description

This paper focuses on a specific variant of the VRP. This consists of defining routes for a heterogeneous fleet of vehicles with limited capacity in order to supply clients with multiple types of products, which may not be compatible with each other. It is also given that each demand point must be serviced within a predefined time window. In what follows we better explain the details of the problem under study.

- *Time Windows:* Every demand point has a predefined time window that is represented with an earliest delivery time and a latest delivery time. Every client must be served within its time window. If a vehicle arrives to a point before the *earliest delivery time*, it has to wait until that time to serve the client. A vehicle cannot arrive after the *latest delivery time*.
- *Multiple Products:* There are P product types and the demand of every client is represented in terms of weight and volume of each product type. Every client demands at least one type of product P.
- *Incompatibility constraint:* Every product type P has its own properties, meaning that there are some types of products that are not compatible with others. Compatibility (C) is defined as a $P \times P$ matrix that represents the compatibility constraints between the P product types, taking the value of 1 if the products can be transported in the same vehicle, and 0 otherwise.
- *Heterogeneous Fleet:* The vehicles used to serve all the demand points are not the same. Some vehicles may have more capacity than others; nevertheless, every vehicle is capable of transporting every type of product. This capacity is expressed in terms of weight and volume.

The objective is to find a route for each vehicle that serves all the customers with minimum cost (expressed in terms of travel distance). A route starts from the depot, visits every client assigned to it and then returns to the depot. These routes must respect the time window for each customer, the vehicle capacity in terms of weight and volume, and the compatibility constraint associated with each type of product.

An instance of the HVRPTWMPIC is defined as follows. $G = (V, E)$ is a directed and fully connected graph where V is the set of N nodes representing the customers, with node 0 being the depot and E is the set of edges connecting them. D is an $N \times N$ matrix representing the distance traveled from customer i to customer j. T is an $N \times N$ matrix which specifies travel times between customers. TW is an $N \times 3$ matrix that contains earliest delivery time (e_i), latest delivery time (l_i) and estimated service time (s_i) at point i. A and B are $N \times P$ matrices that specifies the demand of every type of product in terms of weight and volume, respectively, for every customer. C is a $P \times P$ matrix that represents the compatibility constraints between the P product types, taking the value of 1 if the products can be transported in the same vehicle, and 0 otherwise. Finally, F is an $M \times 2$ matrix that contains the capacity in terms of weight and volume for the M vehicles.

A solution to the optimization problem is an $N \times N \times M$ matrix X, with element $X_{ijk} = 1$ if customer j is served immediately after customer i by vehicle k or $X_{ijk} = 0$ otherwise.

The objectives are the minimization of the routing cost and the vehicle fleet size. An objective function is defined as follows:

$$F = \sum_{k=1}^{M} \xi_k \sum_{j=1}^{N} X_{0jk} + \sum_{i=0}^{N} \sum_{j=0}^{N} \sum_{k=1}^{M} X_{ijk} D_{ij} \tag{1}$$

where ξ_k is the fixed cost of vehicle k, D_{ij} is the cost (represented as distance) of going from node i to j. The fixed cost ξ_k is defined as:

$$\xi_k = \frac{\sum_{i=0}^{N} \sum_{j=0}^{N} D_{ij}}{N^2} \times 1.2 \tag{2}$$

This objective function allows to control routing cost and fleet size, giving solutions that can be easily applied in a real situation, using less vehicles.

4   Two-Pheromone Trail Ant Colony System Approach

A very similar VRP variant was previously studied by De la Cruz and Paternina in 2013 [13]. In that paper, a VRP with heterogeneous fleet, time windows and multiple products was studied and solved via two-pheromone ACS approach with tabu search. This paper adapts the procedure proposed by De la Cruz and Paternina [13] to the HVRPTWMPIC, using a colony of cooperative agents to build feasible solutions for the HVRPTWMPIC with the objective of minimizing the total distance traveled and the fleet size.

The procedure starts with the initialization of the parameters and routes for every agent. Then, every agent builds a feasible solution based on a decision that evaluates desirability of the next customer in the short term (heuristic approach) and long term (weak pheromone trail). When all agents finish their construction, the procedure selects the best solution of the iteration and performs fleet size minimization and additional improvements based on local search procedures. Once the best solution of the iteration is refined, it is compared to the global best solution and the strong pheromone trail is updated, affecting the decisions in the next iterations.

The pseudo-code for the proposed approach is shown in Algorithm 1.

Algorithm 1. Two-pheromone trail ant colony system proposed

---

begin
 Initialize parameter n
 $GlobalBestAgent \longleftarrow \emptyset$
 $GlobalBestDistance \longleftarrow \emptyset$
 $GlobalBestFZ \longleftarrow \emptyset$
 $GlobalBestObj \longleftarrow \emptyset$
 Execute System initialization with Algorithm 2
 for $i = 1, n$ **do**
  $LocalBestAgent \longleftarrow \emptyset$
  $LocalBestDistance \longleftarrow \emptyset$
  $LocalBestFZ \longleftarrow \emptyset$
  $LocalBestObj \longleftarrow \emptyset$
  Execute Iterative construction of feasible solutions with Algorithm 3
  for *Every agent* **do**
   if $FObj_a < LocalBestObj$ **then**
    $LocalBestAgent \longleftarrow a$
    $LocalBestDistance \longleftarrow Dt_a$
    $LocalBestFZ \longleftarrow FZ_a$
    $LocalBestObj \longleftarrow FObj_a$
  Execute Fleet Size Minimization with Algorithm 4
  Execute 2^k Local search with Algorithm 5
  if $LocalBestObj < LocalBestObj$ **then**
   $GlobalBestAgent \longleftarrow LocalBestAgent$
   $GlobalBestDistance \longleftarrow LocalBestDistance$
   $GlobalBestFZ \longleftarrow LocalBestFZ$
   $GlobalBestObj \longleftarrow LocalBestObj$
  for *Every arc (i,j) in GlobalBestAgent route* **do**
   Update Strong pheromone trail following Eq. 7

4.1   System Initialization

System initialization is done at the beginning of the procedure, assigning values to all the parameters used and initializing the routes for every agent. A total of eight parameters are required:

- Number of iterations, $r \in \mathbb{Z}^+$
- Number of agents, $m \in \mathbb{Z}^+$
- Exploration rate, $\omega_0 \in \mathbb{R},\ 0 \leq \omega_0 \leq 1$
- Learning rate, $\alpha \in \mathbb{R},\ 0 \leq \alpha \leq 1$
- Discounting rate, $\gamma \in \mathbb{R},\ 0 \leq \gamma \leq 1$
- Relative importance of pheromones, $\delta \in \mathbb{R},\ 0 \leq \delta \leq 1$
- Relative importance of heuristic, $\beta \in \mathbb{R},\ 0 \leq \beta \leq 1$
- Initial value of the pheromone $Q_0 \in \mathbb{R}, Q_0 \geq 0$.

The pseudo-code for the system initialization is shown in Algorithm 2.

Algorithm 2. System initialization

---

begin
    Initialize parameters $m, \omega_0, \alpha, \gamma, \delta, \beta, Q_0$
    for $i = 0, N$ **do**
        for $j = 0, N$ **do**
            $Q^S(i,j) \longleftarrow Q_0$
            $Q^W(i,j) \longleftarrow Q_0$

    for $i = 1, m$ **do**
        $r_a \longleftarrow 0$ // `Starting point of agent` a
        $J^a = 1, 2, 3, ... N$ // `Set to be serviced customers`
        $v_{ra} \longleftarrow 0$ // `Leaving time of node` r_a
        $Dt_a \longleftarrow 0$ // `Total distance of agent` a
        for $k = 1, M$ **do**
            $R_{ka} \longleftarrow \emptyset$ // `Route of the vehicle` k `of agent` a
        $FZ_a \longleftarrow 1$ // `Fleet size of agent` a
        $a_{ka} \longleftarrow 0$ // `Used capacity of agent` a `in terms of weight`
        $b_{ka} \longleftarrow 0$ // `Used capacity of agent` a `in terms of volume`

4.2   Construction of Feasible Solutions

The construction of a feasible solution for an agent a starts with a decision of *exploring* or *exploiting* its knowledge. To decide between exploitation or exploration, ant algorithms also use an exploration rate $0 \leq \omega_0 \leq 1$, such that each agent generates a random value ω, uniformly distributed inbetween 0 and 1.

When an agent decides to *exploit* the knowledge, it has to select a node based on the following formula:

$$s_a = \arg\max |Q(r_a, s_a)|^\delta |H(r_a, s_a)|^\beta, \ s_a \in J_a(r_a) \tag{3}$$

In this formula, $Q(r_a, s_a)$ represents the long-term desirability stored as the amount of pheromone in the trail (r_a, s_a), $H(r_a, s_a)$ represents the short-term desirability to service a customer s, starting in r for the agent a; δ and β are parameters that weight their relative importance, respectively. Based on [18], the short-term desirability can be expressed as:

$$H(r_a, s_a) = \frac{1}{[D(r_a, s_a) \times (max(v_r + T(r_a, s_a), e_s) - v_r) \times (l_s - v_r)]} \tag{4}$$

With $D(r_a, s_a)$ being the distance between nodes r_a and s_a, v_r is the leaving time for the customer r, $T(r_a, s_a)$ represents the time required to travel from customer r to s, e_s and l_s correspond to the earliest and latest delivery time for customer s (time window).

If the agent decides to *explore*, it selects randomly the next customer s according to the probability distribution:

$$Pr(r_a, s_a) = \frac{|Q(r_a, s_a)|^\delta |H(r_a, s_a)|^\beta}{\sum |Q(r_a, s_a)|^\delta |H(r_a, s_a)|^\beta} \tag{5}$$

Once a customer is selected, it is added to the vehicle route. Then, the information respecting arrival, waiting, and leaving time is updated. Also, vehicle occupation and weak pheromone trail are updated. Finally, the procedure makes the update that guarantees selecting a feasible node in the next iteration. First, it removes the node from the J_a vector that contains the customers that are left to visit, indicating that this client is satisfied. Then, it updates the J_k^a vector that contains the customers that can be visited by that vehicle in the next iteration. In order to do this removal, the procedure checks:

- Time required to move from the client selected, to all other non-visited clients and remove those who have possible arrival times greater than the latest arrival time allowed.
- Demands of every client that in case of being picked, exceeds the vehicle capacity in terms of weight or volume and remove them.
- **Incompatibility constraints.** Removing every client that has a demand for a type of product that cannot be transported in that vehicle, due to the other products already picked.

The pseudo-code is shown in Algorithm 3.

4.3   Weak and Strong Pheromone Trails

This approach works with two pheromone trails, based on [15]. A weak pheromone trail or local trail and a strong pheromone trail or global trail.

Algorithm 3. Construction of feasible solutions

---

begin
  while *All agents have not finished constructing* **do**
    if $|J^a| > 0$ **then**
      $rand \longleftarrow$ Random number
      if $rand \leq \omega_0$ **then**
        $s \longleftarrow$ Select next node using Eq. 3
      else
        $s \longleftarrow$ Select next node using Eq. 5

      for *Every type of product* **do**
        if *Type p can be transported by vehicle k* **then**
          Update vehicle k occupation in terms of weight and volume
          $a_k \longleftarrow a_k + A(s_a, p)$
          $b_k \longleftarrow b_k + B(s_a, p)$

      $Dt_a \longleftarrow Dt_a + D(r_a, s_a)$ // Update total distance
      $u_{sa} \longleftarrow v_{ra} + T(r_a, s_a)$ // Update arrival time
      $w_{sa} \longleftarrow \max(0, TW[s, 1] - u_{sa})$ // Update waiting time
      $v_{sa} \longleftarrow u_{sa} + w_{sa} + TW[s, 3]$ // Update leaving time
      Update Weak pheromone trail Q^W using Eq. 6
      Update J_k^a, customers that can be visited by agent a in vehicle k
      Update J^a, customers that the agent a has not visited
      Update $R_k a$, Route of the vehicle k
      if $J_k^a = \emptyset$ **then**
        if $J^a = \emptyset$ **then**
          Agent finished constructing a feasible solution
        else
          $k \longleftarrow k + 1$, Open a new vehicle

    else
      Go to the next agent

The first is updated by every agent in the construction of a feasible solution while the other one is only updated at the end of an iteration and can only be updated by the agent with the best solution found at that moment.

The updates on the pheromone trails follow the next formulas, for weak and strong trails, respectively.

$$Q^W(r_a, s_a) = Q^W(r_a, s_a) + \alpha * [\gamma * Q_0 - Q^W(r_a, s_a)] \tag{6}$$

$$Q^S(r_a, s_a) = Q^S(r_a, s_a) + \alpha * [\Delta Q - Q^S(r_a, s_a)] \tag{7}$$

where Q^W and Q^S represent the weak and strong pheromone trails and ΔQ is a reinforcement applied to the pheromone trail estimated as the inverse of the best current solution as suggested by [18].

4.4   Local Search Procedures

Once the two-pheromone trail ACS gives a feasible solution, we perform two local search procedures that seek to refine the solution with the least computational resources. These procedures are the Minimizing Fleet Size procedure and the 2-Opt Modified Local Search.

Minimizing Fleet Size. As mentioned in Sect. 3 this approach seeks to minimize the fleet size and the total distance travelled. Once every agent finished constructing feasible solutions, the procedure selects the best solution based on the objective function and proceeds to reduce the fleet size by removing clients from the vehicle with smaller number of customers and assigning them to the best vehicle and position in the route, based on [18] without using an ant colony dedicated to minimize fleet size.

This procedure improves the objective value of the best solution of each iteration, transmitting that knowledge to the global pheromone trail and leading agents of the next iteration into building better solutions. The pseudo-code for minimizing fleet size is shown in Algorithm 4.

Algorithm 4. Fleet size minimization

---

begin
    Sort vehicle list by number of customers in increasing order
    for *k=1, FZ* **do**
        for *Every customer in the vehicle* **do**
            if *Customer can be assigned in other vehicle* **then**
                | Assign customer on the best vehicle and position of the route
            else
                └ Go to the next customer

***2-opt* Modified Local Search.** Regarding time window restrictions, a feasible solution is easier to find if we interchange two directly connected nodes from a tour. Hence, we modify the way in which the 2-opt algorithm selects nodes to be swapped allowing that the number of interchanges be as follows.

$$\frac{N(N-1)}{2} - 1 \tag{8}$$

This number comes from firstly trying to change all nodes that are directly connected, and then, trying to change nodes which are in an incremental span distance that starts with 2. This procedure is applied after the Minimizing Fleet Size procedure. For every vehicle, we perform these swaps and proceed to update the vehicle route if the total distance is reduced due to the swap. The pseudo-code is shown in Algorithm 5.

5   Experiments

This section explains how we selected the parameters for every run of the procedure and how we modified existing instances in order to adapt them to this specific problem. Finally, results of the experiments are shown.

Algorithm 5. 2-opt modified

```
begin
    lspace = 0
    for Every vehicle of the solution do
        nSwaps = N(N − 1)/2 − 1, k = 1, j = 1
        while j ≤ nSwaps do
            if k + 1 + lspace ≤ RouteLength then
                Swap clients k and k + 1 + lspace
                if Swap is not feasible then
                    Revert swap
                k = k + 1
            else
                k = 1
                lspace = lspace + 1
            j = j + 1
```

5.1   Parameter Setting

We defined the parameters of the procedure based on [13] and evaluating values with multiple runs. The selected values are shown in Table 1.

Table 1. Parameters and values used for the experiment

Parameter	Values tried	Selected value
r	100, 1000, 4000, 10000	10000
m	5, 8, 10, 12	10
ω_0	0.05, 0.1 0.2	0.2
α	0.05, 0.1	0.1
γ	0.1, 0.2, 0.5	0.1
δ	1, 2	1
β	1, 2	2
Q_0	$\dfrac{N-1}{\sum_{i=0}^{N}\sum_{j=0}^{N} D_{ij}}$	$\dfrac{N-1}{\sum_{i=0}^{N}\sum_{j=0}^{N} D_{ij}}$

5.2 Test Instances

Solomon's repository [34] contains benchmark problems, with instances for the VRP with time windows. Solomon's instances can be classified into three groups depending on the distribution of the customer locations: Random (R), clustered (C), and the mixture of both (RC).

The proposed procedure is first evaluated in four original Solomon instances: R201, R207, R209 and C101. Computational results and conclusions are shown in Sect. 5.3. Based on instances R201 and R209, De la Cruz and Paternina [13] expanded the VRPTW instances proposed by Solomon, adding a multiple product variant with a heterogeneous fleet of vehicles. They presented two instances in [12], from now on named instances R201-mod and R209-mod. The instance 209-mod is tested using this procedure.

Finally, we propose two new instances based on R201-mod and R209-mod, including compatibility constraints by splitting the given demand into three different products, that are not compatible among each other. These two new instances are named R201I and R209I.

In the seven instances tested, travel costs between the customers are given in terms of the corresponding Euclidean distance and it is assumed that all vehicles perform a speed equals to a unit (having the same values for both the travel times and distances). It also assumes that service time is always the same (product independent on load/unload). Also customer demands and vehicle capacities were redefined in terms of weight, volume and product type denominations.

5.3 Results

Since the ACS approach is stochastic in nature, each instance was evaluated 25 times and the results are shown in the tables below. First we show the results of the original Solomon instances and compare them with the best known solution.

From the results shown in the Tables 2, 3, 4 and 5 we can conclude that this procedure works better in clustered instances compared to randomized instances. Our method was able to find a better solution than the one reported in [13] for instance R207, which also landed benchmarks for instances R201, R209 and

Table 2. Results for 25 runs of Instance C101

C101	Fleet size	Total distance	Error
Best known solution	3	828.94	0%
Worst approach solution	3	829.01	0.008%
Best approach solution	3	828.94	0%
Average approach solution	3	828.95	0.001%
Standard Deviation	0	0.02	
95% CI Upper Bound	3	828.96	0.000%
95% CI Lower Bound	3	828.94	0.002%

Table 3. Results for 25 runs of Instance R201

R201	Fleet size	Total distance	Error
Best known solution	4	1252.37	0%
Worst approach solution	5	1574.83	25.74%
Best approach solution	4	1430.98	14.26%
Average approach solution	4.4	1501.06	19.85%
Standard Deviation	0.5	46.49	
95% CI Upper Bound	4.6	1592.20	12.5%
95% CI Lower Bound	4.2	1409.93	27.13%

Table 4. Results for 25 runs of Instance R207

R207	Fleet size	Total distance	Error
Best known solution	890.61	2	0%
Worst approach solution	4	1229.46	38.04%
Best approach solution	3	1094.17	22.35%
Average approach solution	3.36	1159.94	30.24%
Standard Deviation	0.49		
95% CI Upper Bound	3.55	1241.25	38.24%
95% CI Lower Bound	3.17	1090.05	21.84%

Table 5. Results for 25 runs of Instance R209

R209	Fleet size	Total distance	Error
Best known solution	3	909.86	0%
Worst approach solution	4	1116.33	22.69%
Best approach solution	3	1023.73	12.5%
Average approach solution	3.12	1094.58	28%
Standard Deviation	0.33	44.61	
95% CI Upper Bound	3.25	1182.01	10,69%
95% CI Lower Bound	2.99	1007.15	29,91%

C101. However, Solomon reports a best found solution with 2 vehicles and 890.61 in total traveled distance that we were not able to replicate since the repository does not report any citation nor the solution of customer visits. Also, it is important to know that in instance C101 we were able to achieve the best solution known, using an approach that is originally designed for a problem with more constraints.

Table 6 shows the results given by our algorithm tested on De la Cruz and Paternina [13] instance R209-mod.

Table 6. Results for 25 runs of Instance 209-mod

R209-mod	Fleet size	Total distance
Worst approach solution	3	1195.69
Average approach solution	3.12	1118.15
Standard Deviation	0.33	44.61
95% CI Upper Bound	3.75	1135.63
95% CI Lower Bound	2.48	1100.63

Table 7. Results for 25 runs of Instance 201I

R201I	Fleet size	Total distance
Worst approach solution	12	3233
Best approach solution	9	2995.82
Average approach solution	10	3104.89
Standard Deviation	1.12	70.02
95% CI Upper Bound	12.19	3242.12
95% CI Lower Bound	7.81	2967.66

Table 8. Results for 25 runs of Instance 209I

R209I	Fleet size	Total distance
Worst approach solution	11	2563.61
Best approach solution	9	2347.94
Average approach solution	9.44	2478.24
Standard Deviation	0.96	56.36
95% CI Upper Bound	11.32	2500.33
95% CI Lower Bound	7.56	2456.14

Finally, Tables 7 and 8 show the results for the instances proposed in this paper for comparison in future research.

Based on the experimental results comparison shown in Table 9 and this approach's performance in this instance (Table 5), we conclude the solutions given by this approach coincide in fleet size with the best known solution and the objective function values are competitive with other heuristic approaches. Thus, applying this procedure into the R209I and R209-mod instances would also give competitive results.

Note that the incompatibility constraint generates big changes in the solutions found by the procedure in terms of distance and computational time. This is expected because if a customer requests three types of products and they are not compatible between them, more than one vehicle is necessary to serve all the demand, generating more routing costs and increasing the fleet size.

Table 9. Comparison with other heuristics for Instance 209

Heuristic	Fleet size	Total distance
Best known solution	3	909.16
This approach	3	1039.41
2-opt strategy [27]	4	1210.96
Simulated Annealing [38]	4	1206.58
Iterated Local Search [38]	4	1110.30
Location-based approach [8]	3	1262.80
Tabu Search [31]	4	901.88
Genetic algorithms [36]	5	1097.42
Probabilistic Tabu Search [31]	3	944.64
Constraint programming [33]	3	923.96
MACS [18]	3	921.66
Evolutionary Strategy [22]	3	910.55

This problem can be solved by considering using multi-compartment vehicles that allows to carry two incompatible products in the same vehicle, this approach is suggested for future research.

6   Conclusions and Future Research Guidelines

This paper considered a variant of the vehicle routing problem that has a lot of applications in real practices. This variant consists in the VRP subject to heterogeneous fleet, time windows, multiple products and incompatibility between them. This paper proposes an approach to solve the HVRPTWMPIC based on an ant colony optimization with a two-pheromone trail strategy, based on the procedure by De la Cruz and Paternina [13].

The procedure gives acceptable solutions in relatively short computational time, considering the number of constraints and customers; however, the algorithm can resolve any combination of the constraints enunciated above, e.g. vehicle routing problems with time windows and incompatible products, resulting in a really practical and competitive procedure. It is competitive in terms of capability of application in real life situations.

Since it is a one phased algorithm, the solutions can be improved by adding more phases to refine the feasible solution found, as shown by De la Cruz and Paternina [13]. Local search procedures are suggested for further research. Also, the parameter setting was made based on a literature review and some brief experiments. It is recommended to perform an experimental design in order to determine an optimal set of parameters for the algorithm. It is expected that computational times will increase with larger problem size, meaning that it should be tested in future research using instances with more than 100 customers, and quantify the effect in the computational time and quality of the solution.

Our approach was able to achieve good solutions on known instances from Solomon and, for one of them, yielded the best value reported. This is important, since our approach is built for a different problem and in order to replicate those values, we switched off some constraints.

References

1. Amador-Fontalvo, J.E., Paternina-Arboleda, C.D., Montoya-Torres, J.R.: Solving the heterogeneous vehicle routing problem with time windows and multiple products via a bacterial meta-heuristic. Int. J. Adv. Oper. Manag. **6**(1), 81–100 (2014)
2. Archetti, C., Bianchessi, N., Speranza, M.G.: Branch-and-cut algorithms for the split delivery vehicle routing problem. Int. J. Adv. Oper. Manag. **238**(3), 685–698 (2014)
3. Baker, B.M., Ayechew, M.A.: A genetic algorithm for the vehicle routing problem. Comput. Oper. Res. **30**(5), 787–800 (2003)
4. Ballou, R.: Business Logistics, Supply Chain Management, Planning, Organizing and Controlling the Supply Chain. Prentice Hall, Upper Saddle Rive (2003)
5. Belhaiza, S., M'Hallah, R., Brahim, G.B., Laporte, G.: Three multi-start data-driven evolutionary heuristics for the vehicle routing problem with multiple time windows. J. Heuristics **25**(3), 485–515 (2019)
6. Bell, J.E., McMullen, P.R.: Ant colony optimization techniques for the vehicle routing problem. Adv. Eng. Inform. **18**(1), 41–48 (2004)
7. Bianchessi, N., Drexl, M., Irnich, S.: The split delivery vehicle routing problem with time windows and customer inconvenience constraints. Transp. Sci. **7**(4) (2019). https://pubsonline.informs.org/doi/10.1287/trsc.2018.0862
8. Bramel, J., Simchi-Levi, D.: A location based heuristic for general routing problems. Oper. Res. **43**(4), 649–660 (1995)
9. Bräysy, O., Gendreau, M.: Genetic algorithms for the vehicle routing problem with time windows. Arpakannus **1**, 33–38 (2001)
10. Chiang, W.C., Russell, R.A.: Simulated annealing metaheuristics for the vehicle routing problem with time windows. Ann. Oper. Res. **63**(1), 3–27 (1996)
11. Clarke, G., Wright, J.W.: Scheduling of vehicles from a central depot to a number of delivery points. Oper. Res. **12**(4), 568–581 (1964)
12. De la Cruz, J.: Alternativa heurística de dos fases para el problema de enrutamiento de vehículos con ventanas de tiempo, múltiples productos y flota heterogénea. Barranquilla. Ph.D. thesis, Tesis de maestría (Ingeniería Industrial), Universidad del Norte, 155 p. (2003)
13. De la Cruz, J., Paternina-Arboleda, C.D., Cantillo, V., Montoya-Torres, J.R.: A two-pheromone trail ant colony system-tabu search approach for the heterogeneous vehicle routing problem with time windows and multiple products. J. Heuristics **19**(2), 233–252 (2013)
14. Dantzig, G.B., Ramser, J.H.: The truck dispatching problem. Manag. Sci. **6**(1), 80–91 (1959)
15. Dorigo, M., Gambardella, L.M.: Ant colony system: a cooperative learning approach to the traveling salesman problem. IEEE Trans. Evol. Comput. **1**(1), 53–66 (1997)
16. Flisberg, P., Lidén, B., Rönnqvist, M.: A hybrid method based on linear programming and tabu search for routing of logging trucks. Comput. Oper. Res. **36**(4), 1122–1144 (2009)

17. Gajpal, Y., Abad, P.: An ant colony system (ACS) for vehicle routing problem with simultaneous delivery and pickup. Comput. Oper. Res. **36**(12), 3215–3223 (2009)

18. Gambardella, L.M., Taillard, É., Agazzi, G.: MACS-VRPTW: a multiple colony system for vehicle routing problems with time windows. In: New Ideas in Optimization. Citeseer (1999)

19. Garcia, B.L., Potvin, J.Y., Rousseau, J.M.: A parallel implementation of the tabu search heuristic for vehicle routing problems with time window constraints. Comput. Oper. Res. **21**(9), 1025–1033 (1994)

20. Gelves-Tello, N.A., Mora-Moreno, R.A., Lamos-Díaz, H.: Solving the vehicle routing problem with stochastic demands using spiral optimization. Revista Facultad de Ingeniería **25**(42), 7–19 (2016)

21. Golden, B.L., Raghavan, S., Wasil, E.A.: The Vehicle Routing Problem: Latest Advances and New Challenges. OR/CS Interfaces, vol. 43. Springer, New York (2008). https://doi.org/10.1007/978-0-387-77778-8

22. Homberger, J., Gehring, H.: Two evolutionary metaheuristics for the vehicle routing problem with time windows. INFOR Inf. Syst. Oper. Res. **37**(3), 297–318 (1999)

23. Jeon, G., Leep, H.R., Shim, J.Y.: A vehicle routing problem solved by using a hybrid genetic algorithm. Comput. Ind. Eng. **53**(4), 680–692 (2007)

24. Laporte, G.: What you should know about the vehicle routing problem. Nav. Res. Logist. (NRL) **54**(8), 811–819 (2007)

25. Li, Y., Soleimani, H., Zohal, M.: An improved ant colony optimization algorithm for the multi-depot green vehicle routing problem with multiple objectives. J. Clean. Prod. **227**, 1161–1172 (2019)

26. Lysgaard, J.: The pyramidal capacitated vehicle routing problem. Eur. J. Oper. Res. **205**(1), 59–64 (2010)

27. Osman, I.H., Christofides, N.: Simulated annealing and descent algorithms for capacitated clustering problem. Research Report, Imperial College, University of London (1989)

28. Pellegrini, P., Favaretto, D., Moretti, E.: Multiple ant colony optimization for a rich vehicle routing problem: a case study. In: Apolloni, B., Howlett, R.J., Jain, L. (eds.) KES 2007. LNCS (LNAI), vol. 4693, pp. 627–634. Springer, Heidelberg (2007). https://doi.org/10.1007/978-3-540-74827-4_79

29. Prins, C.: A simple and effective evolutionary algorithm for the vehicle routing problem. Comput. Oper. Res. **31**(12), 1985–2002 (2004)

30. Rieck, J., Zimmermann, J.: Exact solutions to the symmetric and asymmetric vehicle routing problem with simultaneous delivery and pick-up. Bus. Res. **6**(1), 77–92 (2013)

31. Rochat, Y., Taillard, É.D.: Probabilistic diversification and intensification in local search for vehicle routing. J. Heuristics **1**(1), 147–167 (1995)

32. Schulze, J., Fahle, T.: A parallel algorithm for the vehicle routing problem with time window constraints. Ann. Oper. Res. **86**, 585–607 (1999)

33. Shaw, P.: Using constraint programming and local search methods to solve vehicle routing problems. In: Maher, M., Puget, J.-F. (eds.) CP 1998. LNCS, vol. 1520, pp. 417–431. Springer, Heidelberg (1998). https://doi.org/10.1007/3-540-49481-2_30

34. Solomon, M.: VRPTW benchmark problems, optimal solutions and best known solutions (2005). http://w.cba.neu.edu/~msolomon/problems.htm

35. Tan, X., Zhuo, X., Zhang, J.: Ant colony system for optimizing vehicle routing problem with time windows (VRPTW). In: Huang, D.-S., Li, K., Irwin, G.W. (eds.) ICIC 2006. LNCS, vol. 4115, pp. 33–38. Springer, Heidelberg (2006). https://doi.org/10.1007/11816102_4

36. Thangiah, S.R.: Vehicle routing with time windows using genetic algorithms. Cite-seer (1993). https://citeseerx.ist.psu.edu/viewdoc/summary?doi=10.1.1.23.7903& rank=7

37. Thangiah, S.R., Osman, I.H., Sun, T.: Hybrid genetic algorithm, simulated anneal-ing and tabu search methods for vehicle routing problems with time windows. Technical report SRU CpSc-TR-94-27 69, Computer Science Department, Slippery Rock University (1994)

38. Thangiah, S.R., Osman, I.H., Vinayagamoorthy, R., Sun, T.: Algorithms for the vehicle routing problems with time deadlines. Am. J. Math. Manag. Sci. **13**(3–4), 323–355 (1993)

39. Toth, P., Vigo, D.: The Vehicle Routing Problem. SIAM, Philadelphia (2002)

40. Tripathi, M., Kuriger, G., Wan, H.d.: An ant based simulation optimization for vehicle routing problem with stochastic demands. In: Winter Simulation Confer-ence, pp. 2476–2487. Winter Simulation Conference (2009)

41. Vidal, T., Crainic, T.G., Gendreau, M., Lahrichi, N., Rei, W.: A hybrid genetic algorithm for multi-depot and periodic vehicle routing problems. Oper. Res. **60**, 611–624 (2012)

42. Vincent, F.Y., Redi, A.P., Hidayat, Y.A., Wibowo, O.J.: A simulated annealing heuristic for the hybrid vehicle routing problem. Appl. Soft Comput. **53**, 119–132 (2017)

A Greedy Heuristic for the Vehicle Routing Problem with Time Windows, Synchronization Constraints and Heterogeneous Fleet

Luísa Brandão Cavalcanti[(⊠)] [iD] and André Bergsten Mendes[iD]

Departamento de Engenharia Naval e Oceânica, Universidade de São Paulo,
Av Prof. Mello Moraes 2231, Sao Paulo, SP 05508-030, Brazil
luisa.cavalcanti@usp.br

Abstract. This paper focuses on a new variant of the Vehicle Routing Problem with Time Windows and Synchronization constraints (VRPTWSyn), inspired by a routing problem faced by the oil & gas industry. Here, a heterogeneous fleet of Anchor Handling Tug Supply vessels (AHTS) must be assigned to perform tasks at offshore platforms, and most tasks require the simultaneous action of more than two tugs. Instances are proposed, as well as a greedy randomized heuristic that generates feasible solutions to the problem. As shown by the computational experiments reported here, although the complexity of the AHTS-routing problem is higher than the VRPTWSyn, the proposed method could produce a diverse set of feasible solutions, which in the future may be improved by a meta-heuristic-based algorithm.

Keywords: Anchor Handling Tug Supply · Vehicle Routing Problem · Synchronization · Greedy heuristic · Oil & gas

1 Introduction

Recent literature has shed a light on the Vehicle Routing Problem with Time Windows and Synchronization constraints (VRPTWSyn), applied to various contexts [1–3]. This proven NP-hard problem [4] consists of scheduling the execution of tasks at geographically spread clients that may demand the simultaneous presence of two or more vehicles and that must be started within a given time window [5]. For instance, vehicles may represent caregivers performing health related activities at patients' homes, as in [2, 4–8], or employees that provide different types of services at their clients' locations, such as installing electronic equipment and furniture [1, 9, 10].

The VRPTWSyn may also be applied to create routes for Anchor Handling Tug Supply vessels (AHTS) that must execute activities at offshore oil platforms, as described in [11, 12]. These include towing and anchoring drilling rigs at production units, performing maintenance and installation of underwater equipment, pulling shuttle tankers during offloading operations in order to avoid collision, amongst others (for a general overview of offshore logistics problems, we refer to [13]). In such context, differences amongst the available tugs must be considered when assigning them to tasks, since some tasks may demand vessels of higher potency, whereas other

© Springer Nature Switzerland AG 2019
C. Paternina-Arboleda and S. Voß (Eds.): ICCL 2019, LNCS 11756, pp. 265–280, 2019.
https://doi.org/10.1007/978-3-030-31140-7_17

may be accomplished by simpler, cheaper tugs. Moreover, tasks usually must be executed simultaneously by more than two tugs, which has not been successfully covered in the literature, as instances of the VRPTWSyn found in previous works are restricted to at most two vehicles per client, and the percentage of clients with synchronization requirements is low (while most papers adopt 10% [1, 2, 4, 5, 8], [14] uses 20%, [7] uses 30% and [10] proposes instances with up to 50% synchronized visits). These intrinsic characteristics of the oil industry problem add an extra layer of complexity to the VRPTWSyn, resulting in a new variant of the mathematical model that, to the best of the authors' knowledge, has not been discussed in the literature yet, except for previous works of the authors themselves, as in [12].

In this paper, we address the hereby called Vehicle Routing Problem with Time Windows, Synchronization constraints and Heterogeneous fleet (VRPTWSyn-Het), by introducing its optimization model and presenting a greedy heuristic that produces feasible solutions to the problem. To evaluate our solution method's performance, the heuristic is applied to 3 sets of 12 instances with 10, 20 and 30 tasks, and its results are compared with the ones produced by the Gurobi optimization package. It is important to note that the main goal of the proposed procedure is to produce feasible solutions that may be used as a start to local search procedures and meta-heuristics, so methodologies for improving the solution quality are left for future work. Although this may sound like an easy target, our computational experiments show that Gurobi had major difficulties in finding feasible solutions for instances with 20 and 30 tasks, which justifies the need of building an efficient algorithm to find feasible tug schedules, as well as it illustrates how the VRPTWSyn-Het differs from the VRPTWSyn, which has had instances of up to 50 clients solved to optimality by [14].

The structure of this paper is as follows: in Sect. 2, the VRPTWSyn-Het is properly defined and its mathematical optimization model is introduced; then, a greedy heuristic for solving the VRPTWSyn-Het is documented in Sect. 3; and its performance is attested in Sect. 4. Lastly, in Sect. 5, conclusions and recommendations for future work are presented.

2 Problem Definition and Model Formulation

The VRPTWSyn-Het consists of routing a heterogeneous fleet of non-capacitated vehicles (AHTS) to perform activities at geographically spread locations. The fleet is subdivided into classes according to their potency, such that the vehicles from a class c have lower power levels than vehicles of class $c + 1$. The higher the potency, the higher the fuel consumption of the tug, so variable transportation costs are also higher. Each activity demands a minimum number of vehicles of each class and must be started within a time window, defined by a lower and an upper time limitation. Vehicles of the same class are interchangeable between one another and vehicles of any given class can execute all the tasks that can be performed by vehicles of a lower class. In other words, a vehicle of class c that can execute a task i may be replaced by any vehicle of class $c + 1$, even though the resulting variable costs will probably be higher. Tasks may start and end at different locations, as they include transportation of equipment from one platform to another. A task will only be started after all demanded vehicles arrive at its

starting point and, once it is started, all vehicles assigned will be released to perform other tasks just after its conclusion. Each tug r is associated with a class c_r and a release date s^r.

The following mathematical formulation of the VRPTWSyn-Het is based on previous works that addressed the VRPTWSyn, with some changes to accommodate the specificities of the AHTS routing problem. The VRPTWSyn-Het is formulated as a directed graph $G = (V, A)$ composed by a set V of vertices and a set A of arcs. Each task j is represented by a unique node, being the set of task nodes denoted by J. There are two vertices in V for each tug $r \in M$ to represent the start and end of its route, O_r^+ and O_r^-, but the former represents an actual location, whereas the latter is a dummy node used to avoid sub cycles. The set of arcs A comprises all pairwise connections between nodes of J, plus arcs connecting each O_r^+ to all nodes of $J \cup O_r^-$, and all the nodes of J to the nodes O_r^-. Each arc $(i, j) \in A$ represents the movement of a tug from a task/origin node i to a task/origin node j, so a cost c_{ij}^r for vehicle r to transverse arc (i, j) is also defined for each pair of tug and arc. To execute all demanded tasks, a fleet of m tugs, subdivided into l classes, is available. The set of all tugs is denoted by M, the subset of tugs of class c is $M^c \subset M$ and the set of all classes is C.

Each task j must be started at any time within its time window $[a_j, b_j]$, i.e. no earlier than a date a_j and no later than b_j. However, if it is initiated after a given $\hat{b}_j$ $(a_j \leq \hat{b}_j \leq b_j)$, which is the actual desired day for starting the task, a daily financial penalty w_j applies. The duration p_j of task j includes the eventual transportation time from its start to its end location. Finally, the basic mode for executing j is represented by a demand vector $D_j = \{d_j^1, d_j^2 \ldots d_j^1\}$.

Finally, a solution S to the VRPTWSyn-Het must assign a value to each of the following decision variables: x_{ij}^r, binary that equals one if tug r is assigned to execute task j right after finishing task i; s_j^r, continuous variable that equals the date of arrival of tug r at the starting location of task j; $\hat{s}_j$, another continuous variable that indicates the starting time of task j, when all assigned tugs are available at their starting location; and T_j, the resulting delay of task j, measured from its desired starting time $\hat{b}_j$ and within its time window. The objective function $z(S)$ is:

$$\min z(S) = \sum_{j \in J} T_j w_j + \sum_{i \in N} \sum_{j \mid (i,j) \in A} \sum_{r \in M} x_{ij}^r c_{ij}^r \tag{1}$$

Subject to:

$$\sum_{j \in J \cup O_k^-} x_{O_r^+ j}^r = 1 \quad \forall r \in M \tag{2}$$

$$\sum_{j \in J \cup O_r^+} x_{jO_r^-}^r = 1 \quad \forall r \in M \tag{3}$$

$$\sum_{r \in M \mid c_r \geq c} \sum_{(i,j) \in A} x_{ij}^r \geq d_j^c \quad \forall j \in J, \forall c \in C \tag{4}$$

$$\sum_{r \in M} \sum_{(i,j) \in A} x_{ij}^r \geq \sum_{c \in C} d_j^c \quad \forall j \in J \tag{5}$$

$$\sum_{(i,j)\in A} x^r_{ij} = \sum_{(j,k)\in A} x^r_{jk} \quad \forall r \in M, j \in J \tag{6}$$

$$s^r_j \geq s^r + t^r_{O^+_r j} - \left(1 - x^r_{O^+_r j}\right)K \quad \forall r \in M, \forall (O^+_r, j) \in A \tag{7}$$

$$s^r_j \geq \hat{s}_i + p_i + t^r_{ij} - \left(1 - x^r_{ij}\right)K \quad \forall r \in M, \forall (i,j) | i \in J, j \in J, i \neq j \tag{8}$$

$$\hat{s}_j \geq s^r_j \quad \forall j \in J, \forall r \in M \tag{9}$$

$$\hat{s}_j \geq a_j \quad \forall j \in J \tag{10}$$

$$\hat{s}_j \leq b_j \quad \forall j \in J \tag{11}$$

$$T_j \geq \hat{s}_j - \hat{b}_j \quad \forall j \in J \tag{12}$$

$$x^r_{ij} \in \{0, 1\} \quad \forall (i,j) \in A, r \in M \tag{13}$$

$$\hat{s}_j, T_j \in R | \hat{s}_j, T_j \geq 0 \quad \forall j \in J \tag{14}$$

$$s^r_j \in R | s^r_j \geq 0 \quad \forall j \in J, r \in M \tag{15}$$

In Eq. (1), the first part of $z(s)$ is the total penalty costs of delayed tasks, whereas the second part is the total transportation costs of all arcs traveled by the tug fleet. Constraints (2) state that for each tug, a single arc leaving from its origin O^+_r must be selected, similarly to constraints (3), which restrict the selection of a single arc ending at its final destination O^-_r (note that this arc may be from O^+_r to O^-_r for both cases).

Inequalities set by (4) and (5) guarantee the fulfillment of each task's demand vector, by assuring that for each class c and task j, the number of designated tugs of this class and above is greater or equal to the demanded d^c_j tugs; and that for each task the total number of designated tugs is equal to the total amount of demanded tugs, regardless of their classes. Constraints (6) state that all the vessels that travel to a task node j must leave it. The inequalities (7)–(9) force the starting time of each task to be greater or equal to the arrival time of each of the vessels assigned to perform it, whereas constraints (10) and (11) guarantee tasks will be started within their time windows. Constraints (12) calculate the penalized delay of all tasks, and the domain of all decision variables is stablished by expressions (13), (14) and (15).

3   Solution Procedure

To generate feasible solutions to the VRPTWSyn-Het, a greedy heuristic with randomized steps was developed. Before presenting the algorithm of the Constructive Heuristic for the VRPTWSyn-Het (VRPTWSyn-Het_CH), it is important to define a couple of terms that are used throughout this paper.

As any task with synchronization constraints that requires low class tugs may be executed by tugs of higher classes, when one lists all subsets of tugs that are compatible with such tasks, the number of vessels of each class will not be the same in all subsets (although the total amount of tugs must be the same). For instance, if task i requires two class-1 tugs and one of class 2, subsets of compatible tugs may have: (*i*) 2 class-1 tugs and 1 class-2; (ii) 1 of class 1 and 2 of class 2; (iii) none of class 1 and 3 class-2 tugs. To distinguish subsets with different numbers of tugs of each class, we borrow the term *mode* from the scheduling literature, more specifically from the multi-processor scheduling problems [15, 16]. Here, a *mode* is a combination of the number of tugs from each available class, such that, in the previous example, there are three modes for executing task i. We also name the *basic mode* as the combination that uses the least higher-class tugs, which in the previous example is the first enumerated mode (2,1).

Another useful term we borrow from scheduling literature is the *slack time* of a task. In this problem, however, we define the slack time of a task as the amount of time it may be postponed (from its current starting time) without breaking any of the routed tasks' time windows, including its own. The use of slack times is crucial to the viability of the proposed heuristic, because it spares the method from having to check time windows of all tasks that may be affected by the insertion of a new task every time an insertion option is evaluated. As stated elsewhere [14], synchronizations constraints generate complex interconnections between tasks, making routing problems extremely difficult to solve, so to find a way of smoothing calculations related to route changes is an important gap to bridge in this field.

Algorithm 1. VRPTWSyn-Het_CH algorithm

```
VRPTWSyn-Het_CH(J, M, nExp):
  Initialize Routes and it ← 1
  While no solution has been found AND it ≤ nExp:
    S := sorted tasks of set J
    While S is not empty:
      job ← select_task_from(S, jPos)
      S ← S \ {job}
      Modes := all_modes_compatible_with(job)
      Options := ∅
      For each mode in Modes:
        M̃_mode ← set will all subsets of tugs type mode
        For each tug subset m̃ ⊂ M̃_mode:
          For combination of insertion positions in m̃:
            Routes' := insert(job, m̃, positions in m̃, Routes)
            If Routes' feasible: Options ← Options ∪ Routes'

          End loop that exhaustively checks m̃
        End loop that exhaustively checks mode
        If Options is not empty, exit loop
      End loop that iterates through Modes
      If Options is empty: exit while loop
      Else: Routes ← select_option_from(Options, oPos)
    End loop that removes task by task from S
    it ← it + 1
  End while loop
  Return Routes
End
```

The VRPTWSyn-Het_CH algorithm, presented by Algorithm 1, starts by initializing a set *Routes* of m vectors $route_r$, with r varying from 1 to m, and each $route_r$ is initialized with two elements: the initial and final depot nodes. Then, it performs the following procedure until a solution has been found or the maximum number of restarts is reached.

At the beginning of an iteration, demanded tasks are sorted according to relevant characteristics of each task, and stored in set *S*. As later described, we have tested 8 different sorting rules and selected the one with the best performance. Next, the algorithm removes a task from *S*, by randomly picking one of the first *jPos* of *S* (method *select_task_from*), and then checks possibilities of inserting the task into the routes of vessels that are compatible with the task. However, instead of exhaustively testing all the insertion possibilities in all compatible subsets of tugs, the algorithm starts with the basic mode for fulfilling the task and, if there is at least one feasible partial solution that can be generated using such mode, it will not look any further. In case none of the tug subsets for the basic mode may have the task added to their routes, the algorithm will exhaustively check the insertion possibilities for the next mode (second lowest amount of high class vessels used), and so on, until a feasible partial solution can be generated or all modes have been tested.

Hence, when testing a mode, all the vessel subsets of such mode are considered and, for each subset, all the combinations of insertion positions from each route are also checked. To expedite this exhaustive procedure, we recalculate the slack times of all routed tasks every time the set of routes is updated (i.e. have a task added to one or more vessels' routes), so when trying to add a task i before j at a tug's route r, we calculate the new starting time of task j after such insertion and only accept the insertion if the difference from j's current starting time is equal or smaller than j's current slack time. Not only does this guarantee that task j will have its time window respected, but also that all other impacted tasks will have their time windows preserved. One should also note that, since the starting time of any task depends on the arrival of all assigned tugs, it is possible that the insertion of i into r will not impact any of the following tasks, in case tug r was originally waiting for the arrival of another tug to start task j and had enough idle time to accommodate i.

Every time an insertion check leads to a feasible solution, the partial routes generated by such insertion are added to a set *Options*, sorted according to the cost increment caused by the insertion. Once a mode with feasible solutions has been found and exhaustively examined, the algorithm randomly chooses amongst the first *oPos* partial solutions stored in *Options* and updates *Routes* accordingly. Otherwise, if a task may not be inserted in any of the compatible tug subsets (considering all possible modes), the loop that removes task by task from S is finished earlier and the whole procedure is restarted, provided that the maximum number of restarts has not been reached yet.

The constructive heuristic is finished when a complete solution is found (all tasks have been successfully added to the routes of compatible tugs and their time windows are respected) or the maximum number of restarts has been reached. However, as the purpose of the heuristic is to find a feasible solution to all instances, the maximum number of restarts was only used for tuning experiments, with which the values for *jPos* and *oPos* were determined, as explained in the next section.

It is worth mentioning that in previous versions of the heuristic, at the step of checking the insertion possibilities for a task, the heuristic looked for all modes and picked the cheapest option. However, this final version, which restricts the search to the first viable mode, outperformed the original algorithm in terms of quality and computational time.

4   Computational Experiments

Computational experiments conducted on the VRPTWSyn-Het_CH included the definition of sorting rule, parameter-tuning tests, efficiency evaluation and assertion of solution quality and diversity. More specifically:

- Definition of sorting rule: eight sorting rules, based on classical scheduling approaches for minimizing total tardiness (see [17] for more details), were tested using the VRPTWSyn-Het_CH with both *jPos* and *oPos* equal to one (which is equivalent to a non-randomized version of the heuristic, as the first job of the sorted test is always selected as the next job to insert into routes of the partial solution, as well as the insertion positions yielding the cheapest partial solution).

- Parameter-tuning tests: to calibrate *jPos*, the value of *oPos* was fixed as one and the heuristic was tested with *jPos* set from one to ten. With *nExp* reduced to one hundred restarts, the heuristic was executed ten times for each value of *jPos*, then medium performances were compared with each other. Similarly, the heuristic was executed ten times to calibrate the value of *oPos*, which varied from one to twenty-one whereas *jPos* and *nExp* were fixed at one and one hundred, respectively. Such value ranges were obtained with preliminary tests on randomly picked instances.
- Efficiency evaluation: once *jPos* and *oPos* were tuned, the resulting algorithm was run with *nExp* equal to ten thousand restarts, to enable finding a solution to all proposed instances. Again, ten executions were performed for each instance, to evaluate the robustness of the method. The computational time spent to generate a feasible solution was compared to the time Gurobi needed to find a single solution.
- Assertion of solution quality and diversity: even though the main goal of the constructive heuristic was to find feasible solutions to difficult instances of the VRPTWSyn-Het, we also tested it as a way of producing good solutions to the problem. In these tests, the algorithm was applied with just one stop criterium, the maximum number of restarts. Here, two thousand restarts were allowed and, although the best solution found was retrieved by the method, all feasible solutions were also stored, which allowed us to check how diverse the set of produced solutions is. As we intend to apply GRASP to the VRPTWSyn-Het, having a diverse set of initial solutions may be useful to future developments of this research.

All tests reported here were conducted on a PC notebook with Intel Core i7-6500U CPU, 2.50–2.59 GHz and 8 GB RAM, using Microsoft Visual Studio 2015 and Gurobi version 8.1.0. Next, a description of the instances is provided, along with the test results.

4.1   Instances for the VRPTWSyn-Het

Three sets of 12 instances each are proposed for the VRPTWSyn-Het, with 10, 20 and 30 tasks. These instances were derived from [11], with a reduction of the fleet's size and homogenization of the tug characteristics within each tug class. All instances have two classes of tugs and their planning horizon is a month (starting dates of the tasks are within 30 days).

Table 1 shows the main characteristics that describe an instance: number of demanded tasks; number of tasks with synchronization needs; average time length of the tasks; fleet's size and distribution across classes; number of vehicles required per task (on average and the maximum value). Dashed lines mark the transition between the three sets of instances, according to their sizes (total number of demanded tasks).

It should be noted that for each instance a task requires, on average, the simultaneous action of at least 2.4 vessels, but for some of them this figure can go up to 2.8. At most, five tugs are required per task, but most instances have no tasks demanding more than four tugs. The mean percentage of tasks with synchronized needs is 86% across all instances, but in four of them, all tasks have synchronized requirements. Instance 11 has the smallest number of synchronized tasks, with only seven of them (or 70% of the total) requiring more than one tug.

Table 1. Main characteristics of the proposed instances

ID	# tasks (n)	# sync. tasks	Avg. p_j [days]	Fleet's size			Tugs per task	
				Class-1 tugs	Class-2 tugs	Total (m)	Avg #tugs/task	Max. #tugs/task
1	10	8	9.9	8	4	12	2.4	4
2	10	10	9.9	8	4	12	2.6	4
3	10	9	10.1	9	4	13	2.8	4
4	10	9	11.9	8	5	13	2.4	4
5	10	9	12.1	8	4	12	2.6	4
6	10	9	12.0	11	5	16	2.8	4
7	10	10	10.1	8	3	11	2.4	4
8	10	8	9.9	8	4	12	2.6	4
9	10	9	10.1	9	4	13	2.8	4
10	10	8	12.0	8	4	12	2.4	4
11	10	7	12.0	10	4	14	2.6	4
12	10	10	12.0	10	5	15	2.8	4
13	20	15	9.9	16	6	22	2.4	5
14	20	17	9.9	11	9	20	2.6	4
15	20	19	9.9	14	9	23	2.8	4
16	20	15	12.0	20	7	27	2.4	4
17	20	17	12.0	22	7	29	2.6	4
18	20	18	12.1	20	7	27	2.8	4
19	20	16	10.0	11	6	17	2.4	4
20	20	18	9.5	11	5	16	2.6	5
21	20	20	9.1	20	7	27	2.8	4
22	20	15	12.0	14	7	21	2.4	4
23	20	17	12.1	16	6	22	2.6	4
24	20	19	12.0	18	8	26	2.8	4
25	30	24	9.9	14	6	20	2.4	5
26	30	25	10.1	14	7	21	2.6	4
27	30	27	10.1	15	8	23	2.8	4
28	30	26	12.0	15	9	24	2.4	4
29	30	26	12.0	16	9	25	2.6	4
30	30	28	12.1	17	9	26	2.8	4
31	30	24	10.0	14	7	21	2.4	4
32	30	24	10.0	16	6	22	2.6	4
33	30	25	10.0	16	8	24	2.8	4
34	30	24	12.1	16	8	24	2.4	4
35	30	26	12.0	19	9	28	2.6	4
36	30	27	12.0	19	10	29	2.8	4

4.2   Tuning Experiments

The following sorting rules were tested on instances with 10 and 20 tasks, using a non-randomized version of the VRPTWSyn-Het_CH:

1. Non-decreasingly by a_j (lower limit of the task's time window), or by b_j (its upper limit) whenever there is a tie;
2. Non-increasingly by total amount of vessels demanded, or non-decreasingly by the demand of upper-class vessels in case of a tie;
3. Non-decreasingly by b_j, or non-increasingly by p_j (task's duration) for ties;
4. Non-decreasingly by b_j, or non-decreasingly by the demand of upper-class vessels for ties;
5. Non-decreasingly by bj, or non-increasingly by total amount of vessels demanded;
6. Non-increasingly by total demanded (in number of tugs required) multiplied by p_j, untied by b_j (non-decreasingly)
7. Non-increasingly by compatibility degree, untie by duration of time window $(b_j - a_j)$;
8. Non-increasingly by compatibility degree, untie by total demand;

Rules 7 and 8 use the concept of *compatibility degree* of a task, defined next. As a way of measuring how difficult it is to fit a given task into a tug's route, we checked if such task is able to share a tug with each one of the remaining $n - 1$ tasks of J, considering their time windows and the travel time needed, using the speed of the fastest tug available. The compatibility degree of a task is then equal to the number of tasks it can share a route with.

Table 2 shows the average number of backlogged tasks for each instance set and sorting rule, as well as the number of solved instances by each rule. It can be noted that the 7th rule outperformed the others and, therefore, was chosen for the next tests.

Table 2. Results of experiments to define the sorting rule

Result	Instance set (# of tasks)	Sorting rule							
		1	2	3	4	5	6	7	8
Solved instances	n = 10	10	6	9	9	11	10	8	9
	n = 20	5	2	6	6	7	4	6	7
Avg. backlogged tasks	n = 10	1.0	1.2	1.0	1.0	1.0	1.0	1.0	1.0
	n = 20	2.1	2.5	1.8	2.5	1.4	1.8	1.7	2.0

The main results of the parameter tuning experiments are presented by Tables 3 and 4. Table 3 presents the number of solved instances for each value of *jPos*, showing how it varied on each execution performed, whereas Table 4 shows the heuristic's performance for each value of *oPos*. Based on these results, the values of *jPos* and *oPos* were set to 3 and 16, respectively, as these values yielded the biggest number of solved instances, on average (for the ten executions).

Table 3. Calibration of *jPos*

jPos	Avg. unsolved instances	Avg. Restarts needed for solved instances	Avg. Restarts needed considering all instances
1	10.0	0.0	35.6
2	2.8	1.9	11.8
3	**2.3**	**2.6**	**10.7**
4	2.4	2.9	11.3
5	2.6	3.9	12.8
6	3.0	4.1	14.4
7	2.7	5.0	14.1
8	3.3	4.6	15.8
9	3.5	5.3	17.0
10	3.7	5.0	17.4
11	3.9	6.3	19.2

Table 4. Calibration of *oPos*

oPos	Avg. unsolved instances	Avg. Restarts needed for solved instances	Avg. Restarts needed considering all instances
1	13.0	0.0	35.8
2	5.1	2.0	15.8
3	5.3	2.1	16.3
4	5.7	1.9	17.3
5	4.7	4.1	16.6
6	4.1	3.0	13.9
7	3.6	3.7	13.3
8	3.9	3.8	14.1
9	3.8	2.5	12.7
10	3.7	2.4	12.3
11	3.5	3.3	12.6
12	3.9	2.5	12.9
13	3.6	3.0	12.6
14	3.5	2.9	12.3
15	3.7	2.9	12.8
16	**3.4**	**3.0**	**12.1**
17	3.5	2.7	12.0
18	3.6	2.7	12.3
19	3.9	2.4	12.8

4.3   VRPTWSyn-Het_CH Efficiency Tests

Once the VRPTWSyn-Het_CH was tuned, it was executed with the aim of solving 36 proposed instances. As the heuristic has two randomized steps, each instance was

solved 10 times, with the aim of checking the method's robustness. Also, to grasp the difficulty level of these instances, Gurobi was run with a time limitation of 24 h and a solution count limit of one, enabling the measurement of how long it took to find a single feasible solution.

Table 5 shows the main results obtained by the constructive heuristic (denoted as CH), in terms of solution quality and execution time, along with Gurobi's results. For each instance, bold results within the Gurobi column indicate that the optimization package outperformed the heuristic; whereas within CH's column indicate otherwise.

These results indicate that the heuristic produces feasible solutions much faster than Gurobi for instances with 20 to 30 tasks, and the methods performed similarly on small instances of 10 tasks. Also, the VRPTHSyn-Het_HC successfully solved all instances, whereas there were two cases for which Gurobi could not provide a solution within 24 h, one of which was solved by the heuristic in less than 8 min and the other in about 9 s (on average). The worst performance obtained by the VRPTHSyn-Het_HC, considering its ten executions on all 36 instances, was 17.5 min. On the other hand, the average solution quality delivered by the non-optimizing method was worse than the first solution provided by Gurobi, as expected. Also, comparing its performance over ten executions, one can conclude that the solution's quality varies significantly from one execution to the other, which is reflected both on the standard deviation of $z(s)$ and on the difference between best and average solutions' costs.

4.4   Capability of Generating a Diverse Set of Feasible Solutions

These final experiments on the VRPTWSyn-Het_CH aimed at checking if the heuristic was able to find a heterogeneous set of feasible solutions. To this end, *nExp* was fixed at 2,000 restarts and the method was not stopped when a feasible solution was found. Instead, all generated solutions were stored, so the results' cost range could be evaluated.

Table 6 shows the number of solutions generated after 2,000 restarts of the procedure and the iteration that yielded the best solution. The minimum and maximum objective function values amongst the produced solutions are presented, as well as the total time spent and the average time per generated solution. Bold figures indicate better performance as compared to Gurobi's first solution.

These results show that the method could generate a heterogeneous solution set, as the total cost of the schedules varied greatly amongst them. The iteration yielding the best solution varied from one instance to the other and was close to the stop criterium in several cases, which could mean the method would deliver even better solutions if *nExp* was higher. Even though the aim of the CH was to get a feasible solution in short time, it created relatively good solutions (as compared to Gurobi's results) faster than Gurobi, at least for instances of 20 and 30 tasks. However, to prove or discard these hypotheses, more experiments are necessary.

Table 5. Results obtained by the heuristic and comparison with Gurobi's first solution

ID	Solution quality (z(s)) [$]				Computational time [s]			
	CH			*Gurobi*	CH			*Gurobi*
	Best $z(s)$	Mean $z(s)$	Std.dev.	$z(s^G)$	Avg.	Std.dev.		*Exec.time*
1	1,018,849	1,153,235	75,692	*397,100*	**0.26**	0.10		*0.29*
2	**634,117**	1,320,615	471,689	*789,297*	**0.17**	0.06		*0.45*
3	**293,853**	1,383,961	936,821	*1,071,380*	0.20	0.09		*0.16*
4	**941,325**	1,589,366	432,722	*2,273,560*	0.28	0.14		*0.16*
5	**1,655,450**	1,965,072	279,732	*2,506,470*	0.52	0.29		*0.16*
6	**835,080**	2,167,923	717,704	*2,604,400*	0.27	0.15		*0.12*
7	**2,894,217**	4,004,370	942,723	*3,381,290*	**0.13**	0.04		*2.49*
8	**2,262,165**	3,915,317	1,017,668	*2,608,060*	**0.25**	0.13		*3.03*
9	**1,247,058**	2,496,697	952,264	*3,366,270*	**0.16**	0.06		*1.49*
10	**1,595,904**	5,356,688	2,504,757	*2,091,850*	**0.15**	0.05		*0.79*
11	**1,067,153**	3,995,589	1,441,217	*5,615,650*	0.28	0.10		*0.23*
12	**3,935,391**	5,161,747	874,394	*4,685,390*	0.27	0.11		*0.16*
13	3,735,207	5,303,278	1,028,693	*1,953,910*	**0.84**	0.44		*159.28*
14	**3,887,022**	4,874,806	467,879	*4,199,490*	**2.47**	1.06		*422.24*
15	**4,295,783**	5,142,315	764,492	*5,003,350*	**10.71**	3.79		*143.68*
16	**3,464,147**	4,736,351	755,393	*4,319,430*	**0.85**	0.39		*1.75*
17	**2,268,634**	3,582,896	772,067	*5,484,430*	**1.05**	0.48		*7.66*
18	**2,515,237**	3,581,634	711,066	*3,467,740*	**5.02**	3.18		*115.11*
19	8,074,180	8,964,541	767,770	*6,403,480*	**0.96**	0.59		*74,306.70*
20	**6,599,834**	7,446,873	476,774	–	440.16	257.03		*86,400.00*
21	**914,391**	2,001,079	604,243	*3,257,070*	**4.90**	2.30		*517.16*
22	6,581,820	8,349,962	1,311,610	*6,508,570*	**1.46**	0.84		*648.08*
23	**5,541,485**	6,984,207	983,566	–	**9.03**	3.26		*86,400.00*
24	3,723,538	5,262,822	1,048,945	*2,468,210*	**1.20**	0.60		*330.05*
25	**3,682,847**	4,882,373	850,877	*4,845,650*	442.52	302.44		*211.34*
26	**5,704,137**	6,740,694	860,348	*6,316,010*	**337.23**	191.34		*633.43*
27	**3,397,119**	6,426,697	1,536,478	*6,217,920*	**5.59**	2.67		*1,812.00*
28	**6,021,856**	7,285,937	719,566	*8,996,560*	**4.20**	2.54		*1,579.26*
29	**2,927,804**	4,883,168	808,302	*3,112,020*	**3.09**	1.52		*2,652.72*
30	5,760,690	7,205,681	901,976	*5,273,030*	**2.89**	1.22		*232.31*
31	**8,687,666**	10,597,010	1,405,808	*7,780,720*	**0.62**	0.26		*311.05*
32	**6,247,951**	9,029,322	1,724,166	*6,969,420*	**1.58**	0.79		*3,804.42*
33	**5,726,122**	8,349,493	1,176,210	*6,029,180*	**2.14**	1.40		*3,183.62*
34	4,266,441	8,318,461	1,611,908	*3,174,490*	**0.73**	0.32		*680.31*
35	**8,283,730**	11,103,458	1,518,805	*8,631,800*	**2.40**	1.32		*1,032.58*
36	**9,498,005**	12,633,478	1,929,291	*10,591,500*	**2.11**	1.08		*145.13*

Table 6. Diversity analysis of the solutions produced by the VRPTWSyn-Het_CH

ID	Solution count	Iteration of best solution	Min. z(s)	Max. z(s)	Total time [s]	Time per solution [s]
1	1,326	1,456	**227,380**	2,968,402	44.98	0.03
2	1,468	53	**468,712**	4,098,323	38.78	0.03
3	1,513	1,832	**238,643**	4,365,893	44.31	0.03
4	1,534	636	**435,742**	2,311,310	39.19	0.03
5	232	1,457	**1,129,223**	2,591,399	39.32	0.17
6	1,843	151	**211,671**	4,036,078	60.64	0.03
7	1,632	862	**1,317,689**	7,345,716	38.25	0.02
8	1,683	1,573	**1,252,225**	8,278,343	52.30	0.03
9	1,926	817	**324,914**	9,820,846	52.39	0.03
10	1,935	301	**815,982**	10,198,051	38.46	0.02
11	1,117	543	**1,065,496**	8,650,128	59.93	0.05
12	1,876	1,612	**1,748,199**	9,891,738	77.95	0.04
13	1,429	665	2,986,270	8,297,097	289.81	0.20
14	264	98	**3,122,831**	6,752,361	**136.99**	0.52
15	82	1,989	**3,067,867**	7,311,398	177.27	2.16
16	1,901	1,041	**2,627,635**	6,636,653	442.46	0.23
17	1,943	429	**2,130,372**	7,981,841	646.17	0.33
18	308	1,150	**2,337,985**	5,849,192	372.65	1.21
19	485	740	**6,066,304**	13,588,533	**89.68**	0.18
20	1	1,494	6,921,631	6,921,631	**325.55**	325.55
21	2,000	933	**684,117**	5,913,543	2,910.14	1.46
22	401	46	**5,770,438**	11,792,608	**131.70**	0.33
23	147	507	4,746,371	9,277,908	**224.87**	1.53
24	1,930	1,704	2,948,047	9,883,758	378.72	0.20
25	4	464	5,044,479	6,764,789	**155.19**	38.80
26	2	658	7,246,379	7,758,663	**69.49**	34.74
27	221	755	**3,688,258**	11,625,163	380.06	1.72
28	151	858	**5,889,585**	10,606,733	275.24	1.82
29	486	237	**2,683,374**	7,156,007	**346.72**	0.71
30	745	628	**4,137,186**	10,754,740	476.56	0.64
31	1,854	660	**5,716,261**	16,678,259	**309.06**	0.17
32	1,121	1,123	**4,663,924**	15,986,095	**555.77**	0.50
33	1,466	1,382	**3,611,944**	13,158,217	**965.32**	0.66
34	1,974	709	4,433,396	15,840,437	**341.10**	0.17
35	1,637	1,389	**6,615,046**	17,090,015	**818.78**	0.50
36	1,811	1,290	**7,086,348**	19,640,345	1,330.34	0.73

5   Conclusions

This paper discusses a variation of the VRPTWSyn, named VRPTWSyn-Het, in which the available fleet is heterogeneous and there are several modes for fulfilling a task's demand. A constructive heuristic is presented and tested on three groups of instances proposed for the problem, with the aim of generating a set of initial solutions for future application of a meta-heuristic. Computational experiments using Gurobi illustrate the complexity of the VRPTWSyn-Het, as the optimization package was not able to find a feasible solution within reasonable time. On the other hand, the proposed solution method could rapidly generate a feasible solution to all instances, and our tests indicate it is able to get a heterogeneous solution set.

For future work, we intend to apply a meta-heuristic-based method for improving the initial solution produced by this greedy algorithm. Moreover, Gurobi or another optimization package should be used to find optimal values of the instances, or at least better lower bounds. Finally, the proposed methodology should be tested on instances of the VRPTWSyn found in the literature.

Acknowledgments. This study was financed in part by the *Coordenação de Aperfeiçoamento de Pessoal de Nível Superior – Brasil (CAPES)* – Finance Code 001.

References

1. Parragh, S.N., Doerner, K.F.: Solving routing problems with pairwise synchronization constraints. Cent. Eur. J. Oper. Res. **26**, 443–464 (2018). https://doi.org/10.1007/s10100-018-0520-4
2. Bredström, D., Rönnqvist, M.: Combined vehicle routing and scheduling with temporal precedence and synchronization constraints. Eur. J. Oper. Res. **191**, 19–31 (2008). https://doi.org/10.1016/j.ejor.2007.07.033
3. Drexl, M.: Synchronization in vehicle routing–a survey of VRPs with multiple synchronization constraints. Transp. Sci. **46**, 297–316 (2012). https://doi.org/10.1287/trsc.1110.0400
4. Afifi, S., Dang, D.C., Moukrim, A.: Heuristic solutions for the vehicle routing problem with time windows and synchronized visits. Optim. Lett. **10**, 511–525 (2016). https://doi.org/10.1007/s11590-015-0878-3
5. Afifi, S., Dang, D.-C., Moukrim, A.: A simulated annealing algorithm for the vehicle routing problem with time windows and synchronization constraints. In: Nicosia, G., Pardalos, P. (eds.) LION 2013. LNCS, vol. 7997, pp. 259–265. Springer, Heidelberg (2013). https://doi.org/10.1007/978-3-642-44973-4_27
6. Eveborn, P., Flisberg, P., Rönnqvist, M.: Laps Care-an operational system for staff planning of home care. Eur. J. Oper. Res. **171**, 962–976 (2006). https://doi.org/10.1016/j.ejor.2005.01.011
7. Mankowska, D.S., Meisel, F., Bierwirth, C.: The home health care routing and scheduling problem with interdependent services. Health Care Manag. Sci. **17**, 15–30 (2014). https://doi.org/10.1007/s10729-013-9243-1
8. Ait Haddadene, S.R., Labadie, N., Prodhon, C.: A GRASP × ILS for the vehicle routing problem with time windows, synchronization and precedence constraints. Expert Syst. Appl. **66**, 274–294 (2016). https://doi.org/10.1016/j.eswa.2016.09.002

9. Dohn, A., Kolind, E., Clausen, J.: The manpower allocation problem with time windows and job-teaming constraints: a branch-and-price approach. Comput. Oper. Res. **36**, 1145–1157 (2009). https://doi.org/10.1016/j.cor.2007.12.011

10. Hojabri, H., Gendreau, M., Potvin, J.Y., Rousseau, L.M.: Large neighborhood search with constraint programming for a vehicle routing problem with synchronization constraints. Comput. Oper. Res. **92**, 87–97 (2018). https://doi.org/10.1016/j.cor.2017.11.011

11. Mendes, A.B.: Scheduling offshore support fleet under the requirement of multiple vessels per task. PhD, University of Sao Paulo (2007). (in Portuguese). http://www.teses.usp.br/teses/disponiveis/3/3135/tde-14012008-171216/en.php

12. Shyshou, A., Gribkovskaia, I., Barceló, J.: A simulation study of the fleet sizing problem arising in offshore anchor handling operations. Eur. J. Oper. Res. **203**, 230–240 (2010). https://doi.org/10.1016/j.ejor.2009.07.012

13. Seixas, M.P., et al.: A heuristic approach to stowing general cargo into platform supply vessels. J. Oper. Res. Soc. **67**, 148–158 (2016). https://doi.org/10.1057/jors.2015.62

14. Liu, R., Tao, Y., Xie, X.: An adaptive large neighborhood search heuristic for the vehicle routing problem with time windows and synchronized visits. Comput. Oper. Res. **101**, 250–262 (2019). https://doi.org/10.1016/j.cor.2018.08.002

15. Artigues, C., Roubellat, F.: A polynomial activity insertion algorithm in a multi-resource schedule with cumulative constraints and multiple modes. Eur. J. Oper. Res. **127**, 297–316 (2000). https://doi.org/10.1016/S0377-2217(99)00496-8

16. Edis, E.B., Oguz, C., Ozkarahan, I.: Parallel machine scheduling with additional resources: Notation, classification, models and solution methods. Eur. J. Oper. Res. **230**, 449–463 (2013). https://doi.org/10.1016/j.ejor.2013.02.042

17. Pinedo, M.L.: Scheduling: Theory, Algorithms, and Systems. Springer, New York (2012). https://doi.org/10.1007/978-1-4614-2361-4

Network Design and Distribution Problems

A Facility Allocation-Location Approach to Analyze the Capability of the State-Owned Mexican Petroleum Company to Respond to the Closure of Pipelines

Santiago-Omar Caballero-Morales[✉] and José-Luís Martínez-Flores

Postgraduate Department of Logistics and Supply Chain Management,
Universidad Popular Autónoma del Estado de Puebla A.C.,
17 Sur 901, Barrio de Santiago, 72410 Puebla, Mexico
{santiagoomar.caballero,joseluis.martinez01}@upaep.mx

Abstract. In recent years, Mexico has reported significant economic losses caused by the organized theft of gasoline at different stages of the supply chain of its national petroleum company. As measure against these actions, the government ordered the specific closure of distribution pipelines between refineries and storage depots. However, this led to large shortage of gasoline and economic losses in the south central region of Mexico, even though distribution through tanker trucks was increased. Thus, the response capability of the company must be analyzed to quantitatively assess the shortage risk and propose alternatives if such disruption happens again. The present work addresses this task through a facility allocation-location approach where the location of additional storage depots was considered. For this analysis, the most updated data regarding the locations of gasoline storage depots and stations was used. Also, due to the large size of this data, the results of this approach were obtained through the development of a hybrid meta-heuristic. The results support the increase of storage depots to reduce the shortage risk and the service distance. This corroborates the public discussion that distribution of gasoline is restricted by the current infrastructure and that additional storage depots are needed. This infrastructure can also provide more control to reduce gasoline theft.

Keywords: Gasoline stations · Capacitated facility location problem · Location-allocation · Heterogeneous capacity

1 Introduction

Gasoline distribution depends on a complex infrastructure comprised of petroleum refineries, large storage terminals/depots, fueling/gasoline stations,

© Springer Nature Switzerland AG 2019
C. Paternina-Arboleda and S. Voß (Eds.): ICCL 2019, LNCS 11756, pp. 283–296, 2019.
https://doi.org/10.1007/978-3-030-31140-7_18

roads and pipelines. In Mexico, gasoline is mainly produced, imported and distributed by the Mexican state-owned petroleum company PEMEX. Recently, investigations reported significant losses of gasoline due to organized theft at different stages of the supply chain of PEMEX. Fuel worth more than 7.40 billion USD has been stolen since 2016 [4]. Other studies reported this as equivalent to losses of 10 million USD per day [9].

Since 2009, the national energy regulator found that thieves had tapped pipelines roughly every 1.4 km along PEMEX's 14,000 km pipeline network [4]. Thus, to avoid draining of pipelines by these groups, pipelines between refineries and storage depots were closed and fuel was transported mostly through tanker trucks [9]. However, this strategy worsened distribution and led to fuel shortage in different regions of Mexico which caused large economic losses. This was more severe in the states of Hidalgo, Mexico, Jalisco, Michoacan, Guanajuato, Queretaro and Aguascalientes which are important industrial and economic clusters [4]. Just in Guanajuato, it was reported that 84.0% of the gasoline stations were empty [26].

Even though the country has enough gasoline and oil resources, and distribution through tanker trucks was increased, this disruption caused severe shortages at the consumer points [26]. Road and rail transport is to be assessed as a long term solution for fuel distribution as important investments are being performed in these resources [4]. However, it is not clear if these measures are enough to reduce the shortage risk if such disruption happens again. Here it is important to mention that disruption may be caused by other factors besides measures against gasoline theft such as natural disasters and infrastructure maintenance.

The present work researches on this aspect by analyzing the storage capability to reduce the shortage risk with the current infrastructure in the most severely affected region. This is because it has been reported that storage depots can provide up to three days of gasoline supply in some regions [11,27], and pipelines are the main distribution means to supply these facilities. In contrast, the supply capacity of depots in the United States of America is approximately of 90 days [18]. Thus, storage is an important aspect to reduce the shortage risk and improve distribution of gasoline through road and/or rail transportation if closure of pipelines is performed.

Hence, the analysis extends on the evaluation of a capacitated multi-facility allocation-location approach to improve the response capability of PEMEX considering the current and additional storage infrastructure. Specifically, the proposed approach is focused on analyzing the increase of storage depots in strategic geographical locations in the affected region to increase the coverage rate and reduce the service distance between gasoline stations and depots. Having more storage depots can also provide more control to reduce the frequency of gasoline theft. This is based on the public discussion regarding the need of investment in additional storage infrastructure.

For this facility allocation-location approach, a hybrid meta-heuristic based on k-means Clustering (KMC) and Genetic Algorithms (GA) was developed. Due to the characteristics of the already existing infrastructure and new storage depots, the meta-heuristic extends on considering heterogeneous capacities.

It is important to mention that the analysis and results presented in this work are based on available public data which consists of partially described specific and general information in the media and official sources. Thus, the main contribution of the present work lies in the technical aspects of the methodology to analyze this data, the assessment of the proposed approach to improve the response capability of PEMEX (allocation-location approach), and the hybrid meta-heuristic to solve the facility location approach with heterogeneous capacity.

Route planning is not considered due to the characteristics of distribution through tanker trucks, which, based on the demands of gasoline stations, frequently leads to single-station routes. Also, the economic aspects of the facility location approach is not covered by the present work as this is a current topic of analysis and discussion by the government.

The results obtained by the present work provide quantitative support to the reported public need for additional storage infrastructure. However, these results must be considered with caution due to the previously discussed aspects of the available data and the assumptions of the proposed approaches. Finally, as there are no formal works reported in the literature regarding this problem, the present work represents exploratory research which can provide useful reference in the field.

2 Analysis of Current Infrastructure in the Affected Region – Baseline Scenario

As of 2019, there are approximately 12,000 gasoline stations (filling, fueling or gas stations) throughout the country [7]. These are supplied by 77 gasoline storage depots (oil depots/terminals) and six refineries. Figure 1 presents the general locations of these facilities including the gasoline stations within the affected region which consists of the states of Hidalgo, Mexico, Jalisco, Michoacan, Guanajuato, Queretaro and Aguascalientes.

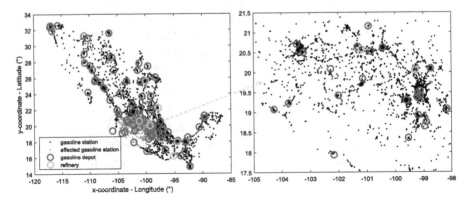

Fig. 1. Location of gasoline facilities within Mexico and the affected South-Central Region.

In the affected states there are approximately 3,959 gasoline stations, 22 gasoline storage depots, and 2 refineries. Available data from PEMEX regarding storing capacity of each depot was considered for the initial analysis of the current infrastructure. As described in [22,23], there are three types of capacity:

- operational capacity: maximum volumes that can be received, stored and delivered considering the design and construction characteristics of the storage system under normal operating conditions;
- contracted capacity: volumes committed through contracts for the provision of services;
- available capacity: volumes to provide services under the commonly used modality (= operational capacity - contracted capacity).

Table 1. Available capacities (liters, l) for each type of gasoline under the commonly used modality of storage depots in the affected region [23].

No.	Storage depot	Magna	Premium	Diesel
1	Añil	1168173	624234	754773
2	Azcapotzalco	16368414	6718386	10492251
3	Barranca Del Muerto	871956	221487	297171
4	Celaya	2749428	473661	1705593
5	Colima	1070229	394797	967038
6	Cuautla	993273	279204	795318
7	Cuernavaca	846834	155184	1671249
8	El Castillo	3412935	1911816	1148775
9	Iguala	299874	245178	816147
10	Irapuato	982779	408153	969423
11	Lazaro Cardenas	2648781	890877	14616393
12	Leon	713274	110823	338193
13	Manzanillo	2266227	659373	1586025
14	Morelia	809469	230709	922677
15	Pachuca	978327	315615	575421
16	Puebla	2510610	612786	1594134
17	Queretaro	1009014	204633	1330035
18	San Juan Ixhuatepec	1046379	162657	1458666
19	Toluca	1310955	261555	532809
20	Uruapan	2471337	225621	2825748
21	Zamora	1693986	46905	243111
22	Zapopan	4345152	1771419	2770416

This work focuses on the available capacity; Table 1 presents the latest capacity data (from barrels into liters, 1 barrel = 159 liters) for the 22 gasoline storage

depots in the affected region [23]. As observed, Azcapotzalco has the largest available capacity for Regular and Premium gasoline (87 and 92 octane, respectively) while Lazaro Cardenas has the largest capacity for Diesel (45 cetane).

Regarding the gasoline stations, it is a common standard that all stations provide all three types of gasoline. Thus, an average capacity of 80,000 to 100,000 l is frequently considered for a gasoline station (exact capacity of each gasoline station is not publicly available). Because distribution of gasoline from the storage depots to the gasoline stations is mainly performed through tanker trucks with average capacity of 30,000 l, each station would require approximately one tanker truck per type of gasoline. From these estimates, the capacity of each depot in terms of tanker trucks per type of gasoline can be estimated as *capacity of depot (l)/capacity of tanker truck (30,000 l)*. Table 2 presents this information based on the capacity data of Table 1.

Table 2. Available capacities (tanker trucks) for each type of gasoline under the common use modality of storage depots in the affected region

No.	Magna	Premium	Diesel	Total	No.	Magna	Premium	Diesel	Total
1	39	21	26	86	12	24	4	12	40
2	546	224	350	1120	13	76	22	53	151
3	30	8	10	48	14	27	8	31	66
4	92	16	57	165	15	33	11	20	64
5	36	14	33	83	16	84	21	54	159
6	34	10	27	71	17	34	7	45	86
7	29	6	56	91	18	35	6	49	90
8	114	64	39	217	19	44	9	18	71
9	10	9	28	47	20	83	8	95	186
10	33	14	33	80	21	57	2	9	68
11	89	30	488	607	22	145	60	93	298
Total	1052	416	1147	2615	Total	642	158	479	1279

From Table 2 the 22 depots can supply up to 1,694, 574 and 1,626 stations with Regular, Premium and Diesel gasoline, respectively (at a rate of one tanker truck per station). Because there are approximately 3,959 stations in the region, this represents $1,694/3,959 = 42.8\%$, $574/3,959 = 14.5\%$, and $1,626/3,959 = 41.1\%$ of the total requirements of Regular, Premium and Diesel, respectively. In general, if each station requires three tanker trucks (one per type of gasoline), the total demand is $3 \times 3,959 = 11,877$ tanker trucks and currently, at once, the available depots can supply $1,694 + 574 + 1,626 = 3,894$ tanker trucks, which represents $3,894/11,877 = 32.8\%$ of the total demand.

In practice, full service is achieved by servicing all stations on a rotating basis with the limited fleet of tanker trucks. Also, the storage depots are supplied daily

by the pipeline and rail networks. Nevertheless, it is important to observe that, if supply to the storage depots is disrupted, the previous estimates can become the effective service rate. And this is the scenario considered in this work.

In order to assess an approach to improve on this scenario, it is important to identify the crucial aspect of limited coverage which leads to increase the shortage risk:

- first, due to the characteristics of demand volumes and tanker trucks' capacities, delivery routes mainly consist of servicing only one station, which restricts the application of route planning strategies;
- second, servicing frequency depends of (1) the distance between depots and stations, and (2) the waiting times at the depots for supply, which have been reported to be up to 15 h. Particularly, distribution and transportation costs depend on the distance between the storage depots and the stations.

While PEMEX implemented measures such as periodically re-opening some pipelines, hiring additional tanker trucks (approximately 3,400) [2] and acquiring 500 new tanker trucks with larger capacity (60,000 l) to serve all the country [20], this is only a partial solution. On the economic aspect, distribution of gasoline through tanker trucks has been reported to be 14 times more expensive than distribution through pipelines [28].

On the other hand, gasoline demand is expected to increase by 136 millions of barrels a day (21,624 millions of liters) within the next decade [24] and limited investment for additional storage infrastructure has been performed. Thus, the improvement strategy to reduce shortage risk should be based on the facility location approach to determine the most suitable location of the additional storage infrastructure. This also can lead to increase servicing frequency and reduce distribution costs associated to transportation distance.

It is important to mention that no benchmark exists with the details of the real assignments/allocations to provide data for comparison purposes. Thus, the estimated service coverage of $3,894/11,877 = 32.8\%$ is considered as baseline metric for assessment of the proposed allocation-location approach.

3   Multi-facility Allocation-Location Approach

The distribution problem associated to PEMEX can be understood as a large-scale problem where approximately 3,959 gasoline stations must be assigned to the closest gasoline depot (22) to ensure minimum transportation time/distance (fast response) while complying with capacity restrictions.

In the logistic field, this problem can be addressed through well-known clustering models such as the Centered Clustering Problem (CCP), the p-Median Problem (PMP), the Capacitated Warehouse Location Problem (CWLP) and the Sum-of-Stars Clustering Problem (SoSCP) [19]. Also, different solving methods have been proposed for these problems due to their NP-hardness. This complexity makes it very difficult or impossible to solve to optimality large-scale problems with exact methods within reasonable time [14].

Hence, for large-scale problems, most of the solving methods are based on meta-heuristics such as Tabu-Search (TS), Nearest Neighborhood Search (NNS) and Genetic Algorithms (GA) [3,10,16]. However, while some solving methods have achieved very competitive results for small, medium and large instances of these problems, their performance is dependent of the size of the instances.

For the present work, the capacitated multi-facility Weber problem (CMFWP) is considered. This is similar to the capacitated PMP with the difference that new facilities are to be established at the locations of minimum distance instead of at median or pre-defined locations [13]. As there are already existing depots and new depots are to be considered, the formulation of the CMFWP considers facilities with heterogeneous capacities. The details of this approach are discussed in the following sections.

3.1   Objective Function

The CMFWP consists in finding the coordinates (x_j, y_j) of p facilities ($j = 1, ..., p$) that minimize the sum of weighted distances between them and n customers ($i = 1, ..., n$) with coordinates (a_i, b_i). In the CMFWP, (x_j, y_j) can take any value within the location space, thus, it can lead to minimum coverage distance [8]. With these definitions, and by considering the formulation of [6], the objective function of the CMFWP can be expressed as:

$$Minimize \sum_{i=1}^{n} \sum_{j=1}^{p} z_{ij} d((x_j, y_j), (a_i, b_i)), \tag{1}$$

where $z_{ij} = 1$ if customer i is served by the facility j ($z_{ij} = 0$ otherwise), and $d((x_j, y_j), (a_i, b_i))$ is the distance between the location of the facility at (x_j, y_j) and the assigned customer i which is located at (a_i, b_i). In order to obtain geographical distances in kilometers, the arc length metric on the spherical Earth was considered [5].

3.2   Restrictions

The optimization of the objective (1) is subject to the following constraints [6]:

$$\sum_{j=1}^{p} z_{ij} = 1, \quad \forall i \in n, \tag{2}$$

$$\sum_{i=1}^{n} z_{ij} = n_j, \quad \forall j \in p, \tag{3}$$

$$\sum_{i=1}^{n} p_i z_{ij} \leq H_j, \quad \forall j \in p, \tag{4}$$

$$x_j, y_j \in \Re, \quad z_{ij} \in \{0, 1\}. \tag{5}$$

In this formulation, (2) and (3) are restrictions that define that each customer is only assigned to one facility and provides the number of customers assigned to each facility, respectively. (4) defines that the total demands of the customers assigned to a facility j must not exceed its capacity H_j. Finally, (5) define the decision variables x_j and y_j as continuous in $\Re$. In common implementations H_j is fixed for all facilities; however, as presented in Tables 1 and 2, at least all existing facilities have different capacities. Thus, the present implementation considers heterogeneous capacity.

Finally, these restrictions consider that all facilities to be located are new. In the present work this is not true as there are already 22 gasoline depots. Thus, the allocation must consider the fixed existing infrastructure (22 depots) and the new infrastructure which is to be determined by the CMFWP. This is to be addressed by the hybrid meta-heuristic which is described in the following section.

3.3 Hybrid Meta-Heuristic

The proposed hybrid meta-heuristic integrates two main meta-heuristics: (a) the capacitated k-means clustering (CKMC) algorithm [15,21,25], and (b) a micro Genetic Algorithm (μGA). The CKMC is used to perform the allocation of stations to each facility while the GA is used to determine the location of minimum distance for each facility considering its allocated stations. While both are well-known meta-heuristics, the present work involves characteristics which are not considered in the standard formulation. These characteristics are the following:

– allocation of stations to facilities with heterogeneous capacity;
– allocation of stations to both, existing and new facilities.

Note that the hybrid meta-heuristic for the adapted CMFWP must simultaneously determine the allocation of stations to new and existing facilities, and the location of new facilities. Figure 2 presents the structure of the meta-heuristic highlighting the new processes to perform the allocation-location task with these characteristics.

As presented in Fig. 2(a) the first step consists in determining a lower bound for the feasible number of depots p (that is, the depots with capacities that can serve the whole tanker truck demand). Once this lower bound is determined, the CKMC which is described in Fig. 2(b), is executed for different number of p depots starting with the lower bound. The minimum distance locations for the depots, as required by the CMFWP, were estimated through the μGA which can lead to similar results than a standard GA but with smaller data arrays (i.e., smaller populations) [17]. Note that the CKMC was adapted to determine just the additional locations because there are already 22 depots which cannot be changed.

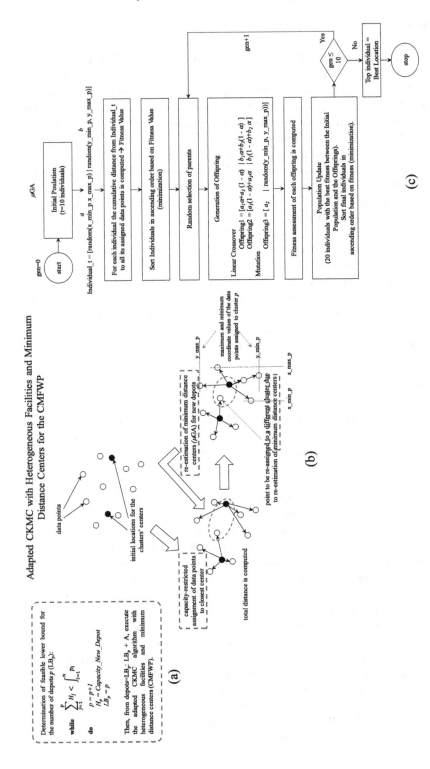

Fig. 2. CKMC and μGA for the estimation of new allocations and additional depots.

4   Results

Implementation of the meta-heuristic was performed with MATLAB R2016a in a HPZ230 Workstation with Intel Zeon CPU at 3.40 GHz with 8 GB RAM. As reference for the capacities of the new storage depot, H_j was estimated as 374 tanker trucks which is equivalent to 30% of the maximum capacity within the existing depots. For all stations, demand was considered as equal to three (one tanker truck per type of gasoline).

Table 3 presents the summarized results regarding the additional storage depots, the total service distance and service coverage obtained with the hybrid meta-heuristic. If 22 depots are considered, the service coverage becomes the baseline estimation of 32.8%. However, if 44 depots are considered (lower bound) then full coverage is achieved.

From this estimation, if the number of facilities is increased, a decrease of service distance can be obtained while maintaining the full service coverage. Just for reference purposes, if allocation of stations is performed solely on the existing 22 depots with the standard minimum distance criterion and no capacity restrictions, a total service distance of approximately 103,921.92 Km is estimated with service coverage of 32.8%. Based on the results presented in Table 3, a service distance of 74,221.98 Km can be achieved with full coverage if 68 depots are considered (22 existing depots plus 46 new depots).

5   Conclusions

In this work, the distribution problem of PEMEX was studied as a consequence of gasoline shortage due to the government's actions against organized gasoline theft. This action made evident that the available infrastructure is not ready to respond quickly to the sudden closure of pipelines between refineries, gasoline depots and stations.

While the actions to compensate this situation consisted of purchasing additional tanker trucks for distribution, the question regarding the main aspect of distribution that worsened gasoline shortage has not been addressed.

Through the analysis presented in this work we identified that, although routing is an important aspect of distribution, it is not the main cause of shortage due to its characteristics: whole tanker trucks are required to serve a gasoline station. In order to reduce the risk of shortage, periodic partial supply of multiple stations is recommended.

On the other hand, the possible main cause of shortage was identified as the limited number of gasoline depots. It has been reported that current depots only have enough gasoline stock for two/three days while in countries such as the USA the average is 90 days. Although this aspect has been discussed in the media, it has not been assessed from the logistic point of view.

As reviewed in Table 3 by increasing the number of depots the service coverage can be increased to 100.0% under the scenario of total pipeline closure while reducing the service distance. This proposal supports the reported projects on

Table 3. Estimated infrastructure by the hybrid meta-heuristic and performance metrics.

Number of depots	Service distance (Km)	Service coverage (%)
22 (existing depots)	103921.92	32.8
44 (lower bound)	123255.85	100.0
45 (23 new depots)	116793.48	100.0
46 (24 new depots)	113689.72	100.0
47 (25 new depots)	112155.62	100.0
48 (26 new depots)	103110.61	100.0
49 (27 new depots)	104497.62	100.0
50 (28 new depots)	101646.07	100.0
51 (29 new depots)	99534.14	100.0
52 (30 new depots)	97998.38	100.0
53 (31 new depots)	97592.84	100.0
54 (32 new depots)	98464.63	100.0
55 (33 new depots)	95056.78	100.0
56 (34 new depots)	95692.28	100.0
57 (35 new depots)	93448.99	100.0
58 (36 new depots)	90628.81	100.0
59 (37 new depots)	85353.15	100.0
60 (38 new depots)	87481.02	100.0
61 (39 new depots)	86051.29	100.0
62 (40 new depots)	81379.67	100.0
63 (41 new depots)	80718.79	100.0
64 (42 new depots)	81994.44	100.0
65 (43 new depots)	82281.33	100.0
66 (44 new depots)	81131.88	100.0
67 (45 new depots)	76787.34	100.0
68 (46 new depots)	74221.98	100.0
69 (47 new depots)	75424.68	100.0

infrastructure which are focused on increasing the storage and processing facilities for gasoline products [1]. With the energy reform of 2015, the participation of private funding makes achievable these projects.

Particularly, service distance is important to improve transportation times and thus, vehicle availability for fast distribution in the presence of sudden closure of pipelines. Re-supply rate of tanker trucks to serve all stations' demands depends of the distance between the gasoline stations and the gasoline depots. A tanker truck which serves a distant gasoline station would take longer to return

to the depot to subsequently supply another gasoline station. Because transportation through tanker truck is very expensive, up to 14 times more when compared to pipeline distribution [12], the increase of depots to reduce service distance in an important action to take.

Although these results highlight the importance of increasing the current infrastructure, more research is required to provide a more comprehensive analysis of PEMEX's supply chain. As future work, the following topics are considered:

- Integrate information about the supply rates and supply channels (pipelines, tanker trucks, trains) from refineries to gasoline depots. This, to analyze the complete two-echelon supply chain of PEMEX around the country.
- Integrate economic costs to evaluate the investment required for the proposed recommendations.
- Develop a simulation model to evaluate the proposed recommendations in a dynamic environment.

References

1. Arias, A.: Se requieren 250 mil mdp en infraestructura. El Heraldo de México, 30 July 2019. https://heraldodemexico.com.mx/mer-k-2/se-requieren-250-mil-mdp-en-infraestructura/
2. Bárcenas, X., González, S.: Cede Pemex parte de la distribución. El Sol de México, 11 January 2019. https://www.elsoldemexico.com.mx/mexico/sociedad/cede-pemex-parte-de-la-distribucion-rocio-nahle-toluca-transporte-combustible-2907209.html
3. Bernábe-Loranca, M.B., et al.: An approximation method for the P-median problem: a bioinspired tabu search and variable neighborhood search partitioning approach. Int. J. Hybrid Intell. Syst. **13**, 87–98 (2016)
4. Carranza, F., Esposito, A.: Mexico offensive against fuel theft leaves motorists stranded. Reuters, 6 January 2019. https://www.reuters.com/article/us-mexico-oil-theft/mexico-offensive-against-fuel-theft-leaves-motorists-stranded-idUSKCN1P00MR
5. Cazabal-Valencia, L., Caballero-Morales, S.O., Martinez-Flores, J.L.: Logistic model for the facility location problem on ellipsoids. Int. J. Eng. Bus. Manag. **8**, 1–9 (2016)
6. Chaves, A.A., Nogueira-Lorena, L.A.: Hybrid evolutionary algorithm for the capacitated centered clustering problem. Expert Syst. Appl. **38**, 5013–5018 (2011)
7. Cuellar, J.: Ubicación de gasolineras y precios comerciales de gasolina y diesel por estación. Goberno de México: Comisión Reguladora de Energía (CRE), 9 May 2019. https://datos.gob.mx/busca/dataset/ubicacion-de-gasolineras-y-precios-comerciales-de-gasolina-y-diesel-por-estacion
8. Dinler, D., Tural, M.K., Iyigun, C.: Heuristic for a continuous multi-facility location problem with demand regions. Comput. Ind. Eng. **62**, 237–256 (2015)
9. Editorial: Is it the right strategy against fuel theft? El Universal, 9 January 2019. https://www.eluniversal.com.mx/english/it-right-strategy-against-fuel-theft
10. Fleszar, K., Hindi, K.S.: An effective VNS for the capacitated p-median problem. Eur. J. Oper. Res. **191**, 612–622 (2008)

11. García, K.: Logística, talón de Aquiles de apertura gasolinera. El Economista, 23 December 2016). https://www.eleconomista.com.mx/empresas/Logistica-talon-de-Aquiles-de-apertura-gasolinera-20161223-0049.html

12. González, N: Sale más caro transportar el combustible en pipas que por ductos. Dinero en Imagen: Imagen Digital, 7 January 2019. https://www.dineroenimagen.com/economia/sale-mas-caro-transportar-el-combustible-en-pipas-que-por-ductos/105988

13. Hakan, M.: The capacitated multi-facility Weber problem with polyhedral barriers: efficient heuristic methods. Comput. Ind. Eng. **113**, 221–240 (2017)

14. Herda, M.: Combined genetic algorithm for capacitated p-median problem. In: Proceedings of the 16th IEEE International Symposium on Computational Intelligence and Informatics, CINTI 2015, pp. 151–154 (2015)

15. Hung, M.C., Wu, J., Chang, J.H., Yang, D.L.: An efficient k-means clustering algorithm using simple partitioning. J. Inf. Sci. Eng. **21**, 1157–1177 (2005)

16. Jánosíková, L., Herda, M., Haviar, M.: Hybrid genetic algorithms with selective crossover for the capacitated p-median problem. Cent. Eur. J. Oper. Res. **25**(3), 651–664 (2017)

17. Liu, Z.-C., Lin, X.-F., Shi, Y.-J., Teng, H.-F.: A micro genetic algorithm with Cauchy mutation for mechanical optimization design problems. Inf. Technol. J. **10**(9), 1824–1829 (2011)

18. Loredo, D.: CDMX tiene capacidad de almacenar gasolina solo para 43 horas: Energía. El Financiero/BLOOMBERG, 9 January 2019. https://www.elfinanciero.com.mx/economia/cdmx-tiene-capacidad-de-almacenar-gasolina-solo-para-43-horas-energia

19. Lorena, L.A.N., Senne, E.L.F.: A column generation approach to capacitated p-median problems. Comput. Oper. Res. **31**(6), 863–876 (2004)

20. Monroy, J: Pemex sumará 500 pipas de 60,000 litros a su flota: AMLO. El Economista, 16 January 2019. https://www.eleconomista.com.mx/empresas/Operativo-gasolinero-un-exito-Canacar-20190116-0041.html

21. Nasiakou, A., Alamaniotis, M., Tsoukalas, L.: Extending the K-means clustering algorithm to improve the compactness of the clusters. J. Pattern Recogn. Res. **11**(1), 61–73 (2016)

22. Pemex: Capacidad: Capacidad de los Sistemas de Gasolina y Diésel. Pemex, 30 July 2019. https://www.pemex.com/nuestro-negocio/logistica/almacenamiento/Paginas/capacidad.aspx

23. Pemex: Términos y Condiciones: ANEXO 1 DE LA RESOLUCIÓN Núm. RES/1680/2016. Pemex, 30 July 2019. https://www.pemex.com/nuestro-negocio/logistica/almacenamiento/Paginas/Terminos.aspx

24. Redacción Opportimes: Cómo crecerá la demanda de gasolina en México? Opportimes, 30 July 2019. https://www.opportimes.com/como-crecera-la-demanda-de-gasolina-en-mexico/

25. Sahraeian, R., Kaveh, P.: Solving capacitated p-median problem by hybrid k-means clustering and fixed neighborhood search algorithm. In: Proceedings of the International Conference on Industrial Engineering and Operations Management, pp. 1–6 (2010)

26. Sieff, K.: Gas stations in Mexico run out of gas as government cracks down on fuel theft. The Washington Post, 9 January 2019. https://www.washingtonpost.com/world/the_americas/gas-stations-in-mexico-run-out-of-gas-as-government-cracks-down-on-fuel-theft/2019/01/09/a9e71da8-1431-11e9-ab79-30cd4f7926f2_story.html?noredirect=on&utm_term=.3215d80e4f40

27. Sígler, É.: El almacenamiento de gasolina en México da apenas para 3 días. Expansión/CNN, 5 January 2017. https://expansion.mx/empresas/2017/01/05/el-almacenamiento-de-gasolina-en-mexico-apenas-para-3-dias
28. Sigler, E.: Los expertos dicen que el plan de las pipas es "improvisado y de corto plazo". EXPANSION/CNN, 30 July 2019. https://expansion.mx/empresas/2019/01/29/los-expertos-dicen-plan-pipas-improvisado-y-de-corto-plazo

Developing an Effective Decomposition-Based Procedure for Solving the Quadratic Assignment Problem

Mehrdad Amirghasemi[1]($\boxtimes$) (iD) and Reza Zamani[2] (iD)

[1] SMART Infrastructure Facility, Faculty of Engineering and Information Sciences,
University of Wollongong, Wollongong, NSW, Australia
mehrdad@uow.edu.au
[2] School of Computing and Information Technology,
Faculty of Engineering and Information Sciences, University of Wollongong,
Wollongong, NSW, Australia

Abstract. The quadratic assignment problem is involved with designing the best layouts for facilities and has a wide variety of applications in industry. This paper presents a decomposition-based procedure for solving this NP-hard optimization problem. The procedure aims at enhancing an initial solution in five interacting and synergetic layers. In the first layer, the initial solution undergoes the process of local search and becomes local optimal. In the second layer, the linear assignment technique is called and results in a suggestion for an enhancement. The third layer is about decomposing the proposed suggestion into several cycles. Then, in the fourth layer, each cycle is further decomposed into several segments. The number of the segments associated with each cycle is the length of the corresponding cycle, which was calculated in the third layer. Finally, the segments are processed in the fifth layer and potentially enhance the original solution. The employed decomposition has proved to be effective, and comparing the performance of the procedure with that of other state-of-the-art procedures indicates its efficiency.

Keywords: Metaheuristics · Local searches ·
Decomposition-based techniques · Facility layout ·
Quadratic assignment · Linear assignment

1 Introduction

Combinatorial optimization problems are widespread and solution strategies used for one problem may create a novel algorithmic ground for developing solution strategies for other similar problems. One of the heuristic algorithms extremely successful in solving one of these problems, namely the Travelling Salesman Problem (TSP), is the Lin-Kernighan algorithm [21] whose success

© Springer Nature Switzerland AG 2019
C. Paternina-Arboleda and S. Voß (Eds.): ICCL 2019, LNCS 11756, pp. 297–316, 2019.
https://doi.org/10.1007/978-3-030-31140-7_19

has inspired the development of various solution strategies for other problems. The key factor which makes this algorithm effective is the notion of the variable depth neighborhood.

The importance of a variable depth neighborhood local search stems from two related facts which necessitate a balance between the size of the neighborhood and computational time. These two facts are as follows. First, local searches with large neighborhood are very effective but impractical in terms of the required computational times. Second, the typical and computationally practical neighborhoods used in most local searches, which change the positions of only two or three elements in the search, are not very effective.

Two examples of the solution strategies developed based on this effective heuristic algorithm are ejection chain [27] and improvement graph [1], both applied to the quadratic assignment problem. They aim at creating a variable depth neighborhood which strikes a balance between the size of the neighborhood and the demand for computational time, albeit in two different fashions.

Inspired by the success of these algorithms, this paper presents a variable depth neighborhood decomposition-based search for the quadratic assignment problem. Aiming at striking a balance between the size of the neighborhood and the required computational time, the procedure, named NDS (Neighborhood Decomposition-based Search), operates through the interaction of four different layers and employs a linear assignment technique for suggesting the cycles which are decomposed. The four layers work as follows.

In the first layer, the initial solution generated for the problem undergoes the process of local search to become local optimal. In this layer, also a linear assignment technique is used to provide a suggestion for enhancement. The second layer decomposes the suggestion into several cycles, and calculates the length of each cycle; then the third layer decomposes each cycle into several segments. The number of these segments is equal to the length of the cycle and each segment itself is a cycle. The final operations of the procedure are performed in its fourth layer, which is responsible for the enhancement process. This layer (i) processes the segments, (ii) finds the best segment among all, and (iii) enforces the best segment to undergo the process of local search. The concepts of *cycle* and *segment* play key roles in the enhancement process and will be discussed in detail.

The paper is organized as follows. In Sect. 2, the quadratic assignment problem is described and the trend of major progress made in solving this problem is discussed. Section 3 discusses the related work, Sect. 4 is devoted to the NDS, and Sect. 5 provides the results of computational experiments. The concluding remarks are presented in Sect. 6.

2   Problem Formulation

As one of the most difficult problems in the NP-hard class, the Quadratic Assignment Problem (QAP) subsumes some other NP-hard problems like the maximal clique, bin-packing, graph partitioning, and the travelling salesman problem [22].

Assuming that there are n facilities to be allocated to n locations, and that the flow between facilities i and j is f_{ij}, and the distance between locations k and l is d_{kl}, then the goal is to find the location of facility i, for $i = 1, 2, \ldots, n$, shown with $\pi(i)$, so that the sum of all possible distance-flow products is minimized. This goal can be stated with the following objective function.

$$\min_{\pi(i), i=1,2,\ldots,n} \sum_{i=1}^{n} \sum_{j=1}^{n} f_{ij} d_{\pi(i)\pi(j)} \tag{1}$$

The trend of the major progress made in solving this problem has been reviewed in [3,14,22]. Solution strategies are mainly based on construction methods and local searches.

3   Related Work

The solution strategies presented for this problem are in a broad spectrum ranging from cooperative parallel search [17] through ant systems [7,10,28,29] to simulated annealing [4,8,24], and tabu searches [13,18]. A large number of approaches presented to solve the QAP belong to local searches and construction methods, with local searches employed in different metaheuristics to escape local optimality.

Unlike local searches, construction methods are based on allocating facilities to locations, one at a time, until every facility is allocated to a location, with no facility occupying more than one location. Different strategies have been used to conduct construction methods. Effective ant systems presented for the QAP include [7,10,15,23,28,29]. The other solution strategy successfully applied to the QAP comprises limited enumeration techniques. These techniques are truncated exact searches that aim at obtaining good solutions in a limited execution time. The techniques presented in [5,33], and [26] are examples of performing a limited and fruitful enumeration.

In the review of the related work, our main emphasis will be on the methods which rely on the incorporation of various mechanisms in the local searches. These mechanisms are aimed at boosting the power of local searches in exploiting the structure of the problems through effective replacement of facilities in locations. With respect to this trend of boosting, five effective solution strategies can be mentioned.

These strategies include (i) λ-exchange neighborhood [20], (ii) extended concentric tabu [13], (iii) extensive mutation-based mechanism [25], (iv) very large scale neighborhood scheme [1], and (v) ejection chain [27]. Concentric Tabu (CT) and its extended version for solving the QAP have been discussed in [12] and [13], respectively. It is worth mentioning that whereas in CT the tabu list is implicitly managed using the distance between "rings", it can also be managed using various techniques such as reverse elimination method as discussed in [16] and [9]. These approaches have been applied to the Quadratic (Semi-)Assignment Problems in [32] and [11].

The notion of improvement graph [1] and that of ejection chain [27] are involved with extending the classical 2-exchange neighborhood to assemble a broad-spectrum k-exchange neighborhood. In other words, they both improve the multi-exchange mechanism to a mechanism in which only a selected subset of moves in each depth is performed. However, they differ from the multi-exchange mechanism in the sense that in each depth, they assemble progressively larger exchanges based upon limited exchanges made in the preceding depth, allowing searching larger neighborhoods quickly and effectively.

The key point with the ejection chain scheme is that unlike the k-exchange scheme, which in level k contains $O(n^k)$ elements, a potentially best member of level k needs only $O(n)$ elements. In other words, the best member of each level is further expanded in $O(n)$ operations to find the best member of the next level, and this creates a neighborhood in successive swaps forming a loop. The length of such a loop is increased by one in each successive level.

The ejection chain procedure starts by identifying the best possible local move for each facility j. This leads to removing facility j from its current position and placing it in the best possible location, with ejecting the facility which has occupied that location. Then the first level of the ejection chain performs a series of best 2-exchange moves. In other words, the best location for the ejected facility is determined and the facility occupying that location is ejected.

The chain is expanded by iteratively selecting the best location for the ejected facility and making a new facility ejected. Since no facility can be moved twice, at most n nodes can be ejected. To create a complete solution with each move, the ejected facility can always be placed in the only empty location at hand, and therefore a solution can be associated with each move tested. In the ejection chain neighborhood, it is among all these solutions that the best one is selected. The idea of using a linear assignment technique in the QAP has been previously discussed in [2] and [34]. Both of these methods, the same as the NDS, use linear assignment as a facilitator for solving the QAP. The differences between these two procedures and the NDS are as follows.

Whereas [34], as an iterated local search, modifies local optimal solutions and runs a local search on the modified solutions, [2] is not involved with any similar local search. In [34], after creating an initial solution, the N* local search [20] is used and upon making possible initial improvement, the linear assignment technique is employed. In effect, [34] applies N* local search to the result of any suggested interchange.

Moreover, [34] uses Randomized Heuristic (RH) which aims at assigning highly connected facilities to closely located sites, as a heuristic procedure for creating initial solutions. Whereas [2] in search for an initial solution uses the average flow of each facility to other facilities as well as the average distance of each location from other locations.

Also, unlike [2] and [34], the current procedure uses five synergetic layers as well as N-Lambda local search to improve its efficiency. It also decomposes cycles into segments which broadens it search capability. The other difference is that the NDS uses the GREEDY method for generating initial solutions. In this

method, as it will be discussed in Sect. 4, each facility is assigned to the location which incurs the minimal partial cost to the objective function.

The method presented in [1] heuristically enumerates promising neighbors and uses the concept of the improvement graph to evaluate the neighbors efficiently. The neighborhood scheme employed is multi-exchange, which is the generalization of the 2-exchange scheme. A multi-exchange is shown by a cycle sequence of k different facilities as $i_1, i_2, i_3, \ldots, i_{k-1}, i_k, i_1$.

To specify the result of a multi-exchange operation, facility i_1 is moved to the location of facility i_2 and facility i_2 is moved to the location of facility i_3 and this process continues until facility i_{k-1} is moved to the location of facility i_k. Now facility i_k has no location and the previous location of facility i_1 is empty. Hence, by moving facility i_k to this only empty location, a new complete solution is generated. When a multi-exchange operation can improve the objective function, it is called a profitable multi-exchange, and the goal is to identify such exchanges.

The number of dislocated facilities shows the length of the corresponding multi-exchange, and, when this number is k, the neighborhood is shown with k-exchange notation. This, however, does not limit a k-exchange notation to the cases in which only k facilities are dislocated, and includes those with less than k facilities as well. In other words, a k-exchange neighborhood scheme includes all multi-exchange schemes with the length of no more than k. This means that 2-exchange is included in any k-exchange neighborhood scheme.

As the following equation shows, the number of neighbors, S, in a k-exchange neighborhood scheme is very large.

$$S = \sum_{k=2,\ldots,n} \binom{n}{k} (k-1)! \tag{2}$$

That is why this neighborhood scheme has been named as very large-scale neighborhood (VLSN). The authors have handled VLSN with employing an improved version of the concept of improvement graph, first introduced in [30]. In this graph, which is associated with a feasible solution, nodes represent facilities and arc (i, j) corresponds to facility i moving to the location of facility j. In the original applications of this graph, for graph partitioning, the cost of each arc in the graph shows the difference between the objective function of the neighboring solution and that of the seed solution.

Assuming that the objective function is involved with minimization, the original concept transforms the problem of finding an improved neighbor for a seed solution to finding a path with a negative cost from the seed solution to another solution, as its improved neighbor. The procedure enumerates directed paths with increasing length in the improvement graph, pruning paths which are unlikely to guide profitable multi-exchanges. For the corresponding neighborhood scheme to be k-exchange, the length of the path should not be more than k.

4 The NDS

The NDS is a five-layer procedure that enhances an initial solution in five inter-acting layers. Using the famous notion of the improvement graph, the NDS employs a linear assignment technique [19] for suggesting the cycles and uses both a local search and a cycle decomposition mechanism. This cycle mechanism decomposes every of the cycles proposed by the linear assignment procedure to smaller cycles so that every one of these smaller cycles can be applied to the solution in the hope of its enhancement.

The NDS initially needs a method for generating initial solutions. Basically, the initial solutions are constructed by incrementally deciding the best possible location for each facility in a greedy fashion. For this purpose, first, a uniformly random permutation of facilities is generated, and, based on this order, each facility is assigned to the location which incurs the minimal partial cost to the objective function. Once the best location for each facility is determined, it is fixed and added to the partial solution to be considered in the next iteration. This procedure, named GREEDY, can be considered as a modified version of the algorithm presented in [20]. After generating the initial solution, the layers, one after another, perform their operations.

In the second layer, based on the initial solution generated and improved by the local search in the first layer, the linear assignment technique is applied and results in proposing an assignment. In this assignment, facilities are assigned to facilities, and not to locations. The third layer is responsible for finding the cycles existing in the proposed assignment made by the linear assignment technique. For instance, if in the proposed assignment, facility 1 is assigned to facility 5 and facility 5 is assigned to facility 9 and facility 9 is assigned to facility 1, a cycle of the length 3 has been suggested. This cycle indicates that facility 1 should go to the location of facility 5 and facility 5 should go to the location of facility 9, and finally facility 9 should go to the location of facility 1.

An assignment proposed by the linear assignment can have several of these cycles, and all of them are found in the second layer. The second layer also calculates the length of each cycle. For instance, the length of the cycle mentioned is three because three facilities of 1, 5, and 9 are involved. In the third layer, each of these cycles is decomposed into several segments, with each segment itself being a small cycle. The length of the corresponding cycle calculated in the second layer shows the number of these segments. The following numerical example shows how a cycle is decomposed into segments (smaller cycles).

Suppose that a cycle has seven facilities. In this case, all the segments of the cycle with the length of 6, 5, 4, and 3 facilities are determined. In a cycle with the size q, the number of segments with the size k, as far as k is smaller than q, is q. As an example, consider the following cycle with the size of 7.

$$[(1), 4, 5, 8, 7, 6, 9, (1)]$$

This cycle indicates that the facility 1 goes to the location of the facility 4, and the facility 4 goes to the location of the facility 5, etc. Now for $k = 5$,

which is smaller than $q = 7$, the number of segments with the size of 5 is 7, and these segments are as follows: [(1),4,5,8,7,(1)], [(4),5,8,7,6,(4)], [(5),8,7,6,9, (5)], [(8),7,6,9,1,(8)], [(7),6,9,1,4,(7)], [(6),9, 1,4,5, (6)], and [(9),1,4,5,8,(9)]. As is seen, every element of the cycle receives the chance of being the first element of a segment and this describes why the number of segments with any length less than q is q.

After creating a number of segments with different lengths in the fourth layer, the segments are given to the fifth layer for being processed, and it is in the fifth layer that the result of applying any segment to the current solution is evaluated separately. For instance, suppose that the segment [(1),4,5,8,7,(1)] is to be applied to the current solution. This necessitates evaluating the effect of moving the facility 1 to the location of the facility 4, 4 to that of 5, 5 to that of 8, 8 to that of 7, and finally moving the facility 7 to the location of the facility 1, which has already moved to the location of the facility 4. The number of these segments can be large and every one of them should tested.

The input matrix used for the assignment component of the NDS is constructed based on the notion of the improvement graph. For instance, in the cell (2,5) of such a matrix, the cost of inserting the facility 2 in the location of the facility 5 is provided. Figure 1(a) shows how facilities 2 and 5 are both in location "A" and the relocation cost is calculated based on such double facilities in a single location. Based on this input matrix, the assignment procedure creates a solution for the linear assignment problem. Figure 1(b) shows a sample output of the linear assignment procedure proposing a loop 1-2-4-5-1.

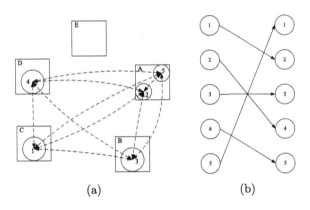

(a)   (b)

Fig. 1. (a) Both facilities 2 and 5 are in location A and the relocation cost is calculated based on such double facilities in a single location, (b) A sample output of the linear assignment procedure proposing the loop: 1-2-4-5-1

Figure 2 shows how the loop proposed by the linear assignment technique is applied to the original problem. As is seen, this loop indicates that facility 1 should go to the location of facility 2 and facility 2 should go to the location of

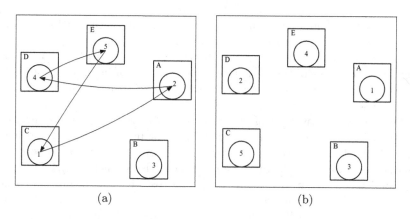

(a)   (b)

Fig. 2. The effect of applying the proposed loop 1-2-4-5-1 to the original solution

```
01 Procedure NDS()
02 {
03        for (int r=0; r<iterations; r++)
04        {
05            Generate an initial solution and call it X;
06            Improve X with N−Lambda;
07            Create the matrix of improvement graph, M_X, associated with X;
08            Set minCycleCost to a large number, and improvement to True;
09            while (improvement)
10            {
11                Make M_X a matrix which has only positive elements;
12                Solve the linear assignment problem associated with M_X;
13                Distinguish all cycles in the linear assignment result;
14                Initialize an array called α with all cycles distinguished;
15                for (each cycle C in α)
16                {
17                    Calculate the length of C and call this length L;
18                    for (i=3; i++; i<=L)
19                        for (each segment s with the length of i in C)
20                        {
21                            cycleCost= the cost of applying s to X;
23                            if (cycleCost<minCycleCost)
24                            {
25                                bestSegment=s;
26                                minCycleCost=cyclCoest;
27                            }
28                        }
29                }
30                apply the cycle associated with the bestSegment to X;
31                Improve X with N−Lambda;
32                if (any improvement has occurred)
33                {
34                    Create the matrix of improvement graph associated with X;
35                    improvement=True;
36                }
37                else  improvement=False;
38            }//end of while (improvement ...
39        } //end of for (int r=0 ...
40        return the best solution obtained;
41 }//end procedure
```

Fig. 3. The C-type pseudocode of the NDS

facility 4 and facility 4 to the location of facility 5, and finally facility 5 to the location of facility 1.

Figure 3 shows a C-type pseudocode for the procedure. As is seen at line 3 of the pseudocode, the procedure is repeated for a given number of times. The process is started by generating an initial solution. Then, at line 6, this solution is improved by a local search based on the λ-exchange neighborhood discussed in the previous section.

After the relevant initialization shown in the pseudocode, the main loop of the NDS starts at line 9. In each iteration, based on the updated, positive improvement graph, M_x, a linear assignment problem is solved and all the cycles proposed by the linear assignment as well as all sub-segments of each cycle is evaluated in the for-loop started at line 15. After evaluating all segments, the best segment is applied to the solution and the resulting solution is further improved with the extended N-Lambda procedure.

If by applying the λ-exchange procedure (Extended N-Lambda), at line 31, the cost of solution X is reduced, the current solution, X, as well as its associated improvement graph are updated, at line 34, and the improvement flag is set to *true* at line 35. Otherwise, a new initial solution is generated, at line 5. The entire process continues for a number of iterations and at the same time if a specified time-limit is reached, it stops.

Since in the NDS, computational bottleneck is involved with applying the segments to the solution one after another, the effectiveness of the module which does this operation can play a critical role in the efficiency of the procedure.

5 Computational Experiments

For testing the procedure, a selection of 76 problem instances, which consist of 53 real-life instances, and 23 randomly generated instances, have been selected from the QAPLIB [6]. Also, the procedure has been run on seven large instances with 100 facilities/locations. As a repository of benchmark instances for the QAP, the QAPLIB has a related website and is updated regularly. It is worth noting that randomly generated instances, shown with the patterns of *tai*a*, *tai*b* and *sko** are among the most difficult problems in the QAPLIB. The NDS has been implemented in C++ and has been compiled via Microsoft Visual C++ compiler on a laptop PC with 2.2 GHz speed.

Before running the NDS on benchmark instances, the method incorporated for generating initial solutions, GREEDY, has been analyzed. For this purpose, computational experiments have been made on hard-to-solve randomly generated instances, comparing the GREEDY with a uniform construction method, shown with UNIFORM. In the uniform method, simply, a uniformly random permutation is generated as the initial solution. As can be seen in Table 1, each method is run 1000 times on each individual instance and average percent deviation from the best known solution, %DEV$_{avg}$, and %STDEV is reported. %STDEV is simply the sample standard deviation estimate divided by the best known solution. As is seen in Table 1, while the uniform method has a total

average deviation of 47.26%, the greedy method has an average deviation of 26.59%. Moreover, assuming a Normal distribution function, $P(X \leq 0) = \Phi(0)$ can be easily calculated for UNIFORM and GREEDY as $4.5 * 10^{-12}$ and $1.2 * 10^{-6}$, respectively. Comparing the construction times in Table 1, we can see that UNIFORM is faster by a factor of $\frac{2.967}{0.083} = 35.7$.

Table 1. Comparing the performance of greedy solution construction with that of uniform method

Instance	UNIFORM		GREEDY	
	%DEV$_{avg}$	%STDEV	%DEV$_{avg}$	%STDEV
tai20a	27.60	3.39	17.61	2.85
tai25a	23.88	2.42	15.32	2.06
tai30a	20.88	1.81	13.55	1.48
tai35a	21.21	1.71	12.50	1.32
tai40a	20.63	1.49	12.19	1.22
tai50a	19.48	1.20	11.14	0.95
tai60a	18.21	1.01	9.96	0.73
tai80a	15.76	0.72	8.24	0.52
tai20b	163.57	51.68	102.87	41.28
tai25b	145.02	26.87	79.52	23.05
tai30b	106.84	17.98	50.64	13.55
tai35b	82.05	11.68	44.00	9.49
tai40b	77.70	8.82	40.38	7.21
tai50b	72.01	7.54	33.62	6.28
tai60b	66.21	6.51	35.04	5.47
tai80b	51.77	3.79	28.28	3.31
sko42	26.93	2.26	17.37	1.97
sko49	24.16	1.91	15.30	1.62
sko56	23.81	1.76	15.06	1.41
sko64	21.28	1.45	13.35	1.18
sko72	20.33	1.21	12.72	1.03
sko81	19.17	1.13	11.67	0.90
sko90	18.43	1.00	11.22	0.81
Average	47.26	6.93	26.59	5.64
Total Time(s)	0.083		2.967	

Having analyzed the initial solution construction method, we now analyze the performance of NDS on benchmark instances. First, NDS is run on the single instance, *lipa60a*. Figure 4(a) shows best-solution-cost versus iteration for the instance *lipa60a* in the first seven initial solutions generated. Figure 4(b)

depicts how, when the number of solutions evaluated increases, solution quality improves.

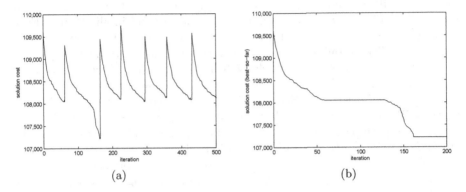

Fig. 4. Iteration/best-solution-cost for the instance *lipa60a* in (a) the first seven restarts and (b) in the first 200 iterations

For removing the effect of the random seed, in line with other procedures, the NDS has been run 10 times for each instance, and each run has been associated with a different random seed. A time-limit of $\frac{n}{10} + \frac{1}{10} * 2^{\frac{n}{10}}$ min for each run has been set, in which n shows the size of the problem instance. This time limit has been set in this way so that it will be close to the average running times of the procedures being compared. Furthermore, if the optimal solution is known for a problem instance in the literature, the procedure stops once the optimal solution is found.

In comparing the results produced by the NDS with the best available result in the literature for each instance, the following performance measures have been presented: (i) best percentage deviation ($\%DEV_{best}$) from the Best Known Solution (BKS) or the optimal solution (OPT), (ii) average percentage deviation ($\%DEV_{avg}$) for 10 runs, (iii) T_{best} and T_{avg}, which show the best and average time in which the best solution for the instance has been found, respectively, and (iv) T_{max}, which shows the maximum allowed running time for each run. All times are in minutes. The rationale behind providing the value of T_{avg} is that, for some instances different runs lead to generating the best solution and the average time needed for producing the best solution is as indicative as the shortest time needed for producing such a solution.

Table 2 presents the detailed results of running the NDS on all 53 real-life instances. Since the optimal value for all real-life instances are known, the optimal cost (OPT) is presented and the column (#OPT) shows the number of runs, out of 10, in which the optimal value has been found. As is seen, except for *chr22b*, and *ste36a*, the optimal value has been found in all 10 runs. Even for *chr22b* and *ste36a*, the optimal value has been found in 9, and 5 trials, respectively. The overall average percent deviation is as small as 0.003%.

The results of experiments on relatively harder randomly generated instances have been shown in Tables 3 and 4. Since for most instances, the optimal value is not known, the Best Known Solution (BKS) and number of times the BKS was found (#BKS) has been reported. In addition, $T_{max} = \frac{n}{10} + \frac{1}{10} * 2^{\frac{n}{10}}$ has also been reported for each instance. Due to complexity of the problem, the experiments are limited to the instances with up to 100 facilities/locations. As can be seen in Table 3, for 13 out of 23 instances, the BKS has been obtained by NDS in a very short amount of time. Furthermore, the overall average deviation from the best known solution is as small as 0.231%.

In order to demonstrate its effectiveness when applied to larger size instances, as shown in Table 4, the NDS has been separately run on problem instances with 100 facilities/locations. For these instances, the time limit has been set to $T_{max} = n/3$ min where $n = 100$. It is worth noting that this time limit is not as large as those set for previous instances. For example, whereas our time limit for problem instances with 90 facilities let the system to work for over 60 min, in this case only less than 34 min has been allowed for each run. As is seen, even with this small time limit, an average deviation of less than 0.2 % from the BKS has been achieved.

To examine its comparative performance, the NDS has been compared with four high-performance well-known methods from the literature. These procedures include the Progressive Adjusting Structural Solver (PASS) [2], the Cooperative Parallel Tabu Search (CPTS) [17] the Diversified Tabu Search (DivTS) [18], and a hybrid algorithm named ACO-GA/LS, which combines Ant Colony Optimization(ACO) with Genetic Algorithm (GA) and Local Search(LS) [31]. PASS has been run under the same setting as NDS. ACO/GA-LS and DivTS have been run on an Intel Pentium 2.0 with 2.0 GHz speed, and an Intel Itanium processor with 1.3 GHz speed, respectively. On the other hand, CPTS, which is a parallel procedure, has been run on 10 Intel Itanium processors, each with 1.3 GHz speed.

Since among the last three procedures, only DivTS has reported the complete results on real-life instances, the performance of DivTS on these intances has been compared with that of the NDS in Table 5. As is seen, both methods find optimal solutions for all real-life instances, NDS and DivTS have a %DEV$_{avg}$ greater than zero for 3 and 2 instances, respectively. Also while the overall average percent deviation for NDS is 0.003, this value for DivTS is 0.013. Furthermore, the NDS is at least 15 times faster than DivTS on average.

Table 2. The result of running NDS on real-life benchmark instances

Instance	OPT	#OPT	Best	%DEV$_{best}$	T$_{best}$	AVG	%DEV$_{avg}$	T$_{avg}$
bur26a	5426670	10	5426670	0.000	0.00	5426670	0.000	0.01
bur26b	3817852	10	3817852	0.000	0.00	3817852	0.000	0.01
bur26c	5426795	10	5426795	0.000	0.00	5426795	0.000	0.01
bur26d	3821225	10	3821225	0.000	0.00	3821225	0.000	0.01
bur26e	5386879	10	5386879	0.000	0.00	5386879	0.000	0.01
bur26f	3782044	10	3782044	0.000	0.00	3782044	0.000	0.00
bur26g	10117172	10	10117172	0.000	0.00	10117172	0.000	0.00
bur26h	7098658	10	7098658	0.000	0.00	7098658	0.000	0.00
Average				0.000	0.00		0.000	0.01
els19	17212548	10	17212548	0.000	0.00	17212548	0.000	0.00
had12	1652	10	1652	0.000	0.00	1652	0.000	0.00
had14	2724	10	2724	0.000	0.00	2724	0.000	0.00
had16	3720	10	3720	0.000	0.00	3720	0.000	0.00
had18	5358	10	5358	0.000	0.00	5358	0.000	0.00
had20	6922	10	6922	0.000	0.00	6922	0.000	0.00
Average				0.000	0.00		0.000	0.00
chr12a	9552	10	9552	0.000	0.00	9552	0.000	0.00
chr12b	9742	10	9742	0.000	0.00	9742	0.000	0.00
chr12c	11156	10	11156	0.000	0.00	11156	0.000	0.00
chr15a	9896	10	9896	0.000	0.00	9896	0.000	0.00
chr15b	7990	10	7990	0.000	0.00	7990	0.000	0.00
chr15c	9504	10	9504	0.000	0.00	9504	0.000	0.00
chr18a	11098	10	11098	0.000	0.00	11098	0.000	0.01
chr18b	1534	10	1534	0.000	0.00	1534	0.000	0.00
chr20a	2192	10	2192	0.000	0.00	2192	0.000	0.12
chr20b	2298	10	2298	0.000	0.00	2298	0.000	0.57
chr20c	14142	10	14142	0.000	0.00	14142	0.000	0.00
chr22a	6156	10	6156	0.000	0.00	6156	0.000	0.35
chr22b	6194	9	6194	0.000	0.14	6197.6	0.058	0.83
chr25a	3796	10	3796	0.000	0.00	3796	0.000	0.68
Average				0.000	0.01		0.004	0.18
kra30a	88900	10	88900	0.000	0.00	88900	0.000	0.01
kra30b	91420	10	91420	0.000	0.00	91420	0.000	0.04
kra32	88700	10	88700	0.000	0.00	88700	0.000	0.02
Average				0.000	0.00		0.000	0.02
ste36a	9526	5	9526	0.000	0.36	9531	0.052	1.78
ste36b	15852	10	15852	0.000	0.00	15852	0.000	0.04
ste36c	8239110	10	8239110	0.000	0.03	8239110	0.000	0.41
Average				0.000	0.13		0.017	0.74

(*continued*)

Table 2. (*contniued*)

Instance	OPT	#OPT	Best	%DEV$_{best}$	T$_{best}$	AVG	%DEV$_{avg}$	T$_{avg}$
esc16a	68	10	68	0.000	0.00	68	0.000	0.00
esc16b	292	10	292	0.000	0.00	292	0.000	0.00
esc16c	160	10	160	0.000	0.00	160	0.000	0.00
esc16d	16	10	16	0.000	0.00	16	0.000	0.00
esc16e	28	10	28	0.000	0.00	28	0.000	0.00
esc16f	0	10	0	0.000	0.00	0	0.000	0.00
esc16g	26	10	26	0.000	0.00	26	0.000	0.00
esc16h	996	10	996	0.000	0.00	996	0.000	0.00
esc16i	14	10	14	0.000	0.00	14	0.000	0.00
esc16j	8	10	8	0.000	0.00	8	0.000	0.00
esc32a	130	10	130	0.000	0.00	130	0.000	0.02
esc32b	168	10	168	0.000	0.00	168	0.000	0.00
esc32c	642	10	642	0.000	0.00	642	0.000	0.00
esc32d	200	10	200	0.000	0.00	200	0.000	0.00
esc32e	2	10	2	0.000	0.00	2	0.000	0.00
esc32g	6	10	6	0.000	0.00	6	0.000	0.00
esc32h	438	10	438	0.000	0.00	438	0.000	0.00
esc64a	116	10	116	0.000	0.00	116	0.000	0.00
esc128	64	10	64	0.000	0.00	64	0.000	0.01
Average				0.000	0.00		0.000	0.00
Overall				0.000	0.001		0.000	0.014

A comparison on the performance of the four procedures on the randomly generated, commonly tested instances has also been shown in Table 6. As is seen, the NDS outperforms the PASS and, in general, both the NDS and PASS have a comparatively better performance on the small instances. This can be partly described by comparing the average running times of other procedures with the maximum allowed running time (T$_{max}$) of the NDS. Whereas CPTS, DivTS and ACO/GA-LS have allowed a total of 602.3, 328.7, and 564.9 min for a single run on all instances, respectively, the NDS has a total maximum allowed running time of only 225.9 min. Moreover, comparing the average running times (T$_{avg}$) on those instances for which a %DEV$_{avg}$ of 0.000 has been obtained, indicates that the NDS is generally faster than the compared methods. In total, the NDS has obtained an overall average percent deviation of 0.348% with an average maximum allowed time of only 16.138 min, supporting the fact that the NDS is competitive with those well-known procedures.

Table 3. The result of running NDS on randomly generated benchmark instances

Instance	BKS	#BKS	Best	%DEV$_{best}$	T$_{best}$	AVG	%DEV$_{avg}$	T$_{avg}$	T$_{max}$
Taillard symmetric problems:									
tai20a	703482	10	703482	0.000	0.00	703482	0.000	0.02	2.40
tai25a	1167256	10	1167256	0.000	0.01	1167256	0.000	0.23	3.07
tai30a	1818146	6	1818146	0.000	1.09	1818264.4	0.007	1.77	3.80
tai35a	2422002	1	2422002	0.000	3.09	2431263.2	0.382	2.42	4.63
tai40a	3139370	0	3156844	0.557	1.49	3161756.4	0.713	2.19	5.60
tai50a	4938796	0	4980630	0.847	0.63	4995391	1.146	3.89	8.20
tai60a	7208572	0	7282038	1.019	5.44	7301371.2	1.287	6.71	12.40
tai80a	13499184	0	13674098	1.296	13.22	13696120.8	1.459	15.70	33.60
Average				0.465	3.12		0.624	4.12	9.21
Taillard asymmetric problems:									
tai20b	122455319	10	122455319	0.000	0.00	122455319	0.000	0.00	2.40
tai25b	344355646	10	344355646	0.000	0.00	344355646	0.000	0.05	3.07
tai30b	637117113	10	637117113	0.000	0.01	637117113	0.000	0.37	3.80
tai35b	283315445	10	283315445	0.000	0.02	283315445	0.000	0.62	4.63
tai40b	637250948	10	637250948	0.000	0.05	637250948	0.000	0.29	5.60
tai50b	458821517	1	458821517	0.000	3.84	459004373.5	0.040	4.70	8.20
tai60b	608215054	0	608228537	0.002	3.29	608423633.3	0.034	4.77	12.40
tai80b	818415043	0	819032458	0.075	28.58	819839123.6	0.174	14.66	33.60
Average				0.010	4.47		0.031	3.18	9.21
Skorin-Kapov problems:									
sko42	15812	10	15812	0.000	0.01	15812	0.000	0.28	6.04
sko49	23386	10	23386	0.000	0.07	23386	0.000	3.05	7.89
sko56	34458	0	34462	0.012	2.11	34465.4	0.021	5.05	10.45
sko64	48498	5	48498	0.000	0.76	48501.8	0.008	7.96	14.84
sko72	66256	0	66272	0.024	9.88	66300.4	0.067	9.86	21.90
sko81	90998	0	91008	0.011	27.65	91076.4	0.086	17.62	35.54
sko90	115534	0	115574	0.035	58.05	115626.4	0.080	30.36	60.20
Average				0.012	14.08		0.037	10.60	22.41
Overall				0.162	7.22		0.231	5.97	13.61

Table 4. The result of running NDS on instances with $n = 100$

instance	BKS	#BKS	Best	%DEV$_{best}$	T$_{best}$	AVG	%DEV$_{avg}$	T$_{avg}$	T$_{max}$
Instances with size 100									
tai100a	21052466	0	21232880	0.857	15.684	21257331.2	0.973	16.963	33.33
sko100a	152002	0	152090	0.058	18.483	152206.8	0.135	12.664	33.33
sko100b	153890	0	154002	0.073	1.160	154136.6	0.160	19.356	33.33
sko100c	147862	0	147918	0.038	5.053	148030.2	0.114	19.004	33.33
sko100d	149576	0	149740	0.110	26.599	149853.8	0.186	19.734	33.33
sko100e	149150	0	149216	0.044	23.713	149329.8	0.121	14.087	33.33
sko100f	149036	0	149262	0.152	21.115	149384.4	0.234	16.435	33.33
Average				0.190	15.972		0.275	16.892	33.333

Table 5. Comparing the performance of NDS with that of DivTS on real-life instances.

	NDS		DivTS	
Instances	%DEV$_{avg}$	T$_{avg}$	%DEV$_{avg}$	T$_{avg}$
bur26a-h	0.000	0.01	0.000	0.52
els19	0.000	0.00	0.000	0.19
had12-20	0.000	0.00	0.000	0.32
chr12a-25a	0.004	0.18	0.093	0.50
kra30a-32	0.000	0.02	0.000	1.27
ste36a-c	0.017	0.74	0.000	2.26
esc16a-128	0.000	0.00	0.000	10.06
Average	0.003	0.137	0.013	2.159

Table 6. Comparing the performance of NDS with that of other procedures on randomly-generated instances.

	NDS			PASS		CPTS		DivTS		ACO-GA/LS	
Instance	%DEV$_{avg}$	T$_{avg}$	T$_{max}$	%DEV$_{avg}$	T$_{avg}$	%DEV$_{avg}$	T$_{avg}$	%DEV$_{avg}$	T$_{avg}$	%DEV$_{avg}$	T$_{avg}$
Taillard Instances:											
tai20a	0.000	0.0	2.4	0.000	0.2	0.000	0.1	0.000	0.2	–	–
tai25a	0.000	0.2	3.1	0.041	0.7	0.000	0.3	0.000	0.6	–	–
tai30a	0.007	1.8	3.8	0.181	1.6	0.000	1.6	0.000	1.3	0.341	1.4
tai40a	0.713	2.2	5.6	0.851	8.4	0.148	3.5	0.222	5.2	0.593	13.1
tai50a	1.146	3.9	8.2	1.324	11.8	0.440	10.3	0.725	10.2	0.901	29.7
tai60a	1.287	6.7	12.4	1.458	9.8	0.476	26.4	0.718	25.7	1.068	58.5
tai80a	1.459	15.7	33.6	1.524	17.9	0.570	94.8	0.753	52.7	1.178	152.2
Average	0.659	4.358	9.867	0.768	7.197	0.233	19.571	0.345	13.700	0.816	50.980
Skorin-Kapov Instances:											
sko42	0.000	0.3	6.0	0.000	1.8	0.000	5.3	0.000	4.0	0.000	0.7
sko49	0.000	3.0	7.9	0.007	5.7	0.000	11.4	0.008	9.6	0.056	7.6
sko56	0.021	5.1	10.5	0.033	9.6	0.000	21.0	0.002	13.2	0.012	9.1
sko64	0.008	8.0	14.8	0.051	15.1	0.000	42.9	0.000	22.0	0.004	17.4
sko72	0.067	9.9	21.9	0.111	15.7	0.000	69.6	0.006	38.0	0.018	70.8
sko81	0.086	17.6	35.5	0.156	11.2	0.000	121.4	0.016	56.4	0.025	112.3
sko90	0.080	30.4	60.2	0.146	12.9	0.000	193.7	0.026	89.6	0.042	92.1
Average	0.037	10.6	22.4	0.072	10.3	0.000	66.5	0.008	33.3	0.022	44.3
Overall	0.348	7.479	16.138	0.420	8.744	0.117	43.021	0.177	23.479	0.419	47.633

6   Concluding Remarks

Despite the significant progress made in the development of systematic search methods for the QAP, the achievement of these methods are still bounded by solving comparatively undersized instances through highly parallel programming environments in comparatively long, sometimes several days, of computation times. Unlike systematic searches which explore search spaces of partial solutions, the NDS explores search spaces of complete solutions.

In the NDS, moves occur from one state to the next through making local changes to the value(s) of one or more component(s) of an encoding. A decoder

has to decode such an encoding and find the value of the objective function. Despite exploring search spaces of complete solutions, the NDS, like systematic search methods, performs limited enumeration for a set of promising solutions.

For the purpose of marginally improving a set of high quality solutions through the repeated calls of a linear assignment technique, the NDS, as a limited enumeration technique, relies on the rule that as soon as a suitable solution is found, it should spend a great deal of time to marginally improve such a solution. To balance speed versus solution quality, by the use of its linear assignment component, the NDS virtually enumerates only those solutions which are fruitful and ignores a large number of potential solutions. In other words, it yields deeper cuts in a virtual search tree to reach promising solutions at the cost of losing insignificant preciseness. Tested on a broad spectrum of benchmark instances, the NDS has proved to be robust and effective.

One reason for the success of the NDS can be the exploitation of the swap neighborhood structure of the QAP. With this respect, Fig. 5 provides a glimpse of the neighborhood structure for the instance $chr25a$ near its optimal solution, positioned in location $(0,0)$. This instance is involved with 25 items and the surface, $z = f(x,y)$, shows the result of swapping item x with item y. Since the QAP is a minimization problem, the solution evaluated for each swap has been decreased from the maximum value of z, indicating that the value of z is obtained by $z = Max(Z) - f(x,y)$. As is seen, this formula causes the original largest value to become zero, and the original smallest value to change into the difference between the original smallest and largest values. In other words, in this 3D figure, original points with larger values appear closer to the surface. As can be expected, for drawing this figure 625, 25 * 25, permutations have been evaluated.

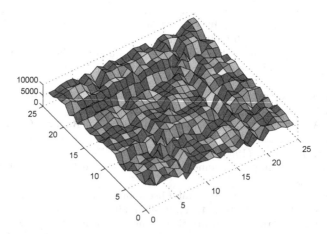

Fig. 5. A neighborhood structure for the instance $chr25a$ near its optimal solution, $z = Max(Z) - Swap(x,y)$

The NDS can be improved in two different directions. First, by injecting controlled randomizing elements into the generation of solutions, the procedure can extend the range of complete solutions it searches, focusing on diversification at the cost of slight decrease in intensification. The less such randomizing elements degrade solution quality, the more superior solutions the procedure may produce. Hence, in the development of such possible controlled randomizing elements, care should be taken that despite diversifying the search, they should be biased towards generating high quality solutions.

Second, since in the NDS, computational bottleneck is involved with applying the segments to the solution consecutively, an efficient parallelization mechanism can facilitate the application of segments to the solutions. For this purpose, separate threads need to calculate the effect of applying a particular segment to the solution. Since the number of segments proposed by the linear assignment technique depends on a parameter controlling the number of segments, in the case of such parallelization, this parameter should increase significantly, contributing both to the speed in which the procedure calculates a solution and the overall quality of the solution produced.

References

1. Ahuja, R., Jha, K., Orlin, J., Sharma, D.: Very large-scale neighborhood search for the quadratic assignment problem. INFORMS J. Comput. **19**(4), 646–657 (2007)
2. Amirghasemi, M., Zamani, R.: An effective structural iterative refinement technique for solving the quadratic assignment problem. In: Cerulli, R., Raiconi, A., Voß, S. (eds.) Computational Logistics, pp. 446–460. Springer International Publishing, Cham (2018)
3. Anstreicher, K.: Recent advances in the solution of quadratic assignment problems. Math. Program. **97**(1), 27–42 (2003)
4. Baykasoğlu, A.: A meta-heuristic algorithm to solve quadratic assignment formulations of cell formation problems without presetting number of cells. J. Intell. Manufact. **15**(6), 753–759 (2004)
5. Burkard, R.E., Bönniger, T.: A heuristic for quadratic boolean programs with applications to quadratic assignment problems. Eur. J. Oper. Res. **13**(4), 374–386 (1983)
6. Burkard, R.E., Karisch, S.E., Rendl, F.: QAPLIB-a quadratic assignment problem library. J. Glob. Optim. **10**(4), 391–403 (1997)
7. Colorni, A., Dorigo, M., Maffioli, F., Maniezzo, V., Righini, G., Trubian, M.: Heuristics from nature for hard combinatorial optimization problems. Int. Trans. Oper. Res. **3**(1), 1–21 (1996)
8. Connolly, D.: An improved annealing scheme for the QAP. Eur. J. Oper. Res. **46**(1), 93–100 (1990)
9. Dammeyer, F., Voß, S.: Dynamic tabu list management using the reverse elimination method. Ann. Oper. Res. **41**(2), 29–46 (1993)
10. Demirel, N., Toksarı, M.: Optimization of the quadratic assignment problem using an ant colony algorithm. Appl. Math. Comput. **183**(1), 427–435 (2006)
11. Domschke, W., Forst, P., Voß, S.: Tabu search techniques for the quadratic semi-assignment problem. In: Fandel, G., Gulledge, T., Jones, A. (eds.) New Directions for Operations Research in Manufacturing, pp. 389–405. Springer, Heidelberg (1992). https://doi.org/10.1007/978-3-642-77537-6_23

12. Drezner, Z.: Heuristic algorithms for the solution of the quadratic assignment problem. J. Appl. Math. Decis. Sci. **6**, 163–173 (2002)
13. Drezner, Z.: The extended concentric tabu for the quadratic assignment problem. Eur. J. Oper. Res. **160**(2), 416–422 (2005)
14. Drezner, Z., Hahn, P., Taillard, E.: Recent advances for the quadratic assignment problem with special emphasis on instances that are difficult for meta-heuristic methods. Ann. Oper. Res. **139**(1), 65–94 (2005)
15. Gambardella, L., Taillard, E., Dorigo, M.: Ant colonies for the quadratic assignment problem. J. Oper. Res. Soc. **50**(2), 167–176 (1999)
16. Glover, F.: Tabu search - part II. ORSA J. Comput. **2**(1), 4–32 (1990)
17. James, T., Rego, C., Glover, F.: A cooperative parallel tabu search algorithm for the quadratic assignment problem. Eur. J. Oper. Res. **195**(3), 810–826 (2009). https://doi.org/10.1016/j.ejor.2007.06.061
18. James, T., Rego, C., Glover, F.: Multistart tabu search and diversification strategies for the quadratic assignment problem. IEEE Trans. Syst. Man Cybern. Part A Syst. Hum. **39**(3), 579–596 (2009)
19. Lee, Y., Orlin, J.: Quickmatch: a very fast algorithm for the assignment problem. Report, Massachusetts Institute of Technology, Sloan School of Management (Report number: WP#3547-93) (1993)
20. Li, Y., Pardalos, P., Resende, M.: A greedy randomized adaptive search procedure for the quadratic assignment problem. In: Pardalos, P., Wolkowicz, H. (eds.) Quadratic Assignment and Related Problems, DIMACS Series on Discrete Mathematics and Theoretical Computer Science, vol. 16, pp. 237–261 (1994)
21. Lin, S., Kernighan, B.: An effective heuristic algorithm for the traveling salesman problem. Oper. Res. **21**, 443–452 (1973)
22. Loiola, E.M., de Abreu, N.M.M., Boaventura-Netto, P.O., Hahn, P., Querido, T.: A survey for the quadratic assignment problem. Eur. J. Oper. Res. **176**(2), 657–690 (2007). https://doi.org/10.1016/j.ejor.2005.09.032
23. Maniezzo, V., Colorni, A.: The ant system applied to the quadratic assignment problem. IEEE Trans. Knowl. Data Eng. **11**(5), 769–778 (1999)
24. Mavridou, T., Pardalos, P.: Simulated annealing and genetic algorithms for the facility layout problem: A survey. Comput. Optim. Appl. **7**(1), 111–126 (1997)
25. Misevicius, A.: A tabu search algorithm for the quadratic assignment problem. Comput. Optim. Appl. **30**(1), 95–111 (2005)
26. Nissen, V., Paul, H.: A modification of threshold accepting and its application to the quadratic assignment problem. OR Spectr. **17**(2), 205–210 (1995)
27. Rego, C., James, T., Glover, F.: An ejection chain algorithm for the quadratic assignment problem. Networks **56**(3), 188–206 (2010)
28. Solimanpur, M., Vrat, P., Shankar, R.: Ant colony optimization algorithm to the inter-cell layout problem in cellular manufacturing. Eur. J. Oper. Res. **157**(3), 592–606 (2004)
29. Talbi, E.G., Roux, O., Fonlupt, C., Robillard, D.: Parallel ant colonies for the quadratic assignment problem. Future Gener. Comput. Syst. **17**(4), 441–449 (2001). https://doi.org/10.1016/S0167-739X(99)00124-7
30. Thompson, P., Orlin, J.: The theory of cyclic transfers. Technical report, Operations Research Center Report, MIT, Cambridge, MA, oR 200–89 (1989)
31. Tseng, L.Y., Chen, S.C.: A hybrid metaheuristic for the resource-constrained project scheduling problem. Eur. J. Oper. Res. **175**(2), 707–721 (2006)

32. Voß, S.: Solving quadratic assignment problems using the reverse elimination method. In: Nash, S.G., Sofer, A., Stewart, W.R., Wasil, E.A. (eds.) The Impact of Emerging Technologies on Computer Science and Operations Research. Operations Research/Computer Science Interfaces Series, vol. 4. Springer, Boston (1995). https://doi.org/10.1007/978-1-4615-2223-2_14

33. West, D.: Algorithm 608: approximate solution of the quadratic assignment problem. ACM Trans. Math. Softw. (TOMS) 9(4), 461–466 (1983)

34. Zamani, R., Amirghasemi, M.: A self-adaptive nature-inspired procedure for solving the quadratic assignment problem (2019, submitted)

A Stochastic, Multi-Commodity Multi-Period Inventory-Location Problem: Modeling and Solving an Industrial Application

Mauricio Orozco-Fontalvo[1]($\boxtimes$) (iD), Víctor Cantillo[2] (iD),
and Pablo A. Miranda[3] (iD)

[1] Department of Civil Engineering, Universidad Militar Nueva Granada,
Carrera, 58#55-66 Bogotá, Colombia
mauricio.orozco@unimilitar.edu.co
[2] Department of Civil and Environmental Engineering, Universidad del Norte,
Km 5 Vía Puerto Colombia, Barranquilla, Colombia
victor.cantillo@uninorte.edu.co
[3] Department of Engineering Sciences, Faculty of Engineering,
Universidad Andres Bello, Campus Viña Del Mar, Santiago, Chile
pablo.miranda@pucv.cl

Abstract. This paper addresses the real-world supply chain network design problem with a strategic multi-commodity and multi-period inventory-location problem with stochastic demands. The proposed methodology involves a complex non-linear, non-convex, mixed integer programming model, which allows for the optimization of warehouse location, demand zone's assignment, and manufacturing settings while minimizing the fixed costs of a distribution center (DC), along with the transportation and inventory costs in a multi-commodity, multi-period scenario. In addition, a genetic algorithm is implemented to obtain near-optimal solutions at competitive times. We applied the model to a real-world industrial case of a Colombian rolled steel manufacturing company, where a new, optimized supply chain distribution network is required to serve customers at a national level. The proposed approach provides a practical solution to optimize their distribution network, achieving significant cost reductions for the company.

Keywords: Supply chain network design · Inventory location problems · Facility location problems · Genetic algorithms · Safety stock · Cycle stock · Explicit enumeration · Stochastic modelling

1 Introduction and Problem Statement

Logistics costs have a high impact on a product's final price, and also on the firm's revenue, with an average cost contribution of between 10 and 20%. However, in Latin America, due to its complex geography, infrastructure, and technological gaps, this percentage is higher than in the United States or the European Union, where logistics costs do not exceed 12% of the final product price [23]. For instance, the average ratio between logistics costs and product price in Latin America is 14.7%, while in

© Springer Nature Switzerland AG 2019
C. Paternina-Arboleda and S. Voß (Eds.): ICCL 2019, LNCS 11756, pp. 317–331, 2019.
https://doi.org/10.1007/978-3-030-31140-7_20

Colombia it is 14.9% [4], which is one of the highest values in the region. Among logistics costs, transportation and distribution (37%) and storage (20%) are the costliest. This is a marked tendency in Latin American countries, where storage and transportation costs are almost 60% of total logistics costs. These facts denote an excellent opportunity for improving operational efficiency and reducing costs.

The presence of high logistics costs has become a critical issue for large enterprises, particularly for low-density-value goods, that is, those with low price-weight ratios (i.e. coal, steel, corn, clinker), where supply chain optimization is one of the best ways to reduce the final product cost while increasing the customer service level. A usual and recommended strategy for addressing these shortcomings is to broaden or reconfigure the distribution network, to achieve an improved and quicker response for its retailers, while reducing transportation costs by consolidating cargo to take advantage of economies of scale. However, this strategic decision implies a high capital investment (if new distribution centers need to be installed) and a thorough economic analysis.

In this work, an inventory location model (ILM) is proposed and applied to a laminated steel company in Colombia, allowing for it to optimize its distribution network and analyze whether it should install more distribution centers (DCs) or not. The current distribution network consists of a production plant and a DC located in the city of Barranquilla (northern Colombia) and a small warehouse in Bogotá (center of Colombia), but all retailers are served directly from the plant, missing scale economies, generating higher logistics costs and offering a low service level. The solution approach involves integrating inventory, transportation and fixed costs into the objective function, considering different product categories and demand periods of the company. It was applied to a real-world industrial problem, developing a sensitivity analysis of several variables of the problem. The proposed model optimizes the last two stages of the supply chain, DC locations, and retailer allocation.

The structure of the paper starts with a literature review, followed by the modeling approach and presentation of the case study and results. Finally, we present the relevant conclusions obtained from this work.

2   Literature Review: Inventory-Location Models

The supply chain network optimization problem includes several strategic decisions such as location, number and type of manufacturing factories or DCs; the set of suppliers to select; the choice of transportation modes; the amount of raw materials and products to produce and distribute to the factories, DCs or customers; and the amount of raw materials and products to be held at each inventory location [25]. According to the literature on supply chain management [2, 14, 22] the distribution of supply chain network decisions can be divided into three levels: strategic, tactical, and operational. The first defines the number of facilities and their locations; the second defines the inventory location and inventory availability policies, and the third includes routing, transportation flow and level of inventory at each DC.

In particular, strategic decisions may have long-term effects. These decisions include the location, number and size of each warehouse or manufacturing facility [15].

In most cases, these issues imply a significant amount of capital, hence, it is necessary to make these decisions based on the results of a validated facility location model [10].

Usually, executive directors of a company make strategic decisions, while lower-level managers make tactical and operational decisions. Accordingly, strategic decisions are addressed by specific strategic optimization models that do not include tactical and operational level decisions. This fact might generate incompatibilities and sub-optimality. Tactical and operational decisions are relevant in supply chain management (SCM) for inventory or route planning but are not usually integrated with strategic decisions such as facility locations [9]. Some examples of these decisions are inventory control politics, selection of transportation mode/capacity, facility design and management, and vehicle route planning, among others.

Tactical and operational decisions are often made independently due to their level of complexity which includes different variables at each level to make a joint decision. However, several previous studies integrate both strategic and tactical/operational decisions [3, 5, 6, 11–13, 21, 26].

If the facility location problem (FLP) does not consider capacity constraints at the facilities as in the p-median problem, it is always optimal to allocate each retailer to the closest installed DC [18]. However, due to the safety and cycle stock variables, integrating inventory decisions is not always optimal for closest-facility-allocation. Managing stock inventory consist of two critical tasks: (i) determining the necessary number of DCs, and (ii) calculating the amount of inventory to keep at each DC. Often these tasks are performed separately, resulting in a degree of sub-optimization [3].

The literature concerning inventories tends to focus on finding an optimal resupply strategy for DCs, given a set of spatially distributed facilities. For instance, Daskin et al. [3] concluded that for these types of models, when fixed costs of ordering are reduced in a significant manner, the number of DCs needed increases. Meanwhile, Jamshidi and Esfahani [7] considered the problem of designing a distribution network, transportation planning and allocation model in a supply chain with simultaneous determination of the optimal allocation for supplying and transportation while reaching a minimum cost. Recently, Perez et al. [19] presented a model for designing a supply chain network that involved the relationship between supplier selection, location decisions, and inventory control policy.

Most of the literature considers a supply chain network with a single product, a single stage location, and a single planning horizon. Such simplification does not represent the reality of modern companies, where the increasing variety of products with different sizes, forms, weights or shipping precautions increase logistics times and costs. More advanced models consider the multi-product facility location problem. Also, with the purpose of evaluating the modifications that present themselves to the parameters over time, the multi-period location problem considers a planning horizon divided into several time periods [9]. This study considers both multi-product and multi-period parameters in the model to provide a better approximation to reality.

The major issue with these kinds of approaches is that even those that have a linear programming formulation (basic FLP) result in an NP-hard problem [17]. This issue escalates, even more, when the inventory decisions are included, turning it into a mixed nonlinear programming problem. In response to this, different approaches have multiplied by going from exact methods to heuristics, metaheuristics, and hybrid heuristics.

Over the last couple of years, a new focus on the resolution based on the creation of hybrid heuristics has been developed. Kaya and Urek [8] used three hybrid heuristic models among three widely-used heuristics like tabu search, genetic algorithm and simulated annealing with variable neighbor searches also known as TBVNS, GAVNS, SAVNS, obtaining the best results with the TBVNS approach regarding profitability. Soleimani and Kannan [24] developed a hybrid algorithm between particle swarms and genetic algorithms obtaining much better results than only using genetic algorithms.

This paper proposes a facility location model to optimize the supply chain or distribution network for addressing a real-world industrial case in Colombia, the model integrates several variables addressed in the recent relevant literature. In particular, the model integrates both safety and cycle inventory costs, and multi-period and multi-product decision variables. The problem was solved using both heuristic and meta-heuristic approaches, which makes it different than most of the papers found in literature where models are applied using simulated or non-real data. This case study made it possible to prove consistency between the model and reality, closing the gap found regarding this topic concerning theory and practice.

3 Modeling Approach

This study proposes a mixed integer nonlinear programming model, considering stochastic demands, DC locations, zone demand assignments, a number of planning periods (different demand levels), and different types of products, based on the latest research mentioned in the previous section. The proposed model involves DC operational and fixed costs, transportation costs both between the production plant and the DC, and between the DC and the retailer, and safety and cycle stock costs for each DC, meaning it considers the main variables affecting distribution networks, which differentiates it from other models. The decision variables are the number of locations of proposed DCs and their varieties and the inventory cost of each facility [16].

Integrating both safety and cycle stock in the model may (unlike traditional FLP) generate an optimal solution where retailers may not be assigned to the nearest DC that can serve the demand. Considering the inventory may result in remote allocations due to the reduction of costs by consolidating products into a single DC.

The objective function contains four terms. The first and second terms are the primary FLP costs which are the operational and fixed costs (first term) related to the DC setting, as well as transportation costs (second term) between DCs and customers. The third term refers to safety stock costs and the last to the cycle stock inventory and order costs, both cost terms at the DCs. Inventory costs are not deterministic but are the average values of the distribution (expected values).

The proposed model has five sets of restrictions. Restrictions (1) guarantee that all retailers will be served; while the set (2) prevent the model from assigning a retailer to a non-existing DC. Meanwhile, (3) makes the demand of each DC for each product category and demand period equal to the aggregate requirements of the retailers assigned to the DC so it can supply them. On the other hand, (4) states that the variance of a DC will be equal to the sum of the variances of its retailers. Finally, set (5) guarantees the

binary values by limiting X_i and Y_{ij} to 0 or 1. The main model assumption is that the variance of the demand of each retailer is independent.

Notation:

F_i: Fixed and operational costs of a DC installed in the location i.

N: Number of potential DCs to be installed.

M: Number of retailers.

X_i: Binary variable that has the value of one if a DC is installed in the location i and zero otherwise.

Y_{ij}: Binary variable that has the value of one if retailer j is serviced by DC i and zero otherwise.

C_{ij}: Total transportation cost from DC i to retailer j.

C_{ij}: is given by: $(Tc_{ij} + Pw) \cdot \varphi_j$.

Tc_{ij} is the transportation cost per unit from DC i to retailer j, Pw is the transportation cost from the production plant to the DC and φj is the total demand per year of retailer j. The demand of each DC is the sum of the requests of the retailers allocated to it.

K: Number of product categories.

k: Product category

HC_i^k: Holding cost associated with each product category in location i.

LT_i^k: Lead time to supply DC i, for each product k.

V_{itk}: DC i variance of daily demand, for each product k and each demand period t.

v_j^{tk}: Retailer j variance of daily demand, for each product k and each demand period t.

Q_i: Lot size for location i.

OC_i: Fixed cost per order in location i.

D_i^{tk}: Average daily demand of DC i, for each product k and each demand period t. The demand of each DC is the sum of the demand of the retailers allocated to it.

d_j^{tk}: Average daily demand of retailer j, for each product k and each demand period t.

r_t: Number of days for each demand period.

T: Number of demand periods.

t: Demand period.

The stochastic, multi-commodity and multi-period inventory-location model proposed is:

$$Min \sum_{i=1}^{N} F_i \cdot X_i + \sum_{i=1}^{N} \sum_{j=1}^{M} Y_{ij} \cdot C_{ij} + \sum_{t=1}^{T} \sum_{i=1}^{N} \sum_{k=1}^{K} r_t \cdot HC_i^k \cdot Z \cdot \sqrt{LT_i^k} \cdot \sqrt{V_i^{kt}}$$

$$+ \sum_{t=1}^{T} \sum_{i=1}^{N} r_t \cdot \left(\sum_{k=1}^{K} \left(HC_i^k \cdot \frac{Q_i^{kt}}{2} \right) + OC_i \cdot \frac{\left(\sum_{k=1}^{K} D_i^{kt} \right)}{Q_i^{kt}} \right)$$

$$(1) \ \sum_{i=1}^{N} y_{ij} = 1 \qquad\qquad \forall i = 1, \ldots, M$$

$$(2) \ Y_{ij} \leq X_i \qquad\qquad \forall i = 1, .., N, \ j = 1, .., M$$

$$(3) D_i^{kt} = \sum_{j=1}^{M} Y_{ij} \cdot d_j^{kt} \qquad \forall i, t, k$$

$$(4) \ V_i^{tk} = \sum_{j=1}^{M} Y_{ij} \cdot v_j^{kt} \qquad \forall i, t, k$$

$$(5) \ X_i, Y_{ij} \in \{0, 1\} \qquad \forall i = 1, .., N, \ j = 1, .., M$$

The proposed model is an extension of a classic facility location model, and also of widely-studied inventory location models; therefore, it is also a nonlinear, nonconvex, NP-hard problem. This denotes the high complexity in solving these types of problems, especially with large instances, which is frequent in the real world. The non-linearity, size, and complexity of these kinds of models prohibit researchers from obtaining optimal solutions by exact approaches. Thus, heuristics and metaheuristics are an alternative strategy for solving these problems, providing fast and near-optimal results [11].

Although the model shows inventory levels for the network's warehouse location, these may not be considered the exact levels required for each facility, since the scope of the approach is a strategic decision rather than a tactical one. The products are grouped into categories. Therefore, the resulting levels of the application of the model can be considered as an order of magnitude or as a reference for estimating the costs associated with inventory. To define the necessary inventory for each item, a separate, detailed analysis should be made.

4 Case Study

The model was applied to a laminated steel company in Barranquilla, Colombia. The company has a strategic location for transportation due to its access to the Magdalena River, the Caribbean Sea, and a broad road network that connects Barranquilla with the rest of the country, making it a desirable place for companies to install DCs. Currently the company ships all the products from its production plant, which also serves as DC (DC1) but the transportation is outsourced in an inefficient way. The explanation for this is that truck drivers and transportation companies in the city send their vehicles to the production and plant and get in queue, waiting their turn to be loaded, this means, the company ignores which company or truck will carry their product until the moment the vehicle is getting loaded.

The base scenario had twenty-eight (28) retailers (coded as zones), seven (7) potential DCs, twelve (12) product categories and two (2) different demand periods. The problem was solved using two approaches: a genetic algorithm (GA), and an explicit enumeration procedure, the problem could not be solved using GAMS software due to

its capacity, so both methods were coded in Matlab and solved using an Asus ROG with core i7 processor and 8 Gb RAM, taking 1 min and 10 s to solve it, respectively.

4.1 Data

The company used for the study has more than two thousand retailers located within every department of the country. To reduce complexity, we grouped these retailers into 28 zones per their location, possible routes, and demand characteristics. For the location of potential DC installation sites, we developed a geographical analysis considering network characteristics, accessibility, multi-modal transportation suitability, distance to major retailers, truck restrictions, toll booths, and fixed costs, among others.

The laminated steel company provided transportation fee data for each of its contractor's type of vehicles; retailers demand by type of product; fixed costs of existing DCs; and the desired inventory reliability. Freight charges vary widely, ranging from 8 USD/ton/km for retailers located in cities near DC1 (Northern region of Colombia) to 100 USD/ton/km for retailers located in the Amazon region (Fig. 1) with an average of 40 USD/ton/km for all the zones/retailers. To consider different types of trucks, we established truck mixes from the company's shipping database and defined the truck type used according to the retailer distance from DC1 as shown in Table 1.

Table 1. Truck mixes.

Distance (km)	Truck capacity	
	10 ton	34 ton
<200	90%	10%
200 < D < 500	53%	47%
>500	10%	95%

The company had more than two thousand different products that were grouped into twelve categories: Gutters, Claddings, Steel strappings, Hot Rolled pickled shells, Hot rolled unpickled shells, Galvanized shells, Master 100, Metaldeck, profiles, Coils, Zinc Roofing, Painted Coils.

Holding costs vary due to product price. The range was from 0.002 USD/day to 0.0006 USD/day for the twelve categories created. These values are low because steel is a low-density value good. Finally, fixed costs were obtained from market research in the cities of potential DC locations, focusing on suitable warehouses for each case and its costs, then the square meter cost for each location was calculated.

Figure 1 shows the geographical location of potential DCs and distribution zones in Colombia. Most of the potential DCs are in the central region of the country, where most of the economic activities and the inhabitants of Colombia are. DC1, DC5, and DC6 have ports on the Magdalena River, the largest, most important river in the country, these were included to consider intermodal distribution.

Fig. 1. Potential DC locations and zones.

4.2   Explicit Enumeration

Initially, a reduced explicit enumeration was used to give the model a practical solution approach, as the complete review requires more computational and analysis time, plus intensive database management, considering the combinatorial problem. Given the dimension of the problem, the solution area was limited by discarding unreasonable results to reduce its complexity. For analysis purposes, ten scenarios were considered. The results suggested that the optimal solution was to open two DCs; DC1 and DC5 with a global cost of 8.25 million dollars per year.

The cost of each scenario according to the results are shown in Table 2. The scenario with the lowest total cost is scenario 9 which includes the installation of both DC1 and DC5, in Barranquilla and Puerto Berrio, respectively. This scenario was planned to supply DC5 from DC1 (which is both the production plant and a distribution center) by the Magdalena River and then supply the retailers by road transportation. For the best scenario (Scenario 9) the distribution network would be as is shown in Fig. 2, where the shaded area denotes the retailers served by DC1 and DC5 respectively.

The results show the importance of multimodal transport, allocating all zones/retailers south of DC5 to DC5 and north all the way to DC1. This way, DC1 supplies DC5 by the river while retailers are supplied by road.

In order to analyze the influence of the inventory costs in strategic decision-making, a sensitivity analysis was made, modeling the same scenario with products with a holding cost of 50% of its value.

The results allocate all zones/retailers to DC1, which means only this DC should be installed. This shows that high holding costs imply higher inventory costs and because it is necessary to have a safety stock for each DC, the impact of the high holding cost in the objective function is higher than the savings from economies of scale if several DCs were installed. Therefore, in this case, the best option as the model suggests, is to install only one DC.

Table 2. Results.

Scenario	DC location	Total cost (Million USD/year)
Scenario 1	DC1-DC3	$8.79
Scenario 2	DC1-DC6	$8.35
Scenario 3	DC1-DC2	$8.40
Scenario 4	DC1-DC2-DC3	$9.09
Scenario 5	DC1-DC2-DC4	$9.20
Scenario 6	DC1-DC2-DC3-DC4	$9.45
Scenario 7	DC1-DC7	$9.33
Scenario 8	DC1-DC4	$9.27
Scenario 9	DC1-DC5	$8.25
Scenario 10	DC1-DC2-DC5	$8.40

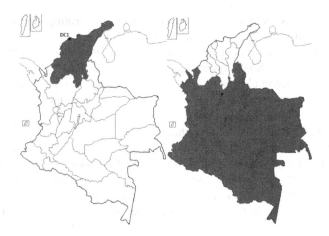

Fig. 2. Allocation results best scenario.

This scenario shows the importance of scale economy in transportation, because in the high holding cost scenario having only one DC, it must supply each retailer directly, unlike the first result where we have two DCs and freight consolidation which results in lower transportation costs. It is evident why the model suggests installing only one DC if the safety stock costs are observed, as these costs would increase for each new DC to be installed.

4.3   Genetic Algorithm

A binary coding was applied, where each gene represents a DC, with the value of 1 if the DC should be installed and 0 otherwise. In this problem, the DC's capacity constraint was relaxed, so it is not necessary to code the retailers inside each chromosome; instead, an internal algorithm allocates each retailer optimally to the DC with the lowest cost.

The parameters of the proposed GA were: (1) Initial population size = 20, (2) Crossover type = roulette (3) Mutation probability = 5%. If $CT(i,j)$ represents the partial total cost associated with the fact that DC i supplies retailer j, we have the following equation:

$$CT^*(j) = \text{ArgMin}_i CT(i,j); \forall j \tag{1}$$

This function searches the DC with the lowest partial cost for each retailer and makes the respective allocation. The chromosome fitness is given by the following equation:

$$Fitness = \frac{1}{\sum_{j=1}^{M} CT^*(j)} \tag{2}$$

Fitness is the basis for determining the algorithm's crossovers, which will lead to the final solution. The methodology of the model starts by creating twenty (20) random solutions, represented as chromosomes, which are composed of chains. The first chain represents the DC results, and the other chains represent the retailers. At the same time, genes form chains, which take binary values (1 or 0).

The problem statement has twenty-nine (29) chains; the first one represents the DCs to be installed and the rest the retailers. The number of zones of the problem defines the number of the retailer's chains.

The initial population of the first chain is generated by filling the chromosomes with binary values (1 or 0). Then, each retailer (each of the 28 chains remaining of each chromosome) is assigned to the DC with the lowest transportation costs between the available DCs. Each chromosome goes through the same process, obtaining the initial population. The next step is to apply the algorithm's operators, which are selection, crossover and mutation. In this methodology, the chromosome with the highest total cost is the one with the lowest probability of being selected for the crossover and the one with a lower cost is the one with the highest probability. This way we guarantee a low-cost solution to the problem.

The final step consists of generating a random number between i positions. This number determines the gene candidate (the position inside the chromosome) for mutation. This algorithm was created with a mutation probability of 5%. To define whether the gene mutates or not, a random number between 0 and 1 is generated. If the number is lower or equal to 0.005, the gene mutates which means the gene changes its value from 0 to 1 or vice versa if the number is higher than 0.005 the gene will not mutate. The procedure is applied to each chromosome.

This process must obtain a population of 20 chromosomes in each generation. Once all three operators are applied, a new generation with new chromosomes is obtained. The algorithm used in this process creates 40 generations. In the last one, the chromosome with the higher fitness value is chosen and showed as the optimal solution of the problem, indicating which DCs should be installed, the retailer allocation and the costs associated with this solution.

5   Results and Discussion

A sensitivity analysis was made by modifying the key parameters such as holding cost (HC) and transportation costs. We created three possible scenarios. The first is the original problem, in which we applied the algorithm to model the real-world industrial case of the company, with the data provided by the managers and all its supply chain characteristics. In the second, we wanted to prove the impact of ignoring inventory policies in facility locations. In this case, we used the real data but applied the algorithm without safety and cycle inventory. The last scenario, with lower transportation costs and lower holding costs, evaluates the sensitivity that the holding and transportation costs have on facility locations.

For the original problem, the algorithm suggests installing two DCs, DC1 and DC6, supplying DC6 by river directly from the production plant and then supplying retailers by road from there. This result is associated with the use of intermodal transportation. (shaded areas denote the retailers served by each DC). The associated costs of this solution are shown in Table 3 and the retailer's allocation is shown in Fig. 3.

Table 3. GA results (USD/year).

Fixed costs	$1,001,819
Inventory costs	$293,935
Transport costs	$6,917,487
Total costs	$8.213

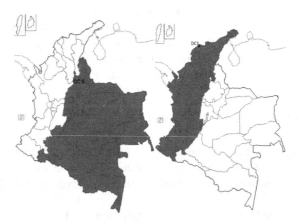

Fig. 3. Retailers' allocation in scenario 1

In the second scenario, ignoring inventory costs, the algorithm suggests installing two DCs, DC1 and DC5. Table 4 presents the associated costs of this result. In this scenario DC1 supplies all the retailers located on the northern coast of the country and DC5 to the rest of the country. In this situation, DC5 receives the products shipped from the production plant by the Magdalena River and distributes them by truck.

Table 4. Results ignoring inventory costs (USD)

Fixed costs	$1,001,819
Inventory costs	–
Transportation costs	$6,786,683
Total costs	$7,788,502

In the last analyzed scenario, we reduced the transportation costs of waterways by 70% and holding costs by 50%; this is a valid scenario given the current project to restore the Magdalena river's navigability. The result suggested installing three DCs (DC1, DC5, and DC6). The associated costs of this configuration are shown in Table 5. DC5 and DC6 are both cities with river ports, basically these results suggest having a main DC (DC1) and serve retailers located in the northern coast from it and ship products through the river to DC 5 and 6 and serve the rest of the country from them.

Table 5. Results for the third scenario

Fixed costs	$1.364.405
Inventory costs	$796.059
Transport costs	$5.466.963
Total costs	$7.627.427

The result of the explicit enumeration methodology suggests installing two DCs (DC1-DC5) with a global cost of 8.2 million dollars, while the GA suggests installing DC1 and DC6 with a global cost of 8.2 million dollars for the same problem. This difference could be a result of the fact that the first methodology is initially based on prefixed matrixes created by the researcher's criteria. It substantially reduces the computational time, but it lowers the algorithm effectiveness.

6   Conclusions

This paper develops and applies an inventory location model to optimize a real distribution network of a laminated steel company located in Colombia. The model considers different manufacturing facilities or factories, inventory policies at DCs for various product categories with stochastic demands, in a multi-period scenario. Unlike some previous research, in which authors apply their models to unreal or simulated data, this approach is applied to a real-world industrial case to study the model's effectiveness and validate its usefulness.

The results show the impact on each logistics cost within overall company performance and both underlying and network configuration. In this case, the transportation costs become the most relevant component for strategic-tactical network configuration decision-making. Nevertheless, building new DCs results in substantial savings when the company consolidates shipment but will increase other costs such as

fixed and inventory costs. It should be clarified that this would happen if they have low-density value goods, like steel. Inventory costs depend directly on the value of the product. On the contrary, in the case of high-density value goods, the inventory costs have a larger value and may turn into the most important cost for facility location decisions. In these cases, it is better to install less DCs than when we are dealing with low-density value goods.

The problem was solved by applying two strategies: explicit enumeration and a genetic algorithm, both methods proved to be valid to solve problems with this size. Therefore, the use of one or the other depends on the researcher or consultant's needs, if there is an important amount of obvious allocation, then explicit enumeration may be used. On the other hand, if there are few, or no apparent allocations and the programming complexity is not relevant, the GA gives a nearly optimum solution.

Different scenarios were evaluated, where some of these consider intermodal transportation processes, by using Colombia's most important waterway, the Magdalena River, complementing the road transportation system. The intermodal network reduces transportation fees by achieving economies of scale, due to the higher capacity of cargo ships compared to trucks, even considering that river transportation requires a higher travel time and therefore higher floating inventory costs, the overall costs of this intermodal network are lower, improving the company's competitiveness, which is an important reason to favor projects that improve the river's navigability.

A sensitivity analysis was performed regarding the holding cost where it shows the importance of making joint inventory-location decisions. When comparing scenarios, the impact of ignoring inventory costs could not be significantly observed due to the low holding costs of laminated steel. However, in the scenario with high holding costs, the model suggests installing only one DC. This means that products with high holding costs or high-density value (e.g. computers, jewelry) should consolidate inventory in as few DCs as possible. For low holding cost products, the inventory cost is not relevant directly, but it has an indirect impact because the quantity of the inventory defines the size of the facility, which factors into fixed costs. Therefore, under these circumstances, it is better to analyze if the fixed costs generated by maintaining multiple facilities are paid back with savings in transportation costs obtained by the freight consolidation, considering an increased service level that they offer to the customers by reducing times of delivery. Naturally, this last concept is hard to quantify. There is still a néed for reviewing each case and companies should utilize a methodology as the analytic hierarchy process [20] or another multi-criteria approach, to study the importance of this service level for the enterprise, to make the correct decisions, such as in [1].

This research is developed under the assumption that the variances between zones were independent. Future research may investigate on how spatial correlation may influence the results. The proposed approach may be enhanced by considering joint production plants and DC location decisions as an extension of the inventory location model.

References

1. Alberto, P.: The logistics of industrial location decisions: an application of the analytic hierarchy process methodology. Int. J. Logist. Res. Appl. **3**(3), 273–289 (2000)
2. Berman, O., Krass, D., Tajbakhsh, M.M.: A coordinated location-inventory model. Eur. J. Oper. Res. **217**(3), 500–508 (2012)
3. Daskin, M.S., Coullard, C.R., Shen, Z.J.M.: An inventory-location model: Formulation, solution algorithm and computational results. Ann. Oper. Res. **110**(14), 83–106 (2002)
4. DNP, Encuesta nacional de logística "Colombia es logística." http://www.colombia-competitiva.gov.co/prensa/2015/Paginas/Colombia-es-Logistica-La-Encuesta-Nacional-de-Logistica-2015.aspx. Accessed 29 July 2019
5. Escalona, P., Ordóñez, F., Marianov, V.: Joint location-inventory problem with differentiated service levels using critical level policy. Transp. Res. Part E Logist. Transp. Rev. **83**, 141–157 (2015)
6. Guerrero, W.J., Prodhon, C., Velasco, N., Amaya, C.A.: Hybrid heuristic for the inventory location-routing problem with deterministic demand. Int. J. Prod. Econ. **146**(1), 359–370 (2013)
7. Jamshidi, R., Esfahani, M.M.S.: A novel hybrid method for supply chain optimization with capacity constraint and shipping option. Int. J. Adv. Manuf. Technol. **67**(5), 1563–1575 (2013)
8. Kaya, O., Urek, B.: A mixed integer nonlinear programming model and heuristic solutions for location, inventory and pricing decisions in a closed loop supply chain. Comput. Oper. Res. **65**, 93–103 (2016)
9. Melo, M.T., Nickel, S., Saldanha-da-Gama, F.: Facility location and supply chain management – a review. Eur. J. Oper. Res. **196**(2), 401–412 (2009)
10. Melo, M.T., Nickel, S., Saldanha-da-Gama, F.: An efficient heuristic approach for a multi-period logistics network redesign problem. TOP **22**(1), 80–108 (2014)
11. Miranda, P.A., Garrido, R.A.: Incorporating inventory control decisions into a strategic distribution network design model with stochastic demand. Transp. Res. Part E Logist. Transp. Rev. **40**(3), 183–207 (2004)
12. Miranda, P.A., Garrido, R.A.: A simultaneous inventory control and facility location model with stochastic capacity constraints. Netw. Spat. Econ. **6**(1), 39–53 (2006)
13. Miranda, P.A., Garrido, R.A., Ceroni, J.A.: e-Work based collaborative optimization approach for strategic logistic network design problem. Comput. Ind. Eng. **57**(1), 3–13 (2009)
14. Mourits, M., Evers, J.J.M.: Distribution network design: An integrated planning support framework. Int. J. Phys. Distrib. Logist. Manag. **25**(5), 43–57 (1995)
15. Olivares-Benitez, E., González-Velarde, J.L., Ríos-Mercado, R.Z.: A supply chain design problem with facility location and bi-objective transportation choices. TOP **20**(3), 729–753 (2012)
16. Orozco-Fontalvo, M., Cantillo, V., Miranda, P.: A meta-heuristic approach to a strategic mixed inventory-location model: Formulation and application. Transp. Res. Procedia **25**, 729–746 (2017)
17. Owen, S.H., Daskin, M.S.: Strategic facility location: a review. Eur. J. Oper. Res. **111**(3), 423–447 (1998)
18. Ozsen, L., Daskin, M.S., Coullard, C.R.: Facility location modeling and inventory management with multisourcing. Transp. Sci. **43**(4), 455–472 (2009)

19. Perez Loaiza, R.E., Olivares-Benitez, E., Miranda Gonzalez, P.A., Guerrero Campanur, A., Martinez Flores, J.L.: Supply chain network design with efficiency, location, and inventory policy using a multiobjective evolutionary algorithm. Trans. Oper. Res. **24**(1–2), 251–275 (2017)
20. Saaty, T.L.: How to make a decision: The analytic hierarchy process. Eur. J. Oper. Res. **48**(1), 9–26 (1990)
21. Shen, Z.-J.M., Coullard, C., Daskin, M.S.: A joint location-inventory model. Transp. Sci. **37**(1), 40–55 (2003)
22. Simchi-Levi, D., Kaminsky, P., Simchi-Levi, E.: Designing and managing the supply chain: concepts, strategies, and case studies, 2nd edn. McGraw-Hill/Irwin, Boston, Mass (2003)
23. Sintec: Transporte, el verdadero reto en latinoamerica y Colombia. http://www.il-latam.com/wp-content/uploads/2018/08/infografia-transporte-Latam-Colombia.pdf. Accessed 29 July 2019
24. Soleimani, H., Kannan, G.: A hybrid particle swarm optimization and genetic algorithm for closed-loop supply chain network design in large-scale networks. Appl. Math. Model. **39**(14), 3990–4012 (2015)
25. Vidal, C.J., Goetschalckx, M.: Strategic production-distribution models: A critical review with emphasis on global supply chain models. Eur. J. Oper. Res. **98**(1), 1–18 (1997)
26. Zhang, Y., Qi, M., Miao, L., Liu, E.: Hybrid metaheuristic solutions to inventory location routing problem. Transp. Res. Part E Logist. Transp. Rev. **70**, 305–323 (2014)

Integrated Production Lot Size and Distribution Planning with Shared Warehouses

Julia Pahl[(✉)]

SDU Engineering Operations Management,
Department of Technology and Innovation, University of Southern Denmark,
Campusvej 55, 5320 Odense, Denmark
julp@iti.sdu.dk

Abstract. Collaboration in supply chains (SCs) is steadily improving especially due to advanced information systems that provide transparency over resource allocation, utilization, and dynamic demands. We are examining the case of shared inventory spaces in SCs where utilization of storage spaces depend on all SC partners. Decisions that need to be made are when to produce, how much of multiple products, and where and how long to store them in order to fulfill known customer demands. Therefore, we take production, inventory control and placement, as well as distribution (logistics) decisions into consideration. Special emphasis is made on setup considerations for detailed production sequencing. We analyze this integrated approach that leads to reduced overall SC costs and extends the literature to the view of shared resources that become increasingly prominent with the developments of Industry 4.0. solutions.

1 Introduction

Tactical production and lot sizing planning including decisions on inventory holding has been one of the classical research areas in operations research since its early days. The same is valid for transportation and logistics problems. The emergence of supply chain management (SCM) in the nineties has urged research and practitioners to consider them together as this can lead to better, i.e., more cost efficient solutions for all SC partners than solving these problems in isolation; see, e.g., Boudia and Prins (2009); Dai and Tseng (2011). In fact, it has been demonstrated that integrated production, inventory, and distribution decision making leads to increased effectiveness, efficiency, and improved resource usage (see Seyedhosseini and Ghoreyshi (2014)). The latter includes less idle times for production equipment that can lead to better return on investment (ROI), but also reduced resource usage regarding (raw) materials and other input factors that might affect the environment, e.g., energy, emissions, etc.; see for instance a discussion and analysis of freight transportation networks employing horizontal collaboration and pooling strategies provided by Sarraj et al. (2013).

© Springer Nature Switzerland AG 2019
C. Paternina-Arboleda and S. Voß (Eds.): ICCL 2019, LNCS 11756, pp. 332–349, 2019.
https://doi.org/10.1007/978-3-030-31140-7_21

As sharing resources increases asset utilization and can decrease idle times and related (opportunity) costs (see Abhishek and Zhang (2016); Darvisch et al. (2016) and the references therein), strategies to streamline material and information flows in SCs while reducing costs focus on inventory holding and warehouse management as well as production operations. Well known concepts like vendor-managed inventory (VMI) are specifically useful in this context as they base on information sharing in between SC partners and collaborative planning approaches (see also Darwish and Odah (2010); Hariga et al. (2014); Mateen et al. (2015); Pandey et al. (2007); Qiu et al. (2019)).

The trend of digitization with the Internet-of-Things, blockchain technology, Industry 4.0 - just to name a few - that permits easy, safe, and secure transmission of information regarding availability of resources in real-time has led to the shared (sharing) economy (see, e.g., Huckle et al. (2016)). Applications like AirBnB or Uber are well known examples in the digital economy (Huckle et al. (2016)) that sneak steadily into (urban) lives. Less well known application possibilities arise in the business to business (B2B) context where companies exchange information and share resources using distributed applications or "Dapps" (Huckle et al. (2016)). These applications use sensor technology and network connectivity to check and facilitate the shared use of vehicles, buildings, warehouses, machines etc. (Huckle et al. (2016)) in a peer-to-peer (P2P) mode. The trend to outsource logistical service to third party logistic companies has increased (Akbari (2018)) and we find different practices in the literature to share spaces. For instance, one way is to rent from developers of shared units of large building warehouse spaces based on square foot, cube, or pallet basis (see, e.g., Klein (2016)) as Amazon does with its "on demand" logistic services. Another way is to pool SC partners that invest, build, and run a warehouse together, so that costs and risks are shared (Klein (2016)). Using new emerging technologies permit private owners to offer their idle warehouse spaces in an "AirBnB" fashion where digital platforms provide an overview of the availability and prices (see PYMNTS (2017)). Nevertheless, barriers of collaboration exist as warehousing logistics is traditionally a personal business where trust the merchandise is stored safely is paramount (see PYMNTS (2017)). However, online platforms with review systems that create transparency like AirBnB can help amend concerns and dispel doubts.

In this paper, we propose an integrated production planning and scheduling inventory distribution problem including shared warehouse spaces. The combination of shared warehouse spaces together with more detailed production operations is a new aspect in production and distribution planning (PDP). It is a first step to develop models and solution algorithms for distributed applications (dapps) for sharing company (privately) owned warehouse spaces with SC members that takes into consideration geographical location opportunities and distances to production and customer sites as well as utilization-dependent pricing which is an interesting new aspect requested by companies and enabled by emerging information and communication technology (ICT) trends enveloped by digitization.

2   Literature Review

The literature on integrating production and inventory routing and distribution decisions has evolved in recent years in the operations research community (see, e.g., Fahimnia et al. (2013); Qiu et al. (2019); Rezaeian et al. (2016); Seyedhosseini and Ghoreyshi (2014)). This is also due to the possibilities given by increased computational power to develop and solve more complex models for SCM and planning. The interested reader is referred to Chen (2004, 2010); Fahimnia et al. (2013) for reviews on the PDP. However, the consideration of shared resources has not equally grown. We find some work that consider sharing resources in production and inventory holding in different perspectives (see Bashiri et al. (2012); Yaghin (2018)), but more research is needed. The literature confirms that it is beneficial to consider production planning and distribution decisions in an integrated manner (see, e.g., Boudia and Prins (2009); Seyedhosseini and Ghoreyshi (2014)).

2.1   Strategic Considerations

Production planning, inventory control including setup and lot size decisions, as well as distribution logistics are addressed on a tactical time level where the network of SC stages is assumed to be fixed (see also Raa et al. (2013)) which is also our focus of this paper. Nevertheless, the incorporation of strategic decisions that regard the overall production-supply network can lead to further enhancements of material and information flows through joint planning. For instance, Bashiri et al. (2012) focus on the network design problem of the production-distribution problem combining the tactical decision level with the strategic decisions on how to expand the network. They further consider limits on capacity as well as utilization rates of facility operations and take into account the option to rent space at a public warehouse. In the network design problem, own warehouses may be opened at a fixed investment setup cost rate while at public warehouses, only the unit holding cost factor accrues.

2.2   Integrated Production and Distribution Planning Approaches

Dai and Tseng (2011) study the combined production and logistics system planning with a special focus on the integration of logistics decisions including, a.o., labor costs, shipping lines, and economies of scale of shipment distance into production lot size decisions for better production control. They formulate a continuous time model for a manufacturer-distribution-center-retailer SC for a single product on a tactical time horizon implying that the logistics configuration is given. The retailer symbolizes a centralized warehouse that holds the aggregate demand information of other retailers. Economies of scale occur for the transportation costs which is influenced by the shipment size and the shipping distance: if more items share a fixed transportation cost, the average costs per item decreases if the shipment size increases until the maximum shipment size is reached. The authors also assume a concave freight rate per unit meaning

that the freight rate increment decreases non-proportional to the increase of the distance, i.e., the transportation cost increase per unit per distance decreases with the increase of the distance. Akbalik and Penz (2011) consider the replenishment side of production and lot size planning in the uncapacitated case where fixed ordering costs occur plus an extra fixed cost per batch ordered. They show that the problem is NP-hard and that it can be transformed, so that it becomes a special case of a lot sizing problem with time-varying batch sizes (see also Akbalik and Rapine (2018)).

Raa et al. (2013) consider the inter-dependencies of multiple production plants from a highly investment cost-intensive production resource which, in the specific case studied, is a mould for injection moulding. By sharing moulds across different production plants, equipment costs can be significantly reduced as moulds are very expensive compared to the value of the products. On the other hand the interdependency highly complicates the planning as the moulds need to be exchanged in between the plants in a time period. They allow for multiple exchanges during a macro time period and show that it is beneficial to share the resource over all plants which also gives a better resource utilization ratio.

Li and Meissner (2010) analyze the optimal investment in capacity. Their model bases on the single item capacitated lot sizing problem (CLSP). Besides inventory holding costs, they consider fixed and variable production costs. Setup costs occur every time production is issues (see also Boudia and Prins (2009)). Due to the single item case, there are no setup changeovers from one product type to another. They also abstract from setup times. The capacity must be decided at the beginning of the considered time horizon and is fixed for the entire planning horizon. It must satisfy the demand while excess capacity is "disposed" (not used) without additional costs.

The capacity adjustment problem is further analyzed by Ou and Feng (2019) for the single item case. Sambasivan and Yahya (2005) consider a CLSP-based lot sizing model with multiple plants with inter-plant transfers that is developed for a real case of a manufacturing company producing steel rolled products. They provide a Lagrangean-based solution approach including a "lot shifting-splitting" routine that provides good solutions to this NP-hard problem (see also Bitran and Yanasse (1982)).

The case of limited lifetimes of items is studied by Qiu et al. (2019); Rezaeian et al. (2016); Seyedhosseini and Ghoreyshi (2014). For instance, Seyedhosseini and Ghoreyshi (2014) propose a model for the PDP for a single perishable product. The objective is to minimize overall costs composed by setup costs, variable production costs, inventory holding costs, and fixed transportation costs for a vehicle in use which are attributed to the number of trips in the planning horizon. Additionally, Rezaeian et al. (2016) incorporate routing decisions of multiple vehicles into a PDP for items with limited lifetimes. A production and inventory routing problem (PRP) for a perishable item is proposed by Qiu et al. (2019) which leads us to further aspects of PDP considered in the literature.

2.3 Further Aspects: Inventory Routing, Marketing Strategies and Uncertainty

There is a vast amount of work that deals with and proposes mathematical models for integrated PDP and PRP. A recent review on the inventory routing problem (IRP) is given by Adulyasak et al. (2015).

The IRP pertains to the distribution planning problem that considers the distribution of products from a central location to multiple geographically dispersed customers by use of a number of definite or indefinite capacitated vehicles (see also Seyedhosseini and Ghoreyshi (2014)). The distribution and routing plan thus needs to respond to the questions when to deliver which products to which customers in which time periods using the vehicles at hand. In this regard, Park et al. (2016) study a combined VMI-IRP model and Neves-Moreira et al. (2019) a large scale problems with delivery time windows. A network flow formulation for the multi-scale routing problem is proposed by Zhang et al. (2017) considering detailed production scheduling.

The consideration of combined production and marketing plans for the tactical planning level has been proposed by Yaghin (2018) for a two-level SC containing a manufacturer and a retailer dealing with multiple items in a dynamic time period setting. The objective is to maximize profits where the retailer is allowed to set different prices during the planning horizon. The model also considers minimal and maximal limitations on warehouse spaces and inventory capacities as well as other resources, e.g., work forces. The resulting model is a non-linear non-convex programming formulation which is tackled by an analytical algorithm based on the convex relaxation of the original problem formulation and solved using standard optimization solvers. Details of the production process are taken into consideration, e.g., machine rates and availability, but setup decisions for changing over from one product to another are neglected.

2.4 Incorporating Simultaneous Lot Sizing and Scheduling

The consideration of positive setup times that can be sequence dependent (SD) has often been neglected in optimization models most probably to reduce complexity and to make them applicable in real-world scenarios in case that they are rather negligible. Nevertheless, in case SD-setup times and costs are significant and have a major impact.

The literature on lot sizing that considers more detailed scheduling information extends the CLSP to consider more detailed setup settings. Models in this regard are, e.g., the discrete lot sizing and scheduling problem (DLSP), the continuous setup and lot sizing problem (CSLP), and the proportional lot sizing and scheduling problem (PLSP) that inherit the assumptions of the CLSP, but treat the time horizons differently in that they split up macro periods in micro periods; see also Pahl and Voß (2010) for a study on these models with limited lifetimes. Models splitting macro periods further into micro periods also referred to as "small bucket" can consider start-up, switch-off, and/or change-over times and costs, and more detailed information about the shop floor, so that feasible

plans can be determined. The considered information can be interpreted as a meta-level in between the tactical and operational level, as detailed scheduling including due dates, machine utilization, makespan, etc. are not accounted for. These hybrid models allow the production of multiple products per time period while preserving setup states across periods by carrying over setup states to one or more periods and, consequently, providing sequencing information of (ordered) items. Such models are the subject of extensive research; for more details see Babaei et al. (2013); Fadaki and Bijari (2016); Karimi-Nasab and Seyedhoseini (2013); Poursabzi et al. (2017); Ramezanian et al. (2012); Rohaninejad et al. (2014); Seeanner (2013). For an excellent review on scheduling with lot-sizing and/or sequence-(in)dependent setup times and costs see Copil et al. (2017).

The CLSP with linked lot sizes and sequence-dependent setup costs and times (CLSD) with zero setup times is proposed and studied by Haase (1996) allowing for continuous lot sizes and preservation of setup states over idle times. His paper provides a local search heuristic including priority rules. A branch-and-bound method with rescheduling possibilities for the CLSD with positive setup times is proposed by Haase and Kimms (2000) and tested using a practical example.

The literature on integrating production, inventory, and distribution planning generally considers generic setup costs that accrue when production is issued, but approaches mostly abstract from detailed lot sizing decisions and thus from machine setups and scheduling. In this paper, we fill this gap and analyze the effects of including details of production operations in a production-distribution framework with shared warehouse capacities.

3   Model Formulation

Our model is a combination of the CLSD and the PDP integrating production, lot sizing, and setup changeover decisions with distribution logistics planning with shared warehouses.

We use the nomenclature given in Table 1 throughout the paper where indices of parameters and variables can vary, but used nomenclature remains consistent.

The material flow is depicted in Fig. 1. We consider multiple manufacturing plants p that can be of the same company/corporation or of distinct companies that delivery their products after assembly to shared warehouses. In Fig. 1, we show two shared warehouses to ease the presentation, but we allow multiple ones in our model.

Shared warehouses have overall inventory space capacity limits IK_w measured in units of products. Inventory holding costs can be different for plants in warehouses to allow for different warehouse customer price schemes, but they do not vary over the regarded time horizon. Demand for a product of a certain plant occurs at the customer/retailer who orders at the warehouse. We abstract from the retailer level in our model assuming direct orders to the warehouse by the customer, but this can be easily added. Plants have to make their decision regarding when to produce which item. For items, SD-setup times and costs as well as and production times occur that are limited by production capacities at

Table 1. Variables, parameters and indices used in the basic models

Sets and Indices	
j	Products with indices $j, i, k = 1, \ldots, J$
p	Plants with $p = 1, \ldots, P$
w	Warehouses with $w = 1, \ldots, W$
c	Customers with $c = 1, \ldots, C$
t	macro periods with $t = 1, \ldots, T$
Parameters	
a_{jp}	Process time per unit of product j of plant p
D_{jpct}	Demand of product j at plant p and customer c in period t
h_{jw}	Inventory holding cost factor for product j in warehouse w
IK_{wt}	Available inventory holding space capacity at warehouse w measured in time units for each time period t
pc_{jp}	Production costs of plant p per unit of product j
PK_{pt}	Available production capacity at plant p measured in time units for each time period t
st_{jip}	Setup times for a changeover from product j to i of plant p
sc_{jip}	Sequence-dependent setup costs for a changeover from product j to i at plant p
tcp_{pw}	Transportation cost from plant p to warehouse w
tcc_{wc}	Transportation cost from warehouse w to customer c
TK_{wt}	Available transportation capacity at warehouse w measured in time units for each time period t
$Y0$	Initial setup state
Variables	
$I_{jwt} = 0$	Inventory holding variable of product j in warehouse w in period t
$f_{jt} \geq 0$	Position of product j in period t
U_{jwpt}	Volume of j transported from plan p to warehouse w in period t
$V_{jwct} \geq 0$	Volume of j transported from warehouse w to customer c in period t
$x_{jpt} \geq 0$	Production amount of product i of plant p in period t
y_{jpt} int	Setup state variable for item j in period t
z_{jipt} int	SD-setup variable, setting up from product j to product i in period t

the plants for each time period. Produced items are transported to the shared warehouses. No storage at plants can take place. At the shared warehouses, items of all plants are stored together, i.e., they are not conflicting in terms of included materials or hygiene requirements like some medical equipment. All stored items can serve customer demands independent of their origin, i.e., specific plants. Items that are used to serve customer orders are distributed to the customers where transportation costs occur that depend on the distance of the customer from the respective distributing warehouse(s). At this stage, we do not take into account fleets of vehicles with restricted transportation space and thus routing considerations.

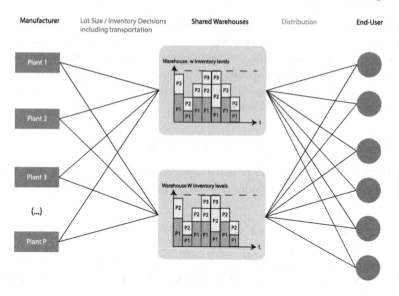

Fig. 1. Material Flow of the Shared Warehouses: Model with multiple plants and customers and two shared warehouses

Regarding the production process, we base our formulation on the CLSD that allows for the preservation of setup states over idle time periods, so that setup times and costs are not overstated which would be the case, e.g., in the CLSP where each time a product is produced, a setup needs to take place even if the machine previously has produced that specific item. The CLSD further includes the following assumptions:

- Production takes place on a single machine restricted in its capacity that produce multiple items
- The planning horizon contains T discrete time periods
- The orders to produce items are given in the beginning of the time period
- Products are delivered in the beginning of the macro period in lots
- Initial and ending inventory of the planning horizon are assumed to be zero
- The initial setup state, i.e., for which item the machine is set up, is given
- The triangle inequalities must be fulfilled

The triangle inequalities derive from graph theory and state that it is not more expensive and time consuming to setup directly from a product j to a product i than setting up from product j to product i via product k independently of the time period, thus

$$z_{jit} \leq z_{jkt} + z_{kit} \qquad \forall\, j, i, k, t \qquad (1)$$
$$sc_{jip} \leq sc_{jkp} + sc_{kip} \qquad \forall\, j, i, k \qquad (2)$$

where z_{jit} are the SD-setup variables and sc_{ji} denote the SD-setup costs.

Additionally, production managers at the plants have to decide how much to put on stock at the warehouses where available capacities also depend upon the decision of the other plants. Depending on the demand at the customers and on available transportation capacities, items can be passed to the customer via the warehouses handling capacities in the same time period t or stored at the warehouses w in case of bottlenecks.

3.1   PDP with Shared Warehouses with SD-Setups

The PDP has been subject to extensive research (see, e.g., Raa et al. (2013)). We extend the basic PDP to consider SD-setup times and costs with the following modifications:

- The inventory holding cost factor at the warehouse w is item-dependent denoted by h_{jw}
- Maximum capacities for production, storage at the warehouses, as well as for handling/transportation to and from plants-warehouses-customers, are time varying
- Ending inventory at the warehouses is assumed to be zero

The complete production lot size scheduling and distribution planning (PLSSDP) with shared warehouses is stated as follows:

$$\min \sum_{j=1}^{J}\sum_{t=1}^{T}\sum_{p=1}^{P}\left[\left(\text{pc}_{jp}\cdot x_{jpt}+\text{sc}_{jip}\cdot z_{jipt}\right)+\sum_{w=1}^{W}\text{tcp}_{pw}\cdot U_{jpwt}\right] \quad (3)$$
$$+\sum_{j=1}^{J}\sum_{t=1}^{T}\sum_{w=1}^{W}\left(h_{jw}\cdot I_{jwt}+\sum_{c=1}^{C}\text{tcc}_{wc}\cdot V_{jwct}\right)$$

subject to:

$$\sum_{j=1}^{J}a_{jp}\cdot x_{jpt}+\sum_{i=1}^{J}\text{st}_{jip}\cdot z_{jipt}\le \text{PK}_{pt} \qquad \forall\, p,t \quad (4)$$

$$a_{jp}\cdot x_{jpt}\le \text{PK}_{pt}\cdot\left(\sum_{i=1}^{J}z_{ijpt}+y_{jp,t-1}\right) \qquad \forall\, j,p,t \quad (5)$$

$$\sum_{j=1}^{J}y_{jpt}=1 \qquad \forall\, p,t \quad (6)$$

$$\sum_{j=1}^{J}(z_{jkpt}+y_{kp,t-1})=\sum_{i=1}^{J}(z_{kipt}+y_{kpt}) \qquad \forall\, k,p,t \quad (7)$$

$$f_{jpt}-f_{ipt}+J\cdot z_{jipt}\le J-1 \qquad \forall\, j,i,t \quad (8)$$

$$x_{jpt} = \sum_{w=1}^{W} U_{jpwt} \qquad \forall\, j,p,t \qquad (9)$$

$$\sum_{p=1}^{P} U_{jpwt} + I_{jw,t-1} = \sum_{c=1}^{C} V_{jwct} + I_{jwt} \qquad \forall\, j,w,t \qquad (10)$$

$$\sum_{w=1}^{W} V_{jwct} = D_{jct} \qquad \forall\, j,c,t \qquad (11)$$

$$\sum_{j=1}^{J}\sum_{p=1}^{P} U_{jpwt} + \sum_{j=1}^{J}\sum_{c=1}^{C} V_{jwct} \le TK_{wt} \qquad \forall\, w,t \qquad (12)$$

$$\sum_{j=1}^{J} I_{jwt} \le IK_{wt} \qquad \forall\, w,t \qquad (13)$$

$$I_{jw0} = I_{jwT} = 0 \qquad \forall\, j,w \qquad (14)$$

$$y_{jp,0} = Y0_{jp} \qquad \forall\, j,p \qquad (15)$$

$$x_{jpt}, z_{jt}, U_{jpwt}, I_{jwt}, V_{jwct}, f_{jpt} \ge 0 \qquad \forall\, j,t,p,w,c \qquad (16)$$

$$z_{jipt}, y_{jpt} = \text{int} \qquad \forall\, j,i,p,t \qquad (17)$$

The objective function given in (3) minimizes the overall SC costs concerning production, SD-setups, transportation from plants to warehouses, inventory holding at warehouses, as well as transportation costs to the customers.

Production at plant p is composed by production and setup times that is restricted by available capacities at the plant in each time period t given in Constraint (4). SD-setups are included by Constraints (5)–(8) where Constraint (5) binds production from above by requiring that production for an item j in time period t can only occur if a setup operation from any product i to product j is executed in time period t or if the setup state for product j is carried over from the previous period $t-1$ to period t. Equation (6) require that the production resource at plant p at the end of period t and, consequently, at the beginning of the next period $t+1$ is set up for a product j. The setup flow condition is given in Eq. (7). The subtour elimination is given in Inequalities (8) where f_{jt} presents some kind of "position" of the product in the production sequence, e.g., the higher the number of the position, the later is the product scheduled for production. Equalities (9) require that all produced units of product j at plant p in time period t are shipped to warehouses w, so that the sum of volumes transported from plant p in period t equals all produced units. Equalities (10) give the flow balance of the warehouses stating that the sum of volumes produced at plants p of items j in time period t plus inventory holding of warehouses w at the previous time period $t-1$ must equal the sum of volumes distributed to customers c regarding items j at warehouse w plus inventory holding at warehouse w of item j in time period t. Equality (11) state that demand of customer orders must be fulfilled by volumes transported from the warehouses. Overall transportation is restricted by available transportation capacity as stated in Constraint

(12). The same is valid for inventory holding space capacitites at warehouses as stated in Constraint (13). Equalities (14) give standard assumptions inherited from the CLSP that initial and ending inventory must be zero. We are aware that this is not realistic and leads to increased production and thus setups at the beginning of the regarded time horizon $t = 1$ as well as less production and inventory holding at the end of the time horizon T. These effects can be eased by assuming positive initial and ending inventories calculated, e.g., by using the economic order quantity (EOQ) formula for overall cost minimal inventory holding based on assumed/predicted future demands and in combination with service level agreements implying safety stocks. Another option is to regard the pipeline inventory holding levels. Pipeline stock is a kind of safety stock that is build because of the difference in time between an order for stock replenishment until it is ready for use in sales or production. This equals the expected demand over the lead time which, in turn, equals the expected demand per day times the length of the lead time measured in, e.g., days (see Muckstadt and Sapra (2010)). Equation (15) sets the initial setup state of the production resource to one specific product j in time period $t = 0$ meaning that a machine has an initial setup state that can be interpreted as the last setup state from previous time horizons, e.g., a previous shift, and defined by parameter $Y0$. The last two restrictions (16) and (17) define and restrict the decision variables.

4   Numerical Study and Discussion

We test our model for correctness with a small numerical example that highlights some insights. We further state the resulting overall SC costs for the cases without detailed information on setups as well as with the consideration of them to highlight the benefit of including them.

Regarding complexity considerations, our model is similar to the proposed model of Raa et al. (2013) that further links setup decisions for plants together as they depend on using the same equipment and thus need to be passed from one plant to another. The insertion of binary decision variables to reflect the setup decisions greatly complicates the solution process. The same is valid for our model where we link setup decisions from macro periods together. As argued in Raa et al. (2013), the basic model without SD setups is already NP-hard and so is its extension. As a result, linear programming (LP) relaxations or heuristic solution approaches need to be developed in order to solve practical problem instances in acceptable time.

For the small example, we consider four products $J = 4$ that are produced at two plants $p = 2$ and stored alternatively in two warehouses $w = 2$ within a planning horizon of $T = 10$ macro periods. Orders of demands occur from three customers $c = 3$. The production capacities of the plants are assumed to be 100 units of time per macro period. Inventory holding capacities as well as transportation capacities are assumed to be 1.000 units for each plant and warehouses at each macro period, so that they do not constitute a bottleneck. The inventory holding costs for each item j in plant $p = 1$ are assumed to be

equal with $h_{j1} = 1$ and for plant $p = 2$, $h_{j1} = 2$. Similarly, we assume that production costs and times are the same for each product in each plant, i.e., $pc_{jp} = 1$ and $a_{jp} = 1$. Table 2 states further cost factors of the model for the case of simple setups viz. non-SD setup times and costs.

Table 2. Setup cost factors for the small numerical example

Products	st_{jp}		sc_{jp}	
	Plant 1	Plant 2	Plant 1	Plant 2
A	5	2	3	3
B	10	4	11	9
C	20	5	21	18
D	30	6	31	30

Demand data for all three customers are stated in Tables 4, 5 and 6 given in Appendix A. The transportation costs factors from to warehouses are as follows: from plant 1 to warehouse 1 it is one monetary units and to warehouse 2 five monetary units occur. From plant 2 to warehouse 1, we have two monetary units and to warehouse 2 three monetary units. For the transportation costs from the warehouses to the customers, we have from warehouse 1 to the three customers the values $1, 3, 15$ while from warehouse 2, costs of $14, 4, 2$ accrue for each transportation.

The small numerical example can be solved on a standard computer in few seconds. It is supposed to present the improvement of the SC costs by considering more detailed operations within the plants. For instance, the PDP with simple setups needs 45 in total while the PLSSDP model gives 20 due to linking subsequent setups in macro periods. Inventory levels for the PDP are in total 204 units while the PLSSDP needs an overall of 20 inventoried items. Thus, the PLSSDP achieves less overall SC costs including detailed production operations as in the case of generic setups in the PDP case.

Table 3. Objective function results for the cost terms

Costs	Model with simple setups	Model with SD setups
Production	1.335	1.335
Setup	591	367
Inventory holding	**217**	**20**
Transportation PW	2.931	2.402
Transportation WC	2.244	2.258
TOTAL	**7.318**	**6.382**

Accordingly, as we can see from Table 3, setup costs decrease. More interestingly, overall inventory holding and related costs significantly decrease in the integrated approach with detailed operations consideration and shared warehouses. This is due to the consideration of linked lot sizes with setup carry-overs that avoid planning setups if a production resource is already set up for a specific item and thus succeed in providing more efficient plans.

5 Conclusions

We proposed a model for the joint PDP with shared warehouses and more detailed production operations planning. We believe that the trend of sharing resources shown already in applications for P2P mode within the overall digitization will become more and more used in the B2B context where sharing production equipment and inventory spaces will significantly increase. Besides advances in ICT that are already under its way in the form of digital distributed applications, this requires planning approaches for production and logistics to extend to the overall SC and use opportunities for sharing resources. Merging plans for production, scheduling, inventory holding, and distribution planning has been confirmed by literature and in practice to be beneficial for all planers in the SC. Nevertheless, planning becomes more complicating with more details to consider. The evolution of ICT and computational power permits more and more to take into account details of SC planning thus avoiding isolated planning and related suboptimal and infeasible plans. We proposed a PDP with shared warehouses including details of production planning and scheduling.

This paper presents a first step in this direction. Future work involves a number of different facets and exciting components, such as, e.g., taking into account the smoothing of utilization of capacities of production and warehousing by, e.g., dynamically setting warehouse prices or grouping plants and customers into priority groups thus applying more marketing considerations into the SC model as we can witness by business to consumer (B2C) applications. Moreover, we aim at considering customer demand specific to certain plants that pertain to different companies, but share warehouses, where overall demand cannot be served from the stock in the warehouses.

An issue that might pass as a detail in theory, but which has a large effect in practice is the already mentioned assumption of zero initial and ending inventory levels which is not realistic and leads to distorted plans with overestimated production operations in the beginning and underestimated production operations in the end of the time horizon. Presumably because these assumptions are practical for developing algorithms for efficiently solving large scale practical instances, they are very often integrated in optimization models for production and inventory planning to the expense of realism of related models. This can be eliminated by assuming realistic inventory levels that can either be specific to the next forecasted planning time horizon taking into account average values determined by the EOQ formula or by estimating pipeline inventory stock levels and the like.

Other interesting aspects that can be taken into consideration are companies' own warehouse spaces versus rented warehouse under certain conditions, e.g., when spaces at the own warehouse are less than expected demands, so that inventory needs to be "outsourced" or when better conditions accrue at the rented warehouse, e.g., reduced prices for holding inventory due to learning curves of the personnel or better temperature conditions in case that considered products are food items that are sensible to warehouse conditions. These aspects are highlighted by target companies that may serve as a test bed for extending the presented model to more realistic settings as well as developing heuristic algorithms for solving large scale practical instances.

A   Appendix

The basic PDP with setup costs and times is given in the following; see also Raa et al. (2013) for a version without setups.

$$
\min \sum_{j=1}^{J} \sum_{t=1}^{T} \sum_{p=1}^{P} \left[\text{pc}_{jp} \cdot x_{jpt} + \text{sc}_{jp} \cdot y_{jpt} + \sum_{w=1}^{W} \text{tcp}_{pw} \cdot U_{jpwt} \right] \tag{18}
$$
$$
+ \sum_{j=1}^{J} \sum_{t=1}^{T} \sum_{w=1}^{W} \left[h_{jw} \cdot I_{jwt} + \sum_{c=1}^{C} \text{tcc}_{wc} \cdot V_{jwct} \right]
$$

subject to:

$$
\sum_{j=1}^{J} (a_{jp} \cdot x_{jpt} + st_{jp} \cdot y_{jpt}) \leq PK_{pt} \qquad \forall\, p, t \tag{19}
$$

$$
a_{jp} \cdot x_{jpt} \leq B \cdot y_{jpt} \qquad \forall\, j, p, t \tag{20}
$$

$$
x_{jpt} = \sum_{w=1}^{W} U_{jpwt} \qquad \forall\, j, p, t \tag{21}
$$

$$
\sum_{p=1}^{P} U_{jpwt} + I_{jw,t-1} = \sum_{c=1}^{C} V_{jwct} + I_{jwt} \qquad \forall\, j, w, t \tag{22}
$$

$$
\sum_{w=1}^{W} V_{jwct} = D_{jct} \qquad \forall\, j, t, c \tag{23}
$$

$$
\sum_{j=1}^{J} \sum_{p=1}^{P} U_{jpwt} + \sum_{j=1}^{J} \sum_{c=1}^{C} V_{jwct} \leq HK_{wt} \qquad \forall\, w, t \tag{24}
$$

$$
\sum_{j=1}^{J} I_{jwt} \leq IK_{wt} \qquad \forall\, w, t \tag{25}
$$

$$
I_{jw0} = 0 \qquad \forall\, j, w \tag{26}
$$

$$
x_{jpt}, U_{jpwt}, I_{jwt}, V_{jwct} \geq 0 \qquad \forall\, j, t, p, w, c \tag{27}
$$

Table 4. Demand Data for the small numerical example for Customer 1

Products	Demand in Periods									
	1	2	3	4	5	6	7	8	9	10
A	30	4	28	29	7	35	37	28	13	26
B	6	2	31	22	11	26	23	27	40	6
C	28	8	32	6	19	33	40	17	8	8
D	21	38	0	36	16	40	18	9	13	25

Table 5. Demand Data for the small numerical example for Customer 2

Products	Demand in Periods									
	1	2	3	4	5	6	7	8	9	10
A	11	4	9	11	5	4	3	7	11	5
B	7	13	1	8	11	0	1	14	3	6
C	1	15	10	10	2	8	4	3	11	7
D	15	1	0	15	3	4	8	10	11	6

Table 6. Demand Data for the small numerical example for Customer 3

Products	Demand in Periods									
	1	2	3	4	5	6	7	8	9	10
A	3	9	4	9	2	5	8	4	5	10
B	3	2	2	5	0	6	4	8	10	9
C	5	8	5	8	8	2	3	5	4	7
D	6	5	7	8	9	7	0	1	3	2

References

Abhishek, V., Zhang, Z.: Business models in the sharing economy: manufacturing durable goods in the presence of peer-to-peer rental markets. SSRN Electron. J. (2016). https://doi.org/10.2139/ssrn.2891908

Adulyasak, Y., Cordeau, J.-F., Jans, R.: The production routing problem: a review of formulations and solution algorithms. Comput. Oper. Res. **55**, 141–152 (2015). https://doi.org/10.1016/j.cor.2014.01.011

Akbalik, A., Penz, B.: Comparison of just-in-time and time window delivery policies for a single-item capacitated lot sizing problem. Int. J. Prod. Res. **49**(9), 2567–2585 (2011). https://doi.org/10.1080/00207543.2010.532921

Akbalik, A., Rapine, C.: Lot sizing problem with multi-mode replenishment and batch delivery. Omega **81**, 123–133 (2018). https://doi.org/10.1016/j.omega.2017.10.005

Akbari, M.: Logistics outsourcing: a structured literature review. Benchmarking Int. J. **25**(5), 1548–1580 (2018). https://doi.org/10.1108/bij-04-2017-0066

Babaei, M., Mohammadi, M., Fatemi Ghomi, S.M.T.: A genetic algorithm for the simultaneous lot sizing and scheduling problem in capacitated flow shop with complex setups and backlogging. Int. J. Adv. Manuf. Technol. **70**(1–4), 125–134 (2013). https://doi.org/10.1007/s00170-013-5252-y

Bashiri, M., Badri, H., Talebi, J.: A new approach to tactical and strategic planning in production–distribution networks. Appl. Math. Model. **36**(4), 1703–1717 (2012). https://doi.org/10.1016/j.apm.2011.09.018

Bitran, G.R., Yanasse, H.H.: Computational complexity of the capacitated lot size problem. Manage. Sci. **28**(10), 1174–1186 (1982). https://doi.org/10.1287/mnsc.28.10.1174

Boudia, M., Prins, C.: A memetic algorithm with dynamic population management for an integrated production-distribution problem. Eur. J. Oper. Res. **195**(3), 703–715 (2009). https://doi.org/10.1016/j.ejor.2007.07.034

Chen, Z.-L.: Integrated production and distribution operations. In: Simchi-Levi, D., Wu, S.D., Shen, Z.J. (eds.) Int. Ser. Oper. Res. Manag. Sci., pp. 711–745. Springer, Boston (2004). https://doi.org/10.1007/978-1-4020-7953-5_17

Chen, Z.-L.: Integrated production and outbound distribution scheduling: review and extensions. Oper. Res. **58**(1), 130–148 (2010). https://doi.org/10.1287/opre.1080.0688

Copil, K., Wörbelauer, M., Meyr, H., Tempelmeier, H.: Simultaneous lotsizing and scheduling problems: a classification and review of models. OR Spectrum **39**(1), 1–64 (2017). https://doi.org/10.1007/s00291-015-0429-4

Dai, H., Tseng, M.M.: Determination of production lot size and DC location in manufacturer DC retailer supply chains. Int. J. Logistics Syst. Manag. **8**(3), 284 (2011). https://doi.org/10.1504/ijlsm.2011.038988

Darvisch, M., Lorrain, H., Coelho, L.C.: A dynamic multi plant lot-sizing and distribution problem. Working document CIRRELT-2016-04, Faculte des sciences de l'administration de l'universite Laval (2016)

Darwish, M.A., Odah, O.M.: Vendor managed inventory model for single-vendor multi-retailer supply chains. Eur. J. Oper. Res. **204**(3), 473–484 (2010). https://doi.org/10.1016/j.ejor.2009.11.023

Saleh Fadaki, M., Bijari, M.: Simultaneous lot-sizing and scheduling problem in flow shop environment with outsourcing. Int. J. Plann. Sched. **2**(3), 252 (2016). https://doi.org/10.1504/ijps.2016.080345

Fahimnia, B., Zanjirani Farahani, R., Marian, R., Luong, L.: A review and critique on integrated production-distribution planning models and techniques. J. Manuf. Syst. **32**(1), 1–19 (2013). https://doi.org/10.1016/j.jmsy.2012.07.005

Haase, K.: Capacitated lot-sizing with sequence dependent setup costs. Oper. Res. Spektrum **18**(1), 51–59 (1996). https://doi.org/10.1007/BF01539882. ISSN 1436–6304

Haase, K., Kimms, A.: Lot-sizing and scheduling with sequence-dependent setup costs and times and efficient rescheduling opportunities. Eur. J. Oper. Res. **66**, 159–169 (2000). https://doi.org/10.1016/S0925-5273(99)00119-X

Hariga, M., Gumus, M., Daghfous, A.: Storage constrained vendor managed inventory models with unequal shipment frequencies. Omega **48**, 94–106 (2014). https://doi.org/10.1016/j.omega.2013.11.003

Huckle, S., Bhattacharya, R., White, M., Beloff, N.: Internet of things, blockchain and shared economy applications. Procedia Comput. Sci. **98**, 461–466 (2016). https://doi.org/10.1016/j.procs.2016.09.074

Karimi-Nasab, M., Seyedhoseini, S.M.: Multi-level lot sizing and job shop scheduling with compressible process times: a cutting plane approach. Eur. J. Oper. Res. **231**(3), 598–616 (2013). https://doi.org/10.1016/j.ejor.2013.06.021

Klein, J.E.. Wat exactly is space sharing? Internet Source, November 2016. https://blog.sior.com/what-exactly-is-space-sharing

Li, H., Meissner, J.: Capacitated dynamic lot sizing with capacity acquisition. Working paper, Lancaster University Management School (2010)

Mateen, A., Kumar Chatterjee, A., Mitra, S.: VMI for single-vendor multi-retailer supply chains under stochastic demand. Comput. Ind. Eng. **79**, 95–102 (2015). https://doi.org/10.1016/j.cie.2014.10.028

Muckstadt, J.A., Sapra, A.: Principles of Inventory Management. Springer, New York (2010). https://doi.org/10.1007/978-0-387-68948-7

Neves-Moreira, F., Almada-Lobo, B., Cordeau, J.-F., Guimarães, L., Jans, R.: Solving a large multi-product production-routing problem with delivery time windows. Omega **86**, 154–172 (2019). https://doi.org/10.1016/j.omega.2018.07.006

Ou, J., Feng, J.: Production lot-sizing with dynamic capacity adjustment. Eur. J. Oper. Res. **272**(1), 261–269 (2019). https://doi.org/10.1016/j.ejor.2018.06.030

Pahl, J., Voß, S.: Discrete lot-sizing and scheduling including deterioration and perishability constraints. In: Dangelmaier, W., Blecken, A., Delius, R., Klöpfer, S. (eds.) IHNS 2010. LNBIP, vol. 46, pp. 345–357. Springer, Heidelberg (2010). https://doi.org/10.1007/978-3-642-12494-5_31

Pandey, A., Masin, M., Prabhu, V.: Adaptive logistic controller for integrated design of distributed supply chains. J. Manuf. Syst. **26**(2), 108–115 (2007). https://doi.org/10.1016/j.jmsy.2007.11.001

Park, Y.-B., Yoo, J.-S., Park, H.-S.: A genetic algorithm for the vendor-managed inventory routing problem with lost sales. Expert Syst. Appl. **53**, 149–159 (2016). https://doi.org/10.1016/j.eswa.2016.01.041

Poursabzi, O., Mohammadi, M., Naderi, B.: An improved model and a heuristic for capacitated lot sizing and scheduling in job shop problems. Scientia Iranica (2017). https://doi.org/10.24200/sci.2017.20016

PYMNTS. Meet "the airbnb of warehouse space". Internet Source, September 2017. https://www.pymnts.com/matchmakers/2017/warehouse-space-matches-customers-with-unused-storage-spaces/

Qiu, Y., Qiao, J., Pardalos, P.M.: Optimal production, replenishment, delivery, routing and inventory management policies for products with perishable inventory. Omega **82**, 193–204 (2019). https://doi.org/10.1016/j.omega.2018.01.006

Raa, B., Dullaert, W., Aghezzaf, E.-H.: A matheuristic for aggregate production-distribution planning with mould sharing. Int. J. Prod. Econ. **145**(1), 29–37 (2013). https://doi.org/10.1016/j.ijpe.2013.01.006

Ramezanian, R., Saidi-Mehrabad, M., Teimoury, E.: A mathematical model for integrating lot-sizing and scheduling problem in capacitated flow shop environments. Int. J. Adv. Manuf. Technol. **66**(1–4), 347–361 (2012). https://doi.org/10.1007/s00170-012-4329-3

Rezaeian, J., Shokoufi, K., Haghayegh, S., Mahdavi, I.: Designing an integrated production/distribution and inventory planning model of fixed-life perishable products. J. Optim. Ind. Eng. **9**(19) (2016). https://doi.org/10.22094/joie.2016.229

Rohaninejad, M., Kheirkhah, A., Fattahi, P.: Simultaneous lot-sizing and scheduling in flexible job shop problems. In. J. Adv. Manuf. Technol. **78**(1–4), 1–18 (2014). https://doi.org/10.1007/s00170-014-6598-5

Sambasivan, M., Yahya, S.: A lagrangean-based heuristic for multi-plant, multi-item, multi-period capacitated lot-sizing problems with inter-plant transfers. Comput. Oper. Res. **32**(3), 537–555 (2005). https://doi.org/10.1016/j.cor.2003.08.002

Sarraj, R., Ballot, E., Pan, S., Hakimi, D., Montreuil, B.: Interconnected logistic networks and protocols: simulation-based efficiency assessment. Int. J. Prod. Res. **52**(11) (2013). https://doi.org/10.1080/00207543.2013.865853

Seeanner, F.: Multi-Stage Simultaneous Lot-Sizing and Scheduling. Springer Fachmedien Wiesbaden (2013). https://doi.org/10.1007/978-3-658-02089-7

Seyedhosseini, S.M., Ghoreyshi, S.M.: An integrated model for production and distribution planning of perishable products with inventory and routing considerations. Math. Probl. Eng. **1–10**, 2014 (2014). https://doi.org/10.1155/2014/475606

Ghasemy-Yaghin, R.: Integrated multi-site aggregate production-pricing planning in a two-echelon supply chain with multiple demand classes. Appl. Math. Model. **53**, 276–295 (2018). https://doi.org/10.1016/j.apm.2017.09.006

Zhang, Q., Sundaramoorthy, A., Grossmann, I.E., Pinto, J.M.: Multiscale production routing in multicommodity supply chains with complex production facilities. Comput. Oper. Res. **79**, 207–222 (2017). https://doi.org/10.1016/j.cor.2016.11.001

Optimization of the p-Hub Median Problem via Artificial Immune Systems

Stephanie Alvarez Fernandez[1]($\boxtimes$) (ID), Gabriel Lins e Nobrega[2] (ID),
and Daniel G. Silva[1] (ID)

[1] Department of Electrical Engineering, University of Brasília, Brasilia, Brazil
stephaniemaf@gmail.com, danielgs@ene.unb.br
[2] Exact Sciences Institute, University of Brasília, Brasilia, Brazil
gabriel.lins97@gmail.com

Abstract. Recent advances in logistics, transportation and in telecommunications offer great opportunities to citizens and organizations in a globally-connected world, but they also arise a vast number of complex challenges that decision makers must face. In this context, a popular optimization problem with practical applications to the design of hub-and-spoke networks is analyzed: the Uncapacitated Single Allocation p-Hub Median Problem (USApHMP) where a fixed number of hubs have unlimited capacity, each non-hub node is allocated to a single hub and the number of hubs is known in advance. An immune inspired metaheuristic is proposed to solve the problem in deterministic scenarios. In order to show its efficiency, a series of computational tests are carried out using small and large size instances from the Australian Post dataset with node sizes up to 200. The results contribute to a deeper understanding of the effectiveness of the employed metaheuristic for solving the USApHMP in small and large networks.

Keywords: Artificial immune systems · p-Hub Median Problem · Metaheuristics · CLONALG

1 Introduction

The design of hub-and-spoke networks addresses decisions about where to allocate or install hubs (facilities), considering customers' or users' demands that must be served, in order to optimize one or more criteria. The "hub" term is used to refer to schools, factories, warehouses, telecommunication antennas, etc., while "customers" refer to neighborhoods, sales units, students, etc. Localization problems are combinatorial by nature, since they consist of selecting from a discrete, finite set of data the best subset that satisfies certain criteria. Many are highly complex and costly from a computational point of view.

There are many variants of the hub localization problem (HLP) [25]: the p-hub median problem, the p-hub center problem, the capacitated/uncapacitated hub location problem, and the hub covering problem. Moreover, HLP may

C. Paternina-Arboleda and S. Voß (Eds.): ICCL 2019, LNCS 11756, pp. 350–362, 2019.
https://doi.org/10.1007/978-3-030-31140-7_22

be classified by the way in which the requested points are assigned or allocated to hubs. In this sense, they can assume one of the two allocation schemes: *(i)* single allocation scheme, where each node must be assigned to exactly one hub node (*i.e.*, all flows from/to each node go only via an assigned hub); and *(ii)* multiple allocation scheme, where nodes are allowed to communicate with more than one hub. Furthermore, different constraints may arise, including capacity restrictions on the traffic volume a hub can concentrate. A straightforward version of this problem is the uncapacitated single allocation p-hub median problem (USApHMP), which assumes that the capacity of each hub is virtually unlimited or, at least, far beyond the expected demand. In these types of configurations, the hubs serve as connection points between two installations, allowing to replace a large amount of direct connections between all pairs of nodes, hence, minimizing the total transportation cost of the network.

Recently, the application of heuristic methods to HLP has received considerable attention by many researchers. On one hand, metaheuristics based on General Variable Neighborhood Search [15], GRASP [26], Scatter Search [4,21], Iterated Local Search [3], among others [16,29,31] have been proposed. Furthermore, recent successful approaches using exact methods are [22,23]. On the other hand, bio-inspired methodologies such as GA have been also proposed in order to solve other variants of the problem such as the uncapacitated single allocation hub location problem, in which the number of hubs is not given in advance [13], or the uncapacitated multiple allocation p-hub median problem, in which one node can be allocated to more than one hub [24]. Furthermore, the application of Artificial Immune Systems (AIS) methodology such as Clonal Selection for solving capacitated HLP and p-hub median problems have been applied [19,20].

Particularly, AIS have been widely adopted for optimization purposes. One can find different approaches of AIS algorithms for solving problems such as optimization of distributed generation in distribution systems [7], optimization of material handling system [18], constrained optimization problems [33], dynamic constrained optimization [34], reverse logistics [10], design problems of energy suply networks [32], distributed wireless networks [28], distribution scheduling problem [27] and many others. The reader is referred to [6] for a recent survey of the application of AIS in optimization.

Regarding the application of bio-inspired metaheuristics for specifically solving the USApHMP, just a few articles were found in the literature. In [17], the authors solved this problem by using two distinct GA approaches. They presented solutions with the Civil Aeronautic Board (CAB) and Australian Post (AP) datasets. Likewise, in [14], the authors considered an approach consisting on the application of Clonal Selection for solving the same problem, however, there is no experimental results to support this outlook.

By relying on the aforementioned explanations, the CLONALG method is proposed to solve the USApHMP. In contrast with a standard GA and other bio-inspired proposals which have been applied to solve other variants of the problem, immune inspired algorithms such as CLONALG present an intrinsic capacity of maintaining diversity among the candidate solutions during the execution, which

can be decisive to increase the probability that the global optimum or a good local optimum can be reached. Although it is known that there is no particular global search strategy with superior performance for a wide class of problems, these arguments support the application of such metaheuristic in the context of USApHMP.

The rest of the paper is organized as follows: Sect. 2 describes the mathematical model of the USApHMP. The proposed methodology is presented in Sect. 3. The experiments and results are provided in Sect. 4. Finally, concluding remarks on this work are provided in Sect. 5.

2   Formal Problem Description

The USApHMP consists of choosing p nodes from a given network of n nodes as hubs, while allocating the remaining $(n-p)$ nodes to them, in order to minimize the total transportation cost over the network. As mentioned before, this problem differs from other HLPs because the number of p hubs is given beforehand. We use the integer programming formulation given in [1,17] which, in turn, is based on the very first model proposed by O'kelly [25] and has been extensively applied for solving small to large instances of the problem. Accordingly, the following notation and formulation is given.

Notation:

- $N = \{1, 2, \cdots, n-1, n\}$: set of n distinct nodes in the network. Each node refers to origin/destination (O/D) points or potential hub location;
- C_{ij}: cost per unit flow from each origin node i to destination node j;
- W_{ij}: amount of flow from i to j;
- $X_{ik} = \begin{cases} 1, & \text{if non-hub node } i \text{ is allocated to a hub node } k \\ 0, & \text{otherwise} \end{cases}$
- $X_{kk} = 1$: implies that the node k is a hub;
- $\mathcal{X}$: cost for collecting flow from the origin non-hub node to its hub;
- τ: cost of transferring the collected flow between the interconnected hubs;
- δ: cost of distributing the flow from the hub to the destination node;

Given the aforementioned notation the problem is formulated as follows:

$$\text{Minimize:} \quad \sum_{i,j,k,l \in N} W_{ij} (\mathcal{X} C_{ik} X_{ik} + \tau C_{kl} X_{ik} X_{jl} + \delta C_{jl} X_{jl}) \tag{1a}$$

$$\text{Subject to:} \quad \sum_{k=1}^{n} X_{kk} = p, \tag{1b}$$

$$\sum_{k=1}^{n} X_{ik} = 1, \forall\, i \in N \tag{1c}$$

$$X_{ik} \leq X_{kk} \quad \forall\, i \in N \tag{1d}$$

$$X_{ik} \in \{0,1\} \quad \forall\, i \in N \tag{1e}$$

Equation 1b gives the exact number of hubs to be selected from the node network. Equation 1c is the single allocation constraint, ensuring that the flow from any origin non-hub node i is sent through one and only one hub node to which the node i is allocated. Equation 1d guarantees that a non-hub node is able to be allocated to only one hub, thus avoiding direct movement between O/D nodes.

3   Methodology

In this section, primarily based on [1], the key principles of AIS and CLONALG method are derived. AIS are a class of algorithms inspired by the mammalian's immune system. In the context of engineering and computing, the application of the immunological principles of decentralization and diversity maintenance are often useful in solving problems with large solutions spaces, such as the USApHMP.

The immune system characteristics have received attention from the research community in order to design new intelligent frameworks. Properties such as: automatic recognition of antigen's characteristics, pattern recognition and memorization capabilities, self-organizing memory, adaptation ability, immune response adaptation, learning from examples, distributed and parallel data processing, multilayer structure and generalization capability [11] have been reproduced for solving optimization problems, like the one of this work, as well as identification of non-linear systems [2, 30] and machine learning related problems [5].

In the context of USApHMP, the cells and their antibodies are analogous to a candidate solution for the optimization problem posed by 1a. Thus, it is possible to immunologically simulate such problem, seeking solutions effectively. With this in mind, an immune-inspired metaheuristic is proposed for solving the USApHMP: the CLONALG algorithm, a classical technique that emulates the clonal selection principle for optimization tasks.

3.1   CLONALG Algorithm

Clonal selection algorithms are inspired by the clonal selection theory, which basically addresses the way the immune system responds to infectious agents. It is result of the work accomplished by [8], which in turn, served as inspiration for CLONALG [9], a popular AIS that emulates the cloning and hypermutation process, key principles of the original theory. The improvement of the candidate solutions pool, in CLONALG, follows the steps of all clonal selection-based algorithm: clone, mutate, select and replace.

The clonal selection theory argues that, when an antigen penetrates the host body, a selection of lymphocytes capable of recognizing it takes place. Lymphocytes with higher antigen reactivity would be selected for clone expansion over those with lower neutralizing capability. Each clone should have a unique antigen receiver, $i.e.$ it should not be present in any other. The idea is that only cells capable of properly recognizing antigens should survive and generate offspring, $i.e.$ the process promotes a maturation of the immune system response to the infection.

Algorithm 1. CLONALG(nodes, pHubs, β, ρ, Ninitial, nC, b)

Ab $\leftarrow$ random($Ninitial$)
while *Stop condition is not met* **do**
    Solve $fit \leftarrow$ affinity(Ab)
    $C \leftarrow$ clone(Ab, nC, β)
    $C^* \leftarrow$ mutate(C, fit, ρ)
    $Fit' \leftarrow$ affinity(C^*)
    $R \leftarrow$ select(C^*, Fit')
    Ab $\leftarrow$ replace(R, random(b))
end
return Ab

Based on the clonal selection principle, the CLONALG has been proposed as a method to be applied both in optimization, as presented in this work, and in machine learning. CLONALG is largely referenced by the study presented in [9], in which the sensitivity to its parameters and its performance in a set of test problems were analyzed. Its pseudocode for solving the USApHMP, based on the description presented in [1], can be found in Algorithm 1. The input parameters are: (i) the number of nodes as well as the number of hubs given by the problem; (ii) the clonal factor β; (iii) parameter ρ which controls the shape of the mutation rate; (iv) the size $Ninitial$ of the antibody[1] pool Ab; (v) the number of clones nC; and (vi) the parameter b for selecting the number of new antibodies that will replace the lowest affinity ones from Ab.

The procedure starts generating a pool of antibodies with fixed size $Ninitial$. The authors of [9] proposed that the generation of those antibodies occurs randomly for a greater diversity within the population. In the case of the USApHMP, every member of Ab represents a solution for the problem, *i.e.* an allocation matrix X. First, the given number of p hubs are randomly chosen following a uniform probability distribution, then, the remaining nodes are allocated to their nearest hubs. Next, every antibody in Ab is evaluated by the affinity/fitness function

$$f^{Ag}(Ab_i), \tag{2}$$

where Ag represents the antigens, and the fitness is the cost function given by Eq. 1a. This way, the Ab solutions with lower fitness also have lower cost, therefore, are the best ones. Following, the amount of clones nC to be generated for each individual is calculated as defined in Eq. 3:

$$nC = \text{round}(\beta \cdot Ninitial), \tag{3}$$

where β is the clonal factor. Then, the affinity maturation process is applied at an α rate proportional to the affinity obtained from Eq. 2, as defined in the following equation:

---

[1] Unlike in nature, there is no distinction for the terms antibodies/lymphocytes/cells in the context of AISs.

Algorithm 2. mutate(Ab, pHubs)

---

while *Stop condition is not met* **do**
    select mode *pHubs* or *nodes*
    if *pHubs* **then**
        oldHub ← select one hub randomly
        newHub ← select one node randomly that is not a hub
        replace *oldHub* with *newHub*
        disconnect all nodes
        reconnect all nodes to closest hub
    else
        node ← select one node randomly
        disconnect *node* from its corresponding hub
        connect *node* to a new randomly selected hub
    end
end
return Ab*

$$\alpha = \rho \cdot fit, \tag{4}$$

with ρ being the normalized affinity in $[0, 1]$ of the correspondent antibody. Differently from the proposal in [9], originally designed for function maximization, note that the mutation rate is proportional to their parent's affinity, *i.e.*, the greater the affinity (the USApHMP cost), the greater the mutation intensity.

Afterwards, the affinity of every clone is calculated and it is selected the best one among the clones and their parents to be kept. Finally, the main loop is concluded where the b worse antibodies are replaced with new randomly generated individuals. The process repeats itself until the stopping criteria is met.

The CLONALG mutation operator is a crucial aspect of the method that directly affects the diversity of the candidate solutions and the exploration of the search space. Recall that an antibody represents an allocation matrix of the hubs and nodes, *i.e.* the matrix X in Eq. 1a, then Algorithm 2 presents the pseudo-code for this operator. The procedure selects between *pHubs* or *node* mode based on the mutation rate α. If the procedure *pHubs* is selected, one of the Ab hubs is chosen in order to be replaced with another node which becomes a hub, such node is randomly selected. Then, it reconnects all nodes to their current closest hub. If the *node* procedure is selected, instead of randomly choosing a hub, a node is selected and its current hub is changed and replaced from among Ab hubs possibilities. By choosing either option, one can expect that candidate solutions are able to escape out of local minima or explore local regions.

Once the set of clones have been generated, each clone is then mutated and an affinity-based selection among the clones and their parent ensures that the solutions generated after the clone step have, on average, higher affinities than those of the early primary response. It is worth stressing that the mutation operator follows the restrictions imposed by the problem formulation, seen in Eq. 1a.

In general, the performance of AIS can be affected by variations on its parameters. CLONALG, particularly, have crucial parameters that control the algorithm exploration capacity: (i) the antibody population size; (ii) the number of clones; (iii) the remainder replacement size; and (iv) the stopping criterion. Understanding the effects of each parameter is crucial for a proper understanding of the algorithm as well as for its refinement. The impact that causes changing the setting of each parameter is summarized in Table 1, as presented in [1].

Table 1. CLONALG parameters

Operator	Function
Antibody population size, $Ninitial$	If excessive, an overloaded number of redundant antibodies can be generated
Number of clones, nC	Excessive number of clones may lead to redundancy in the search process, while a few number of clones may cause ineffective results in the search for a favorable mutation
Remainder replacement size, b	An excessive addition of antibodies reduces the convergence speed but provides a higher search space exploration; on the other hand, adding few antibodies may impair the search for a global optimum
Stopping criterion, $maxIt$	A great number of iterations impacts the execution time. On the other hand, few iterations may lead to poor performance of the algorithm

4 Experiments and Results

In this section, some numerical experiments are considered to check the performance of the proposed CLONALG algorithm to solve the USApHMP. For the sake of examination, the AP dataset is used, its details are described in [12]. All computational experiments were performed on an Intel® i7 7700HQ v4 at 2.8 GHz running the Arch Linux 4.19.4 operating system. The method has been implemented as C application.

As previously mentioned in Sect. 3, the CLONALG parameters adjustment is crucial for a proper performance of the method. Hence, they were defined via a preliminary parameter-selection process, which comprised 10 trials of the algorithm for each possible configuration of $Ninitial \in \{1, 3, 5, 7, 9\}$, $nC \in \{5, 10, 20, 30\}$, $b \in \{0.1, 0.3, 0.5, 0.7, 1\}$ and $maxIt \in \{300, 500, 700, 1000, 1500\}$. In order to minimize runtime and the final cost for each instance of the problem, the trials were executed with two n, p representative scenarios of USApHMP. For instances with "Small" size, those with n up to 50, trials were executed with $n = 10$ and $p = 5$. For instances with "Large" size, those with n ranging from 100 to 200, trial configuration was such that $n = 100$ and $p = 10$. Parameters that showed promising results regarding execution time and cost were chosen and are displayed in Table 2. The algorithm stops if it global solution (when previously known) for that particular instance, if the best solution in the population

remains without changes through 50 iterations for small instances and 200 for large ones, or if $maxIt$ iterations is achieved.

Table 2. CLONALG parameters

Parameter	Small	Large
Ninitial	3	7
nC	5	20
b	1	1
maxIt	300	1000

After setting each parameter values, we have run the algorithm 20 times, in order to collect the information regarding its performance for solving the different USApHMP instances. The quality of each solution is evaluated as a percentage gap between the solution found by CLONALG and the solutions presented in [3] and [17] using

$$Gap\,(Cost, Cost_{BKS}) = 100 \times \left(\frac{Cost - Cost_{BKS}}{Cost_{BKS}} \right). \tag{5}$$

In addition, runtime information has been used to compare the computational burden of the aforementioned methods.

Following, results obtained for the set of instances are summarized in Table 3. The first two columns, "n" and "p" indicate the number of nodes and hubs, respectively. The following two columns "$Cost(1)$" and "$Time(s)$" give the best solutions obtained by BRILS, the metaheuristic proposed by the authors in [3] and which, to the best of our knowledge, provides the best average runtime found in literature regarding the solution of the USApHMP using the AP dataset. The optimal solutions are highlighted in bold, notice that the optimal solutions reported in [3] match with the optimal solutions found by CPLEX solver. The next two columns denoted by "$Cost(2)$" and "$Time(s)$" present the best solution and time reported by the GA method in [17]. This classic technique was chosen for comparison because it is, like CLONALG method, a bio-inspired, population-based, non-deterministic algorithm. Following, the next two columns denoted by "$Cost(3)$" and "$Time(s)$" present the best solution and time obtained when applying our proposed CLONALG algorithm. Finally, the last two columns present the percentage gaps regarding the solutions found by CLONALG and the solutions reported when using BRILS and the GA method, respectively.

By looking at the results presented, we can conclude that the proposed AIS can operate efficiently for solving the presented problem in most of the instances in a very competitive time, the average runtime when applying CLONALG is 7.21 s, which is about 52% of the average time consumed by its similar counterpart, the population-based GA technique [17]. Furthermore, it follows that CLONALG solutions outperform the results reported by the genetic method.

Table 3. Solutions found by CLONALG for the AP instances

Instance		BRILS [3]		GA [17]		CLONALG '19		GAP	
n	p	Cost(1)	Time(s)	Cost(2)	Time(s)	Cost(3)	Time(s)	Cost(3)-(1)	Cost(3)-(2)
10	2	**167493.08**	0.00	167493.06	0.00	167493.065	0.00	0.00	0.00
	3	**136008.14**	0.01	-	-	136008.126	0.00	0.00	-
	4	**112396.08**	0.01	112396.07	0.02	112396.068	0.00	0.00	0.00
	5	**91105.38**	0.01	91105.37	0.03	91105.371	0.00	0.00	0.00
20	2	**172816.69**	0.02	172816.69	0.01	172816.690	0.00	0.00	0.00
	3	**151533.08**	0.02	151533.08	0.02	151533.084	0.00	0.00	0.00
	4	**135624.88**	0.03	135624.88	0.03	135624.884	0.00	0.00	0.00
	5	**123130.09**	0.03	123130.09	0.06	123130.095	0.00	0.00	0.00
25	2	**175541.98**	0.03	175541.98	0.03	175541.978	0.00	0.00	0.00
	3	**155256.32**	0.03	155256.32	0.04	155256.323	0.00	0.00	0.00
	4	**139197.17**	0.03	139197.17	0.07	139197.169	0.00	0.00	0.00
	5	**123574.29**	0.03	123574.29	0.03	123574.289	0.00	0.00	0.00
40	2	**177471.68**	0.04	177471.67	0.05	177471.674	0.00	0.00	0.00
	3	**158830.55**	0.04	158830.54	0.09	158830.545	0.01	0.00	0.00
	4	**143968.88**	0.05	143968.88	0.05	143968.876	0.01	0.00	0.00
	5	**134264.97**	0.05	134264.97	0.18	134264.967	0.01	0.00	0.00
50	2	**178484.29**	0.05	178484.29	0.14	178484.286	0.01	0.00	0.00
	3	**158569.94**	0.05	158569.93	0.13	158569.933	0.02	0.00	0.00
	4	**143378.05**	0.08	143378.05	0.20	143378.046	0.03	0.00	0.00
	5	**132366.96**	0.06	132366.95	0.23	132366.953	0.04	0.00	0.00
100	5	136984.7	0.88	136929.44	2.15	136929.44	0.42	−0.04	0.00
	10	106469.57	0.50	106829.15	11.15	106469.57	4.01	0.00	−0.34
	15	90534.79	0.46	90534.78	13.18	90533.52	7.97	0.00	0.00
	20	80305.1	3.86	80471.84	34.36	80270.96	9.79	−0.04	−0.25
200	5	140062.65	1.08	140450.08	18.34	140062.65	15.68	0.00	−0.28
	10	110147.66	0.39	110648.72	57.34	110147.66	29.84	0.00	−0.45
	15	94695.79	0.87	95857.69	81.92	94495.06	59.18	−0.21	−1.42
	20	85006.05	9.11	86069.21	151.20	85337.42	74.71	0.39	−0.85
Average			0.64		13.74		7.21		

Moreover, from the results in Table 3, one can see that three solutions obtained by CLONALG are better and only one case is worse than the respective ones obtained by BRILS [3]. In average terms, it can be concluded that most of the best solutions obtained by GA and BRILS regarding large instances ($n = \{100, 200\}$) are improved by CLONALG.

In order to better depict the CLONALG performance, Fig. 1 shows the fitness behavior among the population through the generations. The figure was obtained when running the algorithm for solving the problem with $n = 50$ nodes and $p = 5$ hubs. Observe that the method can provide in few iterations solutions with low gap to the optimal value.

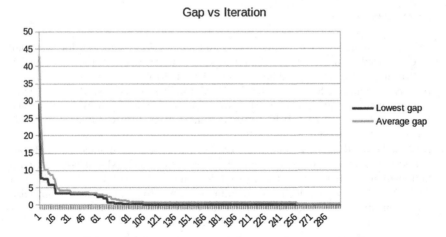

Fig. 1. Simulation results for the network with $n = 50$ nodes and $p = 5$ hubs.

5 Concluding Remarks

In this work we have proposed an Artificial Immune System, the CLONALG optimization algorithm, for solving the USApHMP. The most popular benchmark problem found in the literature, the AP dataset, was considered. A series of tests were carried out, which highlighted in particular the performance of the CLONALG algorithm in the 28 instances that were evaluated. The overall results indicated that the method could outperform its counterpart, the population-based GA metaheuristic, both in terms of quality of solutions and time performance. Furthermore, the algorithm could obtain better solutions than the fast metaheuristic BRILS in 3 out of 4 large instances.

One can conclude that the findings presented in this work add to a growing body of literature on the USApHMP, as they reinforce the effectiveness of AIS as global optimization technique and indicate potential applications of the same method to variations of the USApHMP, where situations with capacity constraints, multiple hub allocation or undefined number of hubs might yield optimization spaces with multiple optima, which positively accounts for the intrinsic population diversity capability of AIS methods.

There are several perspectives for future studies, such as the extension of the experiments for scenarios with more nodes and hubs and a sensitivity analysis of CLONALG with respect to its parameters, which unfortunately were not possible to perform in time for this work. Additionally, studying the potential association between BRILS and CLONALG in order to achieve high quality and, simultaneously, low runtime solutions can lead to interesting new results.

References

1. Alvarez Fernandez, S.: A metaheuristic and simheuristic approach for the p-hub median problem from a telecommunication perspective. Ph.D. thesis, University of Brasília (2018). http://hdl.handle.net/10803/666752
2. Alvarez Fernandez, S., Fantinato, D.G., Montalvao, J., Attux, R., Silva, D.G.: Immune-inspired optimization with autocorrentropy function for blind inversion of Wiener systems. In: 2018 IEEE Congress on Evolutionary Computation (CEC), pp. 1–7. IEEE (2018)
3. Alvarez Fernandez, S., Ferone, D., Juan, A.A., Silva, D., Armas, J.: A 2-stage biased-randomized iterated local search for the uncapacitated single-allocation p-hub median problem. Trans. Emerging Telecommun. Technol. **29**(9), e3418 (2018)
4. Amin-Naseri, M.R., Yazdekhasti, A., Salmasnia, A.: Robust bi-objective optimization of uncapacitated single allocation p-hub median problem using a hybrid heuristic algorithm. Neural Comput. Appl. **29**(9), 511–532 (2018)
5. Aydin, I., Karakose, M., Akin, E.: A multi-objective artificial immune algorithm for parameter optimization in support vector machine. Appl. Soft Comput. **11**(1), 120–129 (2011)
6. Bernardino, H.S., Barbosa, H.J.C.: Artificial immune systems for optimization. In: Chiong, R. (ed.) Nature-Inspired Algorithms for Optimization, pp. 389–411. Springer, Heidelberg (2009). https://doi.org/10.1007/978-3-642-00267-0_14
7. Bhadoria, V.S., Pal, N.S., Shrivastava, V.: Artificial immune system based approach for size and location optimization of distributed generation in distribution system. Int. J. Syst. Assur. Eng. Manag. **10**(3), 339–349 (2019)
8. Burnet, F.M.: Clonal Selection and After. In: Bell, G.I., Perelson, A.S., Pimbley Jr., G.H. (eds.) Theoretical Immunology, pp. 63–85. Marcel Dekker Inc., New York (1978)
9. de Castro, L.N., Von Zuben, F.J.: Learning and optimization using the clonal selection principle. IEEE Trans. Evol. Comput. **6**(3), 239–251 (2002)
10. Diabat, A., Kannan, D., Kaliyan, M., Svetinovic, D.: An optimization model for product returns using genetic algorithms and artificial immune system. Resour. Conserv. Recycl. **74**, 156–169 (2013)
11. Dudek, G.: An artificial immune system for classification with local feature selection. IEEE Trans. Evol. Comput. **16**(6), 847–860 (2012)
12. Ernst, A.T., Krishnamoorthy, M.: Exact and heuristic algorithms for the uncapacitated multiple allocation p-hub median problem. Eur. J. Oper. Res. **104**(1), 100–112 (1998)
13. Filipović, V., Kratica, J., Tošić, D., Dugošija, D.: GA inspired heuristic for uncapacitated single allocation hub location problem. In: Mehnen, J., Köppen, M., Saad, A., Tiwari, A. (eds.) Applications of Soft Computing. Springer, Berlin Heidelberg (2009). https://doi.org/10.1007/978-3-540-89619-7_15
14. Grine, F.Z., Kamach, O., Sefiani, N.: A new efficient metaheuristic for solving the uncapacitated single allocation p-hub median problem. In: 2018 International Colloquium on Logistics and Supply Chain Management (LOGISTIQUA), pp. 69–74. IEEE (2018)
15. Ilić, A., Urošević, D., Brimberg, J., Mladenović, N.: A general variable neighborhood search for solving the uncapacitated single allocation p-hub median problem. Eur. J. Oper. Res. **206**(2), 289–300 (2010)

16. Kratica, J.: An electromagnetism-like metaheuristic for the uncapacitated multiple allocation p-hub median problem. Comput. Ind. Eng. **66**(4), 1015–1024 (2013)

17. Kratica, J., Stanimirović, Z., Tošić, D., Filipović, V.: Two genetic algorithms for solving the uncapacitated single allocation p-hub median problem. Eur. J. Oper. Res. **182**(1), 15–28 (2007)

18. Leung, C.S.K., Lau, H.Y.K.: A hybrid multi-objective ais-based algorithm applied to simulation-based optimization of material handling system. Appl. Soft Comput. **71**, 553–567 (2018)

19. Li, T., Song, R., He, S., Bi, M., Yin, W., Zhang, Y.: Multiperiod hierarchical location problem of transit hub in urban agglomeration area. Mathematical Problems in Engineering 2017 (2017), Article ID 7189060, 15 pages

20. Li, T., Song, R., He, S., et al.: Hierarchical model for regional integrated passenger hub layout. J. Beijing Univ. Technol. **40**(11), 1700–1706 (2014)

21. Martí, R., Corberán, Á., Peiró, J.: Scatter search for an uncapacitated p-hub median problem. Comput. Oper. Res. **58**, 53–66 (2015)

22. Meier, J., Clausen, U.: Solving classical and new single allocation hub location problems on euclidean data. Technical report, Optimization Online (2015)

23. Meier, J.F., Clausen, U., Rostami, B., Buchheim, C.: A compact linearisation of euclidean single allocation hub location problems. Electron. Notes Discrete Math. **52**, 37–44 (2016)

24. Milanović, M.: A new evolutionary based approach for solving the uncapacitated multiple allocation p-hub median problem. In: Gao, X.Z., Gaspar-Cunha, A., Köppen, M., Schaefer, G., Wang, J. (eds.) Soft Computing in Industrial Applications. Springer, Berlin Heidelberg (2010). https://doi.org/10.1007/978-3-642-11282-9_9

25. O'kelly, M.E.: A quadratic integer program for the location of interacting hub facilities. Eur. J. Oper. Res. **32**(3), 393–404 (1987)

26. Peiró, J., Corberán, Á., Martí, R.: Grasp for the uncapacitated r-allocation p-hub median problem. Comput. Oper. Res. **43**, 50–60 (2014)

27. Porselvi, S., Balaji, A., Jawahar, N.: Artificial immune system and particle swarm optimisation algorithms for an integrated production and distribution scheduling problem. Int. J. Logistics Syst. Manag. **30**(1), 31–68 (2018)

28. Rani, R.: Distributed query processing optimization in wireless sensor network using artificial immune system. In: Mishra, B.B., Dehuri, S., Panigrahi, B.K., Nayak, A.K., Mishra, B.S.P., Das, H. (eds.) Computational Intelligence in Sensor Networks, pp. 1–23. Springer, Heidelberg (2019). https://doi.org/10.1007/978-3-662-57277-1_1

29. Rostami, B., Meier, J., Buchheim, C., Clausen, U.: The uncapacitated single allocation p-hub median problem with stepwise cost function. Technical report, Optimization Online (2015)

30. Silva, D.G., Montalvão, J., Attux, R., Coradine, L.C.: An immune-inspired, information-theoretic framework for blind inversion of Wiener systems. Sig. Process. **113**, 18–31 (2015)

31. Sun, X., Dai, W., Zhang, Y., Wandelt, S.: Finding p-hub median locations: An empirical study on problems and solution techniques. J. Adv. Transp. 2017 (2017), Article ID 9387302, 23 pages

32. Wakui, T., Hashiguchi, M., Sawada, K., Yokoyama, R.: Two-stage design optimization based on artificial immune system and mixed-integer linear programming for energy supply networks. Energy **170**, 1228–1248 (2019)

33. Zhang, W., Yen, G.G., He, Z.: Constrained optimization via artificial immune system. IEEE Trans. Cybern. **44**(2), 185–198 (2014)
34. Zhang, Z., Yue, S., Liao, M., Long, F.: Danger theory based artificial immune system solving dynamic constrained single-objective optimization. Soft. Comput. **18**(1), 185–206 (2014)

Selected Topics in Decision Support Systems and ICT Tools

IoT Cargo Weight Tracking System
for Supply Chains

Diego Gomez$^{(\boxtimes)}$, Cesar Viloria, Steven Llerena, and Nel Tinoco

Universidad Del Norte, Barranquilla, Colombia
{dgomez, caviloria}@uninorte.edu.co

Abstract. A cargo weight geo-referenced system for monitoring transportation and delivery tasks, has being developed. Allowing real time weight-tracking reporting, and historic data consultation, transporters can provide a truck with a sensor to control correct cargo delivery. Sensor measures deflection between the vehicle chassis and its rear axle, calculating the cargo weight in a non-intrusive way. A calibration stage must be included after sensor installation, to establish a relation between the vehicle mechanical structure and the sensor electronic output. Data is sent to a cloud database using a GPS/GPRS modem and is gathered by a dynamic web page. The sensor was tested under different weight loads and vehicle speed conditions. A hypothesis test shows that the sensor tolerance, using a 200 kg. load, is less than 10%.

Keywords: IoT · Sensors · Telemetry · Geo-reference · Weight-tracking · Supply chain

1 Introduction

Cargo transportation is the fundamental pillar for the dynamics of countries economy, since it is the basic tool to place the products within reach of the consumer in the national or foreign territory [1, 2, 3]. In Latin America, this activity is being affected in different ways. In the first place, by those people who steal the contents of the transporting trucks which leads to considerable economic losses and to the dissatisfaction of the final consumer since the merchandise is not constantly monitored for a quick action before a possible case of theft. On the other hand, by drivers who make illegal agreements that benefit third parties for the transport of assets not contracted, thus reflecting losses for the owners of the trucks. And finally, the fines for overloads that the owners of the transporting trucks face since they do not know the weight transported by their vehicles continuously [4, 5, 6, 7, 8].

Over the years, telemetry systems have been developed to control the theft of cargo to transport trucks and methods to measure the weight of the cargo transported by different types of vehicles. These systems have been analyzed to build the basis of the present project.

Honeywell International Inc. developed a load control system that is responsible for detecting the presence of cargo inside the container. They use one or more cameras to capture the image of the load to later be stored and processed locally through a device.

© Springer Nature Switzerland AG 2019
C. Paternina-Arboleda and S. Voß (Eds.): ICCL 2019, LNCS 11756, pp. 365–379, 2019.
https://doi.org/10.1007/978-3-030-31140-7_23

The position of the cameras, the viewing angles and the accessories used (flashes, reflectors, etc.) vary according to the space and environment of the container [9].

An anti-theft loading system was also developed in the containers by blocking its doors. It arose due to the easy access of the thieves to the interior of the container since the doors of these use conventional locks on the outside of the doors, thus being easy to intervene by thieves. In this way, the system is responsible for blocking the doors from the inside and it is only possible to unblock it through a remote station via satellite or cell phone [10].

In addition, a mobile computing platform was developed that is equipped with a terminal. Partly due to motion sensors, the terminal is able to determine if a load has been removed from a trailer, and if the trailer is empty. Such determinations are achieved by equipping the terminal with an ultrasonic sensor capable of detecting a load or loads within a trailer. The terminal is also capable of determining whether a load has been placed on or in a trailer. When desired, through an information manager, such determinations and load measurements can be transmitted automatically and continuously. The terminal of the mobile computing platform is also operatively connected to one or more data processors controlled by the information manager in the entire SPS (Satellite Position System) that is able to provide the location through reports indicating the coordinates where the trailer is found, also automatically and continuously [11].

The need to develop a vehicular telemetry system to support the dispatch of merchandise has been found. This system seeks to reduce the costs that owners of cargo transport vehicles have due to the theft of merchandise and/or fines for over-charging. Likewise, it allows to optimize merchandise dispatching by transport companies. This system allows the user to monitor in real time the weight of the vehicle's load and its proper georeferencing in order to carry out a follow-up for the safety of both the vehicle and its merchandise. This is possible thanks to the sensing of the weight inside the vehicle and the sending of the geographical coordinates of its location together with the weight by means of a GSM/GPRS Modem using the UDP protocol. The information is received in a database hosted in an AWS cloud to be managed by a web application hosted in the same cloud. In the web platform of the vehicular telemetry system one can mark the route to be carried out by the vehicle and the expected weight of the load to be delivered for each of the dispatch points. Any anomaly in the weight of the load is reported immediately. For a better management of the dispatches, the platform has historical data of the routes of the vehicles and the weight of the load.

Tests were performed to calculate the maximum tolerance allowed in weight sensing. These tests consisted of comparing weight patterns with the data given by the system. It was concluded to have a maximum tolerance of 5% with a maximum weight of 250 kg at rest.

2 Methodology

The system is composed of a sensing and processing stage that captures and transmits a signal proportional to the weight of a vehicle's load (see Fig. 1). Then, it transmits that signal to a GSM/GPRS module for sending it, with the geographical coordinates of the location of a vehicle, to a database of a server in the cloud. Thus, in the visualization stage the received information is captured to show it through a web application. Finally, the modem is powered by the vehicle's battery; this battery also charges the internal modem battery. This simultaneously feeds the processing unit and, therefore, this unit feeds the sensors.

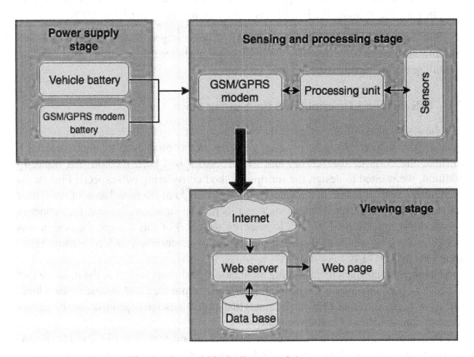

Fig. 1. General block diagram of the system.

The solution has been designed so that it is capable of:

- Calculate the weight of the load transported by a vehicle in a non-intrusive way with a maximum allowed tolerance.
- Implement a wireless communication protocol so that the system constantly shows the route of a vehicle and the weight of its load in a web page.
- Implement a data transport protocol that allows a low data consumption.
- The microcontroller must allow serial communication through a UART (Universal Asynchronous Receiver Transmitter) port, has at least two analog inputs and ensures the best possible resolution.

- The user can mark the dispatch points with the expected weight in each point so that the system can manage the fulfillment of the merchandise loading/unloading route.
- It should be possible to observe the history of the vehicle's routes with the weight at each point where the coordinates are sent for certain time intervals selected by the user.
- Tests must be performed with the vehicle at rest and in motion (three different speeds) for the calculation of the maximum allowed tolerance.
- The modem must send the speed of the vehicle to validate the results; in addition, it must be robust to be suitable for commercial applications.
- A statistical analysis of the weight deviation measured by the system after calibration with respect to heavy patterns on a commercial scale must be performed. In this way, we validate the hypothesis with a confidence interval of 90%.

The technological platform is composed of three stages: Sensing, Processing and Communication. In each stage, the design requirements for the selection of the different technologies are considered.

2.1   Selection and Description of the Sensing Stage

A review was made of the different methods used to measure the weight of a vehicle's load on the market [12, 13, 14]. An analysis of the procedure to implement each method, the costs of the sensors and the accessories of each method was done. In addition, we wanted to design the sensing method considering two aspects: First, to be as unobtrusive as possible in the vehicle, that is to say, that the installation of the sensor does not require to affect its mechanics and its normal structure. Second, the economy versus other conventional sensors to measure weight. For this reason, the design was based on the Technoton method that makes use of its commercial load sensors reference GNOM DP [14].

The method consists in generating an analog signal proportional to the distance that lowers or raises the chassis of the vehicle when it is mounted and disassembles a load. In this way, an arm that can move an angular potentiometer dependent on the movement of the vehicle's chassis was made.

Two arms were used and therefore two potentiometers (see Fig. 2). The potentiometers were installed on the right and left sides of the rear axle of the vehicle. One end of the arm is installed on the knob of the potentiometer, while the other end on the chassis of the vehicle.

Points A and B correspond to the rear axle and vehicle chassis, respectively.

The potentiometers are powered by the selected microcontroller with 3.3 V and a current limiting resistor is used that acts in the event of a short circuit. Thus, when the chassis falls due to the weight of the load, the potentiometer will rotate and change its resistance, which causes a voltage change. This voltage signal reaches the microcontroller to be subsequently transmitted to the GSM/GPRS modem.

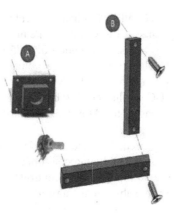

Fig. 2. Weight meter model.

2.2 Selection and Description of the Processing Stage

Three microcontrollers of the market were compared to select the one that could accomplish the indicated requirement in the best way. Table 1 gives a comparison between the different microcontrollers.

Table 1. Comparison of microcontrollers [15, 16, 17].

Name	UART	Clock	Operation range	Memory	Integral voltage reference
PIC16F506	No	20 MHz	2–5.5 V	1.5kB Flash. 67B SRAM	Yes
AT-mega328	Yes	16 MHz	1.8–5.5 V	32kB Flash. 2kB SVRAM	Yes
PIC16LF1554	Yes	32 MHz	1.8–3.6 V	7kB Flash. 256B SRAM	No

When comparing the three microcontrollers, it is observed that all meet the requirement of having at least two analog inputs. But the PIC16F506 does not allow serial communication through the UART port with another device to transmit data, so it is discarded. On the other hand, it is necessary that the microcontroller detect changes for the lowest weights (lowest distances), i.e., that its ADC (Analog-to-digital converter) manages the best possible resolution because the distance that the chassis of the vehicle goes down and up is of the order of millimeters (mm).

Being the resolution of the ADC of the 10-bit PIC16LF1554. The minimum change (resolution) is calculated with the following equation:

$$Resolution = \frac{Operation\,Range}{2^n - 1}$$

This microcontroller does not allow to change its reference voltage from the programming, thus, its working range is 1.8 V and n is the number of bits of the resolution of the ADC. By replacing the values, a resolution of 1.75 mV is obtained. As for the Atmega328, it does allow to change its reference voltage from its programming to achieve a minimum working range of 0 to 1.1 V.

In this way, having the ADC resolution of 10 bits, using the equation, the resolution is equal to 1.07 mV. For this reason, the Atmega328 is chosen for the processing stage since it meets all the requirements.

Once the Atmega328 was chosen, the Arduino UNO open source development card was used, since it contains the integrated microcontroller. In this way, the free Arduino programming interface is used so that the microcontroller receives the analog signals coming from the sensors and its subsequent transmission to the GSM/GPRS modem via serial.

To obtain the weight, a calibration was necessary. That is, use weight patterns and mount them on the vehicle to know the total voltage (sum of the voltages of both potentiometers) that is generated for each weight. Having the relation Total Voltage - Weight, we proceed to use an interpolation polynomial that characterizes the vehicle through the points obtained in the calibration. We used the Newton Interpolation Algorithm [18] developed in the Matlab tool to generate the polynomial that joins the calibration points and that is later introduced to the sniffer of the web application to calculate the weight.

A polynomial was generated for both increasing and decreasing loads, due to the fact that the vehicle does not return to its initial point after placing and removing the same load. In the characterization of the suspension system of the vehicle used, where a hysteresis is reflected, both equations are entered into the sniffer of the website to calculate the weight, either when mounting or removing load (see Fig. 3).

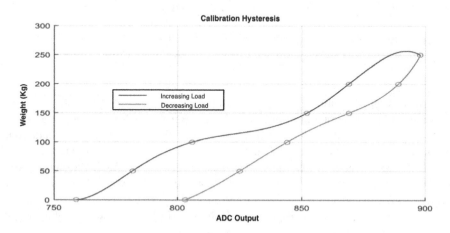

Fig. 3. Calibration polynomials.

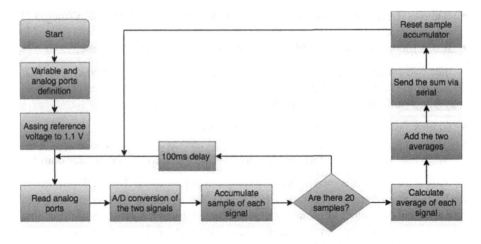

Fig. 4. Flow diagram of the high-level design of the processing stage.

Finally, Fig. 4 shows the flow diagram of the high-level design of the processing stage.

2.3   Selection and Description of the Communication Stage

It was decided to implement a GSM/GPRS Modem as it offers the advantage of having greater coverage because it is linked to cellular telephone networks. Two GSM/GPRS modems were selected for comparison: the SIM900 and the Syrus modem.

The SIM900 module works in the bands of 850/900/1800/1900 MHz and a SIM card with a data plan is needed so that the module can send the information through the APN of the network operator to an IP address of a server. The configuration of the module is done through AT commands incorporated in the ARDUINO source code (Microchip, 2008).

The GSM/GPRS SYRUS modem contains internally GPS and GSM antennas. Adaptable to different network and communication protocols such as TCP/IP, UDP, SMS, SMPP. With an internal battery with a duration of up to 10 h and ease to configure it for sending time, distance, course changes, speed, movement, shock, acceleration, timers, among others. Adaptable for connection to external devices through the RS232 protocol. Its configuration is also done through AT commands. In addition, like the SIM900 module, it uses a sim card with a data plan to send information through the cellular network [18].

When detailing each option, the GSM/GPRS Syrus Modem was selected for the following reasons:

- It includes the variable speed in sending of information, since this is very useful for the validation of the tests.
- It is a more robust module, i.e., with greater protection against external factors of the environment, which implies greater security and adaptability to the market.

- It allows the implementation of the RS232 protocol to connect it to the ATmega328 microcontroller via serial.
- It is a Modem that is available at the Universidad del Norte for its use.

The two common transport protocols used are UDP (User Data Program) and TCP (Transmission Control Protocol). Being the following points, the characteristics to highlight to differentiate the two protocols:

- UDP is faster than TCP because it does not check for errors per data packet.
- UDP is lighter because there is no ordering of messages, verification connections, etc. While TCP requires three packets to establish a connection before transmitting.

This shows the difference in data consumption per protocol. The UDP protocol is the best choice of transport protocol because for the system it is more important to transmit with speed than to guarantee the fact that all the bytes arrive. In addition, it is required to transmit less heavy packages since a sim card will be used with limited data plans that require periodic reloading.

The modem was programmed using AT commands to receive the weight via serial and to send this data, along with the geographic coordinates and speed to an IP address. For programming, the GSM/GPRS Syrus modem manual was consulted and the Hyperterminal software was used to load the command lines to the modem. Within the modem programming parameters are defined as user ID, APN of the telephony network, IP address and port of the cloud server, communication protocol (UDP), work mode and events to send the data.

The PHP language was used to develop the sniffer that will receive the ID of the vehicle, the geographic coordinates of the location of the same along with the time stamp of the modem, in addition to the speed of the vehicle and the digital signal of the microcontroller to calculate the weight through a previously obtained calibration equation. Thus, it will save in a database the information sent by the GSM/GPRS Modem Syrus. In addition, when the GSM/GPRS modem loses signal, the sniffer is able to identify the message and discard it when these erroneous coordinates are sent since the latitude and longitude are filled with 0 by the modem. Figure 5 shows the flow diagram of the high-level design of the communication stage.

2.4   General Description of the Hardware

The GSM/GPRS Syrus modem manual was consulted to identify and select the indicated protocol so that it receives the ADC signal from the ATmega328 microcontroller. The protocol selected is the RS232 because it is an easy implementation protocol and adapts to the modem used.

The connection between the GSM/GPRS Syrus modem and the microcontroller is provided through a TTL/RS232 converter and a DB9 Female-Female cable (see Fig. 6). In this way, the modem receives via serial the output signal of the microcontroller ADC, which is labeled in a message with the geographic coordinates and sent to the cloud through its GPRS module.

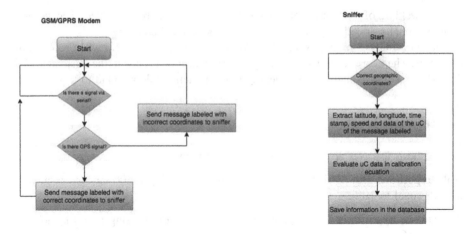

Fig. 5. Flow diagram of the high-level design of the communication stage.

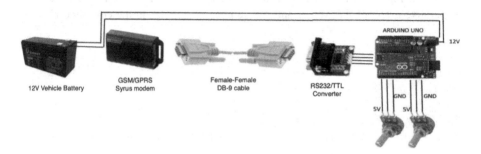

Fig. 6. General outline of the hardware.

2.5   Description of the Viewing Stage

For the visualization of the system, a web application was developed. For the development of the same, the JavaScript language was used to take advantage of the Google APIs to be able to show the location in real time of the vehicle on the Google map. In addition, JavaScript allows the update of the data continuously on the website, such as date, time and weight as the modem is sending information to the cloud. Finally, it allows the marking of routes on the map for weight monitoring.

In addition, the web page allows queries of historical SQL statements through PHP. The query can be done either by time intervals, by location or by combining the queries. In terms of time intervals, the user types the dates to consult and the web application draws the entire route of the vehicle on the map in that time window. By location, the user selects the area on the map where he wants to know if it was the vehicle. Finally, one can combine both queries for better information filtering. Finally, with HTML (HyperText Markup) and CSS (Cascading Style Sheets) the environment of the website was developed.

The welding of a potentiometer without hysteresis must be guaranteed because the system cannot respond to problems related to mechanical coupling.

The GSM/GPRS modem saves the coordinates and it sends them when recovering the signal. The measurement system may present inaccuracies due to the mismatches of the vehicle's chassis and it is not clear how the system could respond to this.

3   Results

A hypothesis test was applied starting from the following:

Null hypothesis H0: Weight measurement tolerance using a 200 kg load is greater than or equal to 10%.

Alternative hypothesis H1: Weight measurement tolerance using a 200 kg load is less than 10%.

Figure 7 shows the route marked to monitor the weight of the merchandise. On the other hand, Fig. 8 shows the historical consulted of the routes and Fig. 9 presents a comparative graph of the weight and the number of samples in the consulted history.

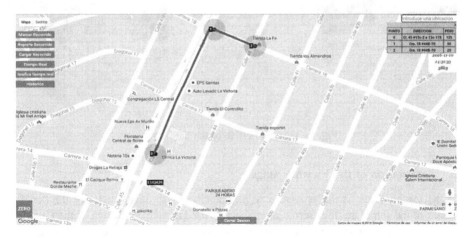

Fig. 7. Route marked to monitor the weight.

The weight of the load transported by a vehicle at rest and in movement was measured at three (3) different speeds: 15, 20, and 30 km/h. Weighing patterns that have previously been weighed by a scale were used. In this way, it is sought to compare known weights with the value thrown by the sensor. When the vehicle is at rest, accuracy tests are performed for increasing and decreasing load in order to calculate the maximum tolerance allowed. That is, we increase the load in steps of 25 kg until reaching the maximum load. In the same way, we decrease the maximum load in the same steps. For tests with the vehicle in motion, the same route will be made for different loads at the three different speeds each. In this way, the average of the

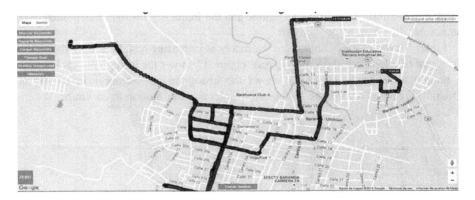

Fig. 8. History consulted.

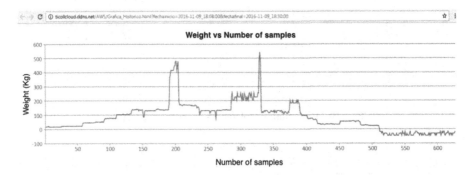

Fig. 9. Comparison of weight and number of samples in the consulted historical.

measurements will be calculated for each load in a time window, this average will be compared with the expected weight (patterns) to calculate the tolerance in movement.

To determine the total number of experiments, the full factorial method is used, given by the following equation:

$$NTE = r \times \prod_{i=1}^{k} n_i$$

Where r is the number of repetitions, k the number of factors and *ni* the number of levels by factors. In this case k = 1 (a single independent variable that is the speed), *ni* = 4 (at rest, 15, 20, 35 km/h) and a number of repetitions equal to 20 was chosen. In this way, *NTE* = 20 × 1 × 4 = 80. Thus, 80 experiments were chosen for the test.

A significance level of 0.05 was established to have a confidence interval of 95% to check if the data had a known distribution, specifically a Normal Distribution. This is in order to perform a statistical analysis from the distributions already established. Through the StatGraphics software, the goodness of fit test is performed with the

Kolmogorov-Smirnov statistical technique and the frequency histogram is plotted over the normal distribution for each test.

Figure 10 shows the histogram of tolerances at increasing rest. Figure 11 shows the histogram of tolerances at decreasing rest. Figure 12 shows the histogram of tolerances in motion at 10 km/h. Figure 13 shows the histogram of tolerances in motion at 20 km/h. Figure 14 shows the histogram of tolerances in motion at 35 km/h.

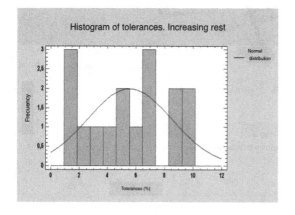

Standart Deviation	2,98673003
Sample mean	4,67568436
P value	0.9331

Fig. 10. Histogram of tolerances at increasing rest.

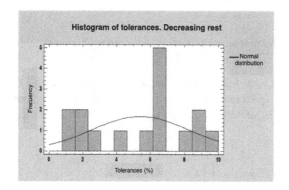

Standart Deviation	3,1490385
Sample mean	4,57726955
P Value	0.5845

Fig. 11. Tolerance histogram at decreasing rest.

Since the P values of all tests are greater than 0.05, the idea that all data fit a normal distribution with a 95% confidence level cannot be rejected.

Based on the requirement of a confidence level of 90%, $\alpha = 0.1$ is selected. The model to be used for the hypotheses is one tail. In this model, the region to the left of the Z values for the level of significance rejects the null hypothesis.

It is established that the null hypothesis is rejected for the increasing and decreasing tests and it is accepted for the other tests with a confidence level of 90%.

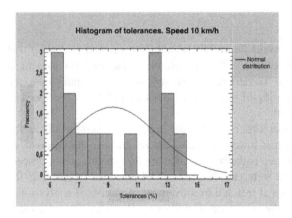

Fig. 12. Histogram of tolerances in movement at 10 km/h.

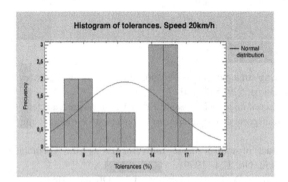

Fig. 13. Histogram of tolerances in movement at 20 km/h.

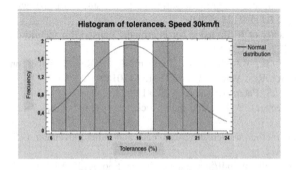

Fig. 14. Histogram of tolerances in movement at 35 km/h.

4   Conclusions

A prototype of a system that measures the weight of a maximum load of 200 kg was developed with an allowable tolerance, either at rest or in motion. The weight is measured more accurately with no movement, in this way, in the web application, the function that performs the monitoring of the weights, must consider that the speed of the GSM/GPRS modem is 0 km/h.

In the calibration stage, it was observed that the suspension system affects the measurement, since when the load is lowered, it does not return to its initial point. In this way, it was necessary to use two equations to calculate the weight, one of descent and another of rise. The tests that were carried out were of precision to calculate the maximum tolerances allowed, both at rest and in movement. A statistical analysis was carried out to accept or reject the proposed hypothesis: The maximum tolerance allowed for a maximum load of 200 kg is greater than 10%. Thus, the Normal probability distribution was applied and the following conclusion was reached: At rest, the hypothesis is rejected, while in speed it is accepted.

References

1. Márquez, L., Cantillo, V.: Evaluating strategic freight transport corridors including external costs. Transp. Planning Technol. **36**(6), 529–546 (2013)
2. Jahir, A., Gutiérrez, O.: El ciclo económico del transporte de carga terrestre carretero en ColombiaEl ciclo económico del transporte de carga terrestre carretero en Colombia. Criterio Libre **18**, 125–154 (2013)
3. Ramírez-Giraldo, M.T.: The impact of transportation infrastructure on the Colombian economy. Borradores de Economíam, No. 124 (1999)
4. De la Torre, E., Martner, C., Quintero, E.M., Martínez, J.L., Olivares Benítez, E.: Herramienta para la evaluación del riesgo de robo en el autotransporte de carga. Nova Scientia **7**(13), 438–469 (2015)
5. Ekwall, D., Lantz, B.: Cargo theft at non-secure parking locations. Int. J. Retail Distrib. Manag. **43**(3), 204–220 (2015)
6. Ghisolfi, V., Ribeiro, G.M., Chaves, G., Orrico Filho, R.D., Hoffmann, I., Perim, L.R.: Evaluating impacts of overweight in road freight transportation: a case study in Brazil with system dynamics. Sustainability **11**(11), 3128 (2019)
7. Hricová, R., Straka, M.: Opportunity of RFID using for intermodal transport in security of goods. Appl. Mech. Mat. **718**, 162–167 (2015)
8. Shah, R., Sharma, Y., Mathew, B., Kateshiya, V., Parmarupal, J.: Review paper on overloading effect. Int. J. Adv. Sci. Res. Manag. **1**(4), 131–134 (2016)
9. Lyall, R., et al.: Cargo sensing. U.S. Patent Application No. 13/923, 259
10. Louis, J.R., Hultberg, M., Seeber, W.J.: Anti-theft locking system. U.S. Patent No. 8,493,193. 23 July 2013
11. Faus, J., Parisi, M.D. System and method for sensing cargo loads and trailer movement. U.S. Patent No. 8,179,286. 15 May 2012
12. Kineo. https://www.basculasprecicell.com/pdf/Sistema_de_Pesaje_Fijo_R125_KINEO.pdf. Accessed 11 June 2019
13. Nordmeyer, D.L.: Weight sensing system and method for vehicles with non-fluid springs. U. S. Patent No. 7,398,668. 15 July 2008

14. Technoton.   https://old.jv-technoton.com/data/dc/GNOM_manual_de_instrucciones.pdf. Accessed 11 June 2019
15. Atmel AVR: DATASHEET ATMEGA (2010)
16. Microchip: DATASHEET PIC16F506 (2007)
17. Microchip: Datasheet PIC16LF1554 (2008)
18. Hernández, S., et al.: Interpolación polinomial. XIV Workshop de Investigadores en Ciencias de la Computación (2012)

Developing Logistic Software Platforms: E-Market Place, a Case Study

Wilson Nieto Bernal[1] (ID), Miguel A. Jimenez-Barros[2](✉) (ID),
Daladier Jabba Molinares[1] (ID), and Carlos D. Paternina-Arboleda[1] (ID)

[1] Universidad del Norte, Barranquilla, Colombia
{wnieto,djabba,cpaterni}@uninorte.edu.co
[2] Universidad de La Costa, Barranquilla, Colombia
mjimenez47@cuc.edu.co

Abstract. This paper describes a framework for software development with emphasis on logistics platforms. Specifically, a case study is presented regarding the Logport program in Colombia's Caribbean coast. The software development is based on the merging of modeling, design, and agile development techniques aimed towards software production. The agile methodologies discussed herein include Scrum, XP, and Crystal among others, and specific verification and validation aspects of CMMI 1.3 are evaluated. Furthermore, the integration of emerging technologies including elastic computing in cloud computing that allows scaled integration, security, load balancing, process speed, and high concurrency among other features are discussed. Finally, the proposed solutions cover actual and emergent needs in a B2B or B2C electronic commerce dynamic environment where suppliers and clients offer and demand transport, storage, and customs services. In this environment, their goal is to ensure there is added value to their logistics processes starting at the inputs and all the way to the outputs of the commercial business processes. Colombia's Caribbean region is one such multimodal and multiport environment in which there is constant demand for low cost, time and storage optimization, and customs reliability for the businesspersons of this region, which serves as a Hub for the American Caribbean.

Keywords: Agile methodology · Enterprise architecture · Software process · Software validation · Cloud computing · Logistics software

1 Introduction

The motivation for this research grows from the fact that the Logistics Performance Index ranks Colombia at position 97th [1]. The importance on reinforcing technology that lead the development at the regional level provides guidelines for focusing on specific topics that derive in improvements on regional competitiveness. Furthermore, the lack of existing modeling frameworks that facilitate trade applied to logistics on emerging countries led to the implementation of this platform. Such structure could be used as a reference in order to improve the implementation of international trade facilitation for the development of our economy.

© Springer Nature Switzerland AG 2019
C. Paternina-Arboleda and S. Voß (Eds.): ICCL 2019, LNCS 11756, pp. 380–396, 2019.
https://doi.org/10.1007/978-3-030-31140-7_24

Regulations in Colombia and many of the countries in our region yield constraints that may inhibit the use of this kind of platforms, especially for those that deal with trade. Furthermore, the use of freight brokering apps and web-based software face many challenges in our country. As an example, our own surveys show that more than 80% of the transport fleet in Colombia is driven by the truck owner, with little access to technology. This makes it difficult for any software platform to perform in an open environment. Although the connectivity is provided through the platform, many drivers still do not carry smart phones to deal with the transport bids.

Most of the products or services can be either bought or contracted through electronic commerce platforms. Business to Business (B2B) and Business to Consumer (B2C) electronic commerce improves the sales or the business that use these platforms [2]; of course, these platforms must be implemented correctly to achieve the businesses goals, with several factors given [3].

As business changes day by day, information systems must also change. For every new requirement, the information system should offer a quick response to the organization, to the users. Agile methodologies are indicated to get these changes running in a short period and to achieve the required quality for the process and the user interface [4]. Some of them such as Scrum, XP, and Crystal have been used in agile software development [5], and as information systems and organizations. They have evolved through time, improving themselves to adapt to new processes and technologies. Although agile methodologies were born for software developing, they must include the Enterprise Architecture (EA) in their operations to achieve the proposed goals. According to [6], EA allows organizations to align information systems with their processes, which includes Business, Application, Data, and Technology Architecture.

To manage the EA, Capability Mature Module Integration (CMMI) is one of the complete process models that allow organization define Product and Service Development (CMMI-DEV), Service Establishment and Management (CMMI-SVC), and Product and Service Acquisition (CMMI-ACQ). CMMI-DEV is a "reference model that provide practices covering project management, process management, systems engineering, hardware engineering, software engineering, and other supporting processes used in development and maintenance" [7]. Appropriating CMMI standards for software and technology implementation projects, allow the quality assurance needed for bringing the organizations high-reliability process and information at the right time. In their research, [6] find Cloud Computing as a high impact technology in EA. Cloud computing allow organizations to consume and pay just the computing resources they need to a service provider [8], such as Amazon Web Services or Microsoft azure, just to mention a few. Organization can contract dedicated or virtual private servers, or elastic services: the latter allow to get as low resources you need for computing and grow as the information systems and technologies require computing power, security, high concurrency and load balancing. [9] highline the importance of Cloud Computing, but at the same time find that the adoption process to this technology is neither easy nor quick, due to several determinants.

The framework construction, named PDLSP, comprises multiple layers of abstraction where all software structures are visualized. The first layer focuses on the strategic modeling while the second layer focuses on the business process model.

The third layer describes the data model and the fourth layer describes the application model. The fifth layer describes the information technology infrastructure while a transversal layer presents validation and verification processes. The seventh, and last layer, describes the delivery channels of the solution. All the layers are depicted in Fig. 1.

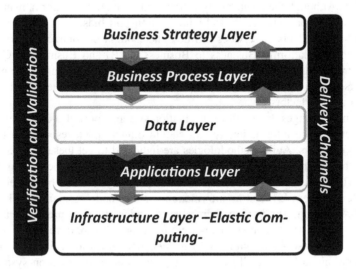

Fig. 1. Methodology structure PDLSP. Made by authors.

2 Research Methodology

A broad methodology is employed herein that begins with the exploration of the available literature and is followed by the identification of software development expertise and information sources regarding the Logport program. This allowed setting up the following investigation phase to build the framework PDLSP.

2.1 Phase 0: State of the Art

The goal for this phase is to evaluate different emerging software development paradigms and highlight their business process evolution. Beginning mainly from MDG (Generation Driver Modeling) paradigms, there are several technologies each with different approaches and techniques to address the development of the IEA. These include MDG SysML, DDS, Zachman[TM] Framework[TM], TOGAF[TM], SOMF[TM], and UPDM[TM] [10, 11, 12, 13].

2.2   Phase 1: Reverse Engineering of Software Products

The goal of this phase is to identify software that offers global solutions regarding loading, storage, and transport logistic support issues. Moreover, the goal is also to identify technology trends and vital components that must be included in the soft-ware platform [14]. All of this is to be accomplished within the context of Colombia's Caribbean region. Some of the platforms that have been identified are discussed next. The first one is AWERY, which features a system to provide an integrated real-time view of core business processes [15]. AWERY is accessible from any place in the world; it is user-friendly; it is fully customizable to different customer's business processes and has been developed in tight cooperation with aviation professionals, among others. The next one is 3plsystems [16]. This software has features such as dispatch and operations, sales portal, customer portal, and carrier selection among others; Amber Road [17] includes features such as vendor and production management, material and product management, risk and quality management, and shipment and logistics management. Finally, Mercatrans [18] whose main feature allows clients and suppliers access to transport service channels [19].

2.3   Phase 2: Strategic Layer Modeling of the E-Market Place Platform

The goal of this phase is to identify and understand the services model/value proposition of the platform to stakeholders (clients and suppliers), understand the geographic implications, identify the persistent data and the process to develop and deploy the value return, and finally to understand how the different actors can access these resources [20].

2.4   Phase 3: Functional and Non-functional Requirements Management of the E-Market Place Platform

The goal for this phase is to establish functional requirements that characterize the computational solution emphasizing management modules of the platform. These are the management of products and services module (including transport, storage, multimodal services, and customs services), the management of the received and completed offers, report module, search module, offer listing, offer selection, offer sending, user and complementary components management of a documental website as a new management, events, blogs and rating [21].

2.5   Phase 4: Architecture of the Solution for the E-Market Place Platform

The first step of this phase is to develop the architecture of the solution over the baseline of the creation of the software. This step is then followed by the deployment of the architecture. In the first step, the components, processes, data, and the detailed requirements for the models are designed through an architecture-oriented software development approach. This allows consolidating the application, the data, and the business process layers of the e-Marketplace platform. The second step allows the

definition of the technologies used for the implementation and deployment of e-Marketplaces. In this case, the final choice helps deploy the platform over an elastic computation model with cloud computing features. In this case, there was integration of PostgreSQL database, Windows Server operation system, and software components PHP based, mainly CakePHP and Foundation frameworks. The result was a totally elastic platform [22].

2.6   Phase 5: Prototype Development of the E-Marketplace Platform

The goal of this phase is to obtain an agile software development process model with a productivity orientation and a focus on software deliverables from the established functional requirements and adding particularities of the associated nonfunctional requirements with the platform performance. In this case FURPS (Functionality, Usability, Reliability, Performance, and Supportability) is framed in ISO 25000:2014 [23].

2.7   Phase 6: Verification and Validation of the E-Market Place Platform

The goal of this phase is to implement the activities associated with the verification and validation of software requirements. In other words, the purpose at this stage is to ensure that the selected solution is correct and fulfills the specified requirements. Verification ensures that "it is correctly built" while validation ensures that "what is built is correct." The activities to be developed include CMMI Specific Goal 1: prepare for verification, establish the verification environment, and establish the verification procedures and criteria. Furthermore, peer reviews are to be included, and the resulting recommendations are to be addressed [24].

2.8   Literature Review

We found, through a literature review, that there are just a few frameworks for logistic platforms. To support this, we made a systematic review and used the phrases "Framework for Developed Software" and "E-market place" to search in scientific databases such as IEEE, Science Direct and Web of Science as on July of 2019. We established the following criteria to select the references: review or research papers. These results are detailed in Table 1. Another search was made including only Open Access research. These results are detailed in Table 2. Finally, we highlight the most relevant papers that had a direct impact on our research.

Table 1. Results of the searches in the scientific databases. Made by authors.

Search terms	IEEE	Science direct	Web of science	Total
Framework for developed software	3475	2864	4345	10684
E-market place	16	156	49	221
Total	**3491**	**3020**	**4394**	**10905**

Table 2. Results of the searches in the scientific databases (Open access). Made by authors.

Search terms	IEEE	Science direct	Web of science	Total
Framework for developed software	110	437	578	1125
E-market place	0	38	4	42
Total	**110**	**475**	**582**	**1147**

Derivate concepts from the literature review

According to Rahman and Suhaimi [25], the concurrent adoption of the engineering software process, and especially frameworks, have been used widely by organizations in software developing. It is fundamental to aboard the quality process of the software, defined as the set of good practices and formal methods to improve the software engineering process. These methodologies are essential to developing unified frameworks.

Mei and Lingjie [26] indicate that frameworks are an option to reuse systems design, considering this as the most effective way in software engineering for developing quality products. Additionally, this eases the developing of automatic tests, which is the tendency to improve the efficiency and effectiveness of software developing. Software engineering, processes, standards, and technologies are increasingly crucial inside quality software processes. A good software developing method can help developers to reduce cycles and reduce costs substantially.

Likewise, Singh et al. [27] point out that to ensure software quality and also to evaluate its reliability, many growth models have been proposed. These models can involve some factors, which includes methodologies, tools, standards, human talent, frameworks, quality metrics, and developing paradigms or approaches. Any change in these factors could end in failures, defects, or errors that increase or decrease software reliability.

Belete et al. [28] describe the design and prototype of a framework named Distributed Model Integration Framework (DMIF) that links the models implemented in different hardware and software platforms. They use distributed computing and service-oriented software developing approaches to use various interoperability aspects. They also integrate automatic conversion of units in semantic mediation using open anthologies, basically for the data structure.

Finally, is important to highlight Siavvas et al. [29] research. The subjectivity that underlies the quality notion does not allow the design and developing of a system accepted worldwide for the evaluation of the quality of the software. Therefore, contemporary research is focused on finding systems capable of developing software quality models and frameworks that can easily fit on the requirements of the users. These frameworks, methodologies, tools, processes, human talent, and also the excellent adoption of standards reduce the uncertainty of software developing, in many formats such as web and mobile platforms or embedded systems.

3 Description of the Framework – PDLSP

The components for PDLSP are described as:

3.1 Business Strategy Layer

This layer comprises a set of models that allow visualizing the business structure as a system that is supported by a Web platform B2B or B2C. The models developed in this layer include key business components: information, business processes or functions, the geographic distribution of the business, organizational structure, process time models, and other strategic elements. For this, object models such as UML class techniques, UML deploy models, BPMN process models, and Norton and Kaplan process map organization models are employed [30].

3.2 Business Process Layer

The goal of this layer is to elaborate and detail business structure having more developed elements as: information entities, business rules, process definition, business geographical distribution definition, responsibility definition, time processes definitions, and time definitions. For these elements the techniques used are relational database, connectivity processes, and Workflow, Master scheduler and business plan models [31].

3.3 Data Layer

The goal of this layer is to elaborate on the data structures associated with the platform. First, it must be identified the information units required to be persistent within the system, elaborate data semantic models, deploy and implement over the database management system, organizational information objects model, logical data Model (high-level Master and Transactional), logical data model (detailed), database design (physical) and repositories [32].

3.4 Applications Layer

The goal of this layer is the visualization of the business processes that will include detailed business processes application models, application architecture, system design, software and software components. The techniques used to achieve this include relational database, processes, components, deployment, web services, service oriented architectures, user interface and data interface models, and middleware software [31].

3.5 Infrastructure Layer

The goal of this layer is the visualization of the necessary components that must be present to support the data and application deployment, network infrastructure, user interface, security systems, and control architectures needed to access all platform services. For this, components must be defined in the specification of physical data

models, system's design, technology and network architecture, server infrastructure, and the concurrent user's estimation of the data volume [33].

3.6   Verification and Validation Layer

The goal of Verification (VER) is to ensure that the selected product meets the specified requirements. The following are a few of the methods used to carry out the verification process: assessment of software architecture, conformity assessment of implementation, functional decomposition-based testing, acceptance testing, and continuous integration (e.g. a flexible approach that identifies integration issues in the initial stages). The goal of Validation (VAL) is to demonstrate that a product and its components fulfill the intended use when placed in the intended environment.

3.7   Delivery Channel Layer

This layer's goal is to model and design the delivery channels of the software solution. To achieve this goal, it is necessary to set the execution environment precisely in regard to if output interfaces match Web interfaces and mobile applications, or are otherwise used in hybrid devices for the input and output data. This layer establishes if the delivery channels are focused on Web interfaces, enterprise sites, or enterprise standardized interfaces.

4   Case Application in the Context of the E-Market Place – Framework PDLSP

The following layers describe the different products to be developed as a result of the application of PDLSP:

4.1   Business Strategy Layer

This layer describes the main business processes (view Fig. 2) and the structured platform for the e-marketplace. In principle, it has a Back-End which is comprised of service processes of transport, storage, and customs documentation and compares data associated with customers and suppliers' business models. To access the core services, users have a Front-End that has been implemented as a web service. Through this, users can access different e-marketplace service platforms.

4.2   Business Process Layer

In this layer, four main components are defined:

Individual and Company-Wide User Sign Up. This process describes how a client signs up in the platform. The sign up must be done as a company (or personal company) and then one or more users who are accredited to access the platform to perform work can be signed up. In terms of a legal certification, documents expedited by the chamber of commerce and other identification documents must be uploaded to the

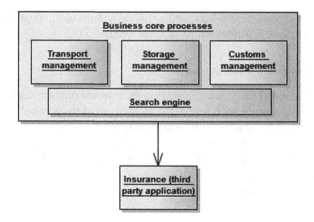

Fig. 2. Business strategy layer. *Made by authors.*

platform to achieve a high security level for the platform users. Once the documents are checked and validated, the company will be made available in the platform to offer and acquire services.

It is very important that the company fills out its information accurately, allowing the platform to execute the internal processes correctly when the B2B or B2C are completed. The accuracy of the information will be important when a user makes a search of the services offered or needed [34, 35].

During the sign up, one or more company users can be setup. First, the user that completes the sign up and registers the company will be set up as administrator. This will allow this user to offer and search for services (view register user in Fig. 3). At the same time, this user can register other users within the company allowing them to have the same security level or a lower security level. A company with more than one user could allocate access for different segments when the company has various transactions at the same time.

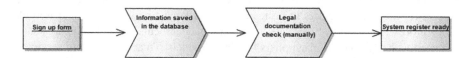

Fig. 3. Register user. *Made by authors.*

User Profile. This process describes the functionality that a specified user can have in the platform. In four general terms, a client can:

- Manage his user (personal information)
- Manage services offered
- Manage offers
- Search for information in the report module that describes all executed transactions.

In this module, the company or client can add users or make changes to the profile information when necessary. In this module, a given user can delete his account if he or she does not want to be registered in the platform anymore.

In this component, the services have a characterization level that allows the supplier to accurately describe his product to the client. This allows his client to search his needs accurately through the search option. At the same time, a client searching for a service has to clarify his needs. It is also possible to make several service connections in different ways through the multimodal services allowing the client to fulfill the logistic chain value of his product.

Once an offer is made, the user can follow it in two ways: as a need-a-service user, having the auction value (in the case of a normal or inverse auction) to give better value, and ask about the service. As a publishing user, to follow the offered value (in the case of a normal or inverse auction) and answer questions made by the users. When a service is published having a unique value, users can ask questions before making an offer, and the owner of the offered service answers it. Once the service is found by a client, the supplier could follow the offer and have access to important information such as the number of users that have visited it, how many offers have been made (in the case of a normal or inverse auction), and how many offers have been finalized for example. Likewise, the client can follow the offers while they are active, inactive, or ended. (View the User Profile Business Process Model in Fig. 4).

Search. This module has a search engine that allows finding keywords of the services that are available in the system and showing a list of corresponding services found. A system user can search for needed services, and the supplier can find services that a client needs to receive. As mentioned in the user profile module, searches are based on matching proper characterizations of the services so that a client can find his need accurately and quickly.

When a user obtains a search result, he or she can select the service needed and verify the information which is matching his/her needs. Moreover, the user can check the info and reputation of the company. (View Fig. 5. Search business process).

Management. The management module allows customers, visitors (non-clients), and system administrators to get help using blogs and forum entries. The blogs allow the administrator to have a constant traceability and to create publications about the upgrades of the platform with specific focus on the clients. The discussion forums are spaces that the users will share to present doubts and suggestions of the platform and in which the users and administrators could participate answering general questions.

Furthermore, administrators can publish news mainly focused in logistics, and not necessary including the platform work. This option is an additional feature of the platform thought as a space where a user can be informed with actual news in an independent segment.

Administrators can manage users, modify their information, delete accounts when security policies are not followed or passwords are forgotten, and add or delete processes. Administrators are also the general managers of the platform (View Fig. 6. Management business process model).

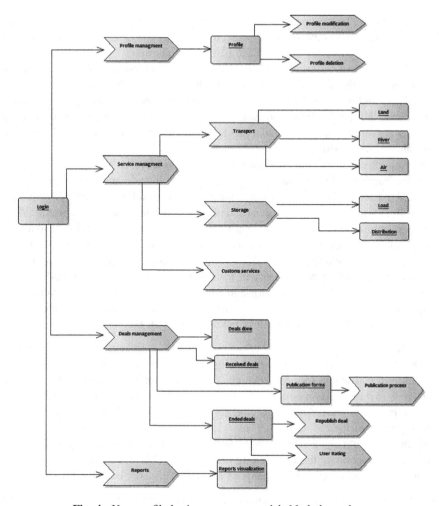

Fig. 4. User profile business process model. *Made by authors.*

Fig. 5. Search business process. *Made by authors.*

4.3   Data Layer: (E-Marketplace)

A database was built in this layer based in the business process designed for the persistence of the information system's data. The relational model of the database includes tables, attributes, and the primary and foreign keys. There are two main goals of this database: Persistence of the data, as mentioned before, and data analysis. Due to

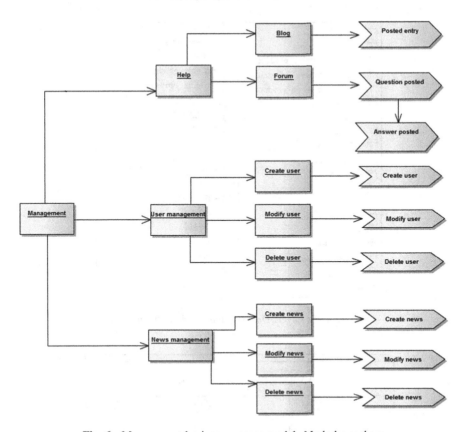

Fig. 6. Management business process model. *Made by authors.*

the importance of the information, there is a transformation of all data and transactions to information, through statistics and representative reports to analyze the market and make decisions. (View Fig. 7. Data layer: Database tables).

4.4   Applications Layer

The application layer represents the model of the platform's components for e-marketplace. It is modeled using the UML notation for describing components, interfaces, structures, and service interfaces under a multilayer model (View Fig. 8. Applications layer).

4.5   Infrastructure Layer

The information system was deployed in an Amazon Web Service that offered all the hardware requirements needed and the Windows Server operating system, where all the third party applications and the e-marketplace are deployed. So all a client has to do is to log in the information system through his personal computer or tablet to access it via

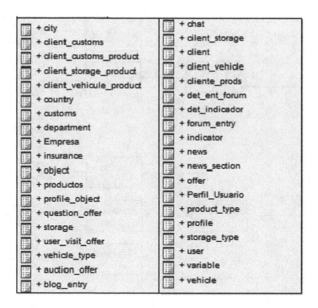

Fig. 7. Data Layer: Database tables. *Made by authors.*

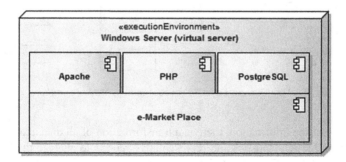

Fig. 8. Applications layer. *Made by authors.*

his internet service through the web server: all the processes are executed in the configured environment. (View Fig. 9. Infrastructure layer)

4.6   Verification and Validation Layer

The verification and validation processes are a part of the comprehensive process management for software quality. In each of the cases, there are a defined set of metrics designated as FURPS – Functionality, Usability, Reliability, Performance and Security (View Fig. 10. Verification and Validation layer).

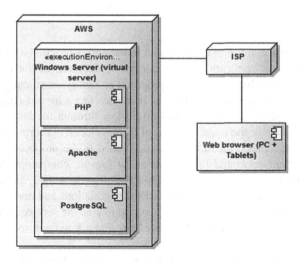

Fig. 9. Infrastructure layer. *Made by authors.*

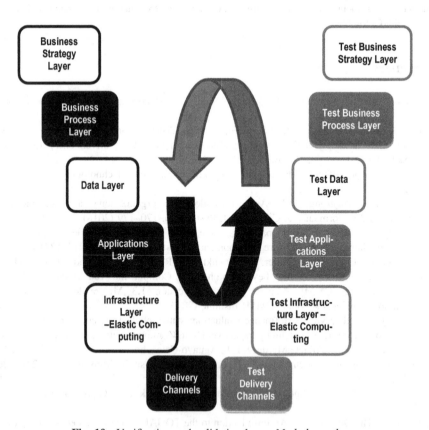

Fig. 10. Verification and validation layer. *Made by authors.*

5 Conclusions

The correct definition and use of the proposed framework lets achieve a successful case of study and business application and a correct deployment of the platform in the proposed infrastructure. By following the proposed phases systematically and using the layers proposed in the framework, business processes are achieved correctly and can cover every required aspect for the information flow through the business processes. This framework will ease the improvements of the developed platform and is pretended to be used for future platforms to be developed based in the proposed work area.

The development presented herein will allow logistic companies that are in the transport, storage, and customs business connect in an agile, secure, and effective way. This will translate into increasing the value of the negotiations and decrease the time of order fulfillment. The logistic chain integration optimizes the services offered towards companies to work in creating new business lines that allow them to group up.

Acknowledgments. This Project was funded using resources from: "Sistema General de Regalías" within the framework of Project LOGPORT and with the support of Government of Atlántico, Government of Bolívar, Government of Sucre, and executed by Universidad del Norte, Barranquilla-Colombia

References

1. Arvis, J., Mustra, M., Panzer, J., Ojala, L., Naula, T.: Connecting to compete, World Bank, mimeo, p. 72, 2014
2. Ranganathan, C., Dhaliwal, J.S., Teo, T.S.H.: Assimilation and diffusion of web technologies in supply-chain management: an examination of key drivers and performance impacts. Int. J. Electron. Commer. **9**(1), 127–161 (2004)
3. Sila, I.: Factors affecting the adoption of B2B E-commerce technologies, vol. 13, no. 2 (2013)
4. Tripp, J.F., Armstrong, D.J.: Agile methodologies: organizational adoption motives, tailoring, and performance. J. Comput. Inf. Syst. **58**(2), 170–179 (2018)
5. Dingsøyr, T., Nerur, S., Balijepally, V., Moe, N.B.: A decade of agile methodologies: towards explaining agile software development. J. Syst. Softw. **85**(6), 1213–1221 (2012)
6. Gampfer, F., Jürgens, A., Müller, M., Buchkremer, R.: Past, current and future trends in enterprise architecture—A view beyond the horizon. Comput. Ind. **100**, 70–84 (2018)
7. CMMI Institute, CMMI ® para Servicios, Versión 1.3 Mejorando procesos para proporcionar mejores servicios CMMI Institute (2013)
8. Duan, Q.: Cloud service performance evaluation: status, challenges, and opportunities – a survey from the system modeling perspective. Digit. Commun. Netw. **3**(2), 101–111 (2017)
9. Palos-Sanchez, P.R., Arenas-Marquez, F.J., Aguayo-Camacho, M.: Cloud computing (SaaS) adoption as a strategic technology: results of an empirical study. Mob. Inf. Syst. **2017**, 20 (2017)
10. Ugavina, N.: MDG Technology For Zachman Framework User Guide. Sparx Syst., 51 (2008)
11. Josey, A.: The Open Group, An Introduction to the TOGAF® Standard, Version 9.2 (2018)
12. Object Management Group, Unified Profile for DoDAF and MODAF (UPDM) (2013)

13. Godinez, M., Hechler, E., Koenig, K., Lockwood, S., Oberhofer, M., Schroeck, M.: The Art of Enterprise Information Architecture: A Systems-Based Approach for Unlocking Business Insight, 1st Editio. IBM Press, Boston (2010)
14. Medina-Dominguez, F., Sanchez-Segura, M.I., Mora-Soto, A., Amescua, A.: Reverse engineering and software products reuse to teach collaborative web portals: a case study with final-year computer science students. IEEE Trans. Educ. **53**(4), 595–607 (2010)
15. AWERY, AWERY. https://awery.aero. Accessed 31 Jul 2019
16. PL Systems Inc., 3PL Systems Inc. http://3plsystems.com. Accessed 31 Jul 2019
17. Road, A.: Amber Road. https://www.amberroad.com. Accessed 31 Jul 2019
18. Mercatrans, "Mercatrans". https://www.mercatrans.com. Accessed 31 Jul 2019
19. Popovic, I., Vrtunski, V., Popovic, M.: Formal verification of distributed transaction management in a SOA based control system. In: Proceedings - 18th IEEE International Conference Work Engineering Computer System ECBS 2011, pp. 206–215 (2011)
20. Marschall, F., Schoemnakers, M.: Towards model-based requirements engineering for web-enabled B2B applications. In: Proceedings - 10th IEEE International Conference Work Engineering Computer System ECBS 2003, pp. 312–320 (2003)
21. Espinosa, E., Junco, A., Ramos, F., Ramirez, J.: One step beyond: making B2B decisions with business activity monitoring and soccer. In: Conference Proceedings - IEEE International Conference System Man Cybernetics, vol. 3, pp. 2084–2089 (2004)
22. Talevski, A., Chang, E., Dillon, T.S.: Reconfigurable web service integration in the extended logistics enterprise. IEEE Trans. Ind. Inform. **1**(2), 74–84 (2005)
23. Siricharoen, W.V.: Applying prototype for software engineering in pilot project of research tracking system. In: 2012 Second International Conference on Digital Information and Communication Technology and it's Applications (DICTAP), pp. 172–176 (2012)
24. Yihua, Z.: The credit mechanism loss with the development of E-commerce. In: 2008 International Conference on Computer Science and Software Engineering, pp. 608–611 (2008)
25. Rahman, A.A., Sahibuddin, S., Ibrahim, S.: A unified framework for software engineering process improvement - a taxonomy comparative analysis. In: 2011 5th Malaysian Conference Software Engineering MySEC 2011, no. December, pp. 153–158 (2011)
26. Yuan, M., Fan, L.: ATS software framework design pattern and application. In: Proceedings - 5th International Conference on Instrumentation and Measurement, Computer, Communication, and Control, IMCCC 2015, pp. 141–146 (2016)
27. Baharuddin, R., Singh, D., Razali, R.: Usability dimensions for mobile applications-a review. Res. J. Appl. Sci. Eng. Technol. **5**(6), 2225–2231 (2013)
28. Belete, G., Voinov, A., Morales, J.: Designing the distributed model integration framework – DMIF. Environ. Model Softw. **94**, 112–126 (2017)
29. Siavvas, M.G., Chatzidimitriou, K.C., Symeonidis, A.L.: QATCH - an adaptive framework for software product quality assessment. Expert Syst. Appl. **86**, 350–366 (2017)
30. Zhimin, W., Zheng, Q.: Mobile agent oriented architecture to build open mobile electronic commerce system. In: Proceedings of IEEE TENCON 2002, pp. 170–175 (2002)
31. Yang, H., Xu, B., Zhou, Y., Zhang, J., Biao, D.: Return in B2C E-Commerce Enterprises. In: 2010 Third International Conference Business Intelligent Finance Engineering, pp. 95–98 (2010)
32. Zhang, W., Kunz, T.: Product line based ontology reuse in context-aware E-business environment. In: Proceedings - IEEE International Conference E-business Engineering ICEBE 2006, pp. 138–145 (2006)
33. Bhattacherjee, A.: Acceptance of E-commerce services: the case of electronic brokerages. IEEE Trans. Syst. Man Cybern. Part A Syst. Hum. **30**(4), 411–420 (2000)

34. He, M., Jennings, N.R., Leung, H.: On agent-mediated electronic commerce. IEEE Trans. Knowl. Data Eng. **15**(4), 985–1003 (2003)
35. Ganguly, D., Chakraborty, S.: E commerce - forward and reverse auction - a managerial tool to succeed over business competitiveness. In: Proceedings 9th ACIS International Conference Software Engineering Artificial Intelligent Networks Parallel/Distributed Computer SNPD 2008 2nd International Work Advantage Internet Technology Applied, pp. 447–452 (2008)

A Decision Support Tool
for Energy-Optimising Railway
Timetables Based on Behavioural Data

Mathias Bejlegaard Madsen[1(✉)], Matthias Villads Hinsch Als[1],
Rune M. Jensen[1], and Sune Edinger Gram[2]

[1] IT University of Copenhagen, Rued Langaards Vej 7, 2300 Copenhagen S, Denmark
{mabm,matt,rmj}@itu.dk
[2] Cubris - A Thales Company, Mileparken 22, 2740 Skovlunde, Denmark
s.gram@cubris.dk

Abstract. Energy-efficient train operation can reduce operating costs
and contribute to a reduction in CO_2 emissions. To utilise the full
potential of energy-efficient driving, energy-efficient timetabling is crucial. To address this problem, we propose a decision support tool to
give timetable planners insight into energy consumption for a given
timetable. The decision support tool uses a recommendation based on
quadratic optimisation of a given timetable. Differently to previous work,
the optimisation uses actual data from the train operation, which is pre-
processed by data reduction, outlier detection, and second-degree regression modelling. With this approach, our results show that the optimised
timetables can save up to 33.07% energy on a single section and up to
6.23% for a complete timetable. Solutions are computed in less than a
microsecond.

Keywords: Energy-efficient train timetabling · Decision support ·
Data mining · Quadratic optimisation

1 Introduction

The threat of climate change urges for energy-efficient solutions to the transportation sector. Even though railway provides one of the most energy-efficient
forms of transportation, there is still significant potential for reduction in energy
consumption [2]. Energy-efficient train operation is one approach that both can
reduce operating costs and contribute to a reduction in CO_2 emissions [8]. To
utilise the full potential of energy-efficient driving, energy-efficient timetabling
is crucial. Timetable planners are experts in the complex process of railway
timetabling, which has to conform to several known internal and external factors. What timetable planners seldom know is how changes to a timetable affect
the total energy consumption. For that reason, insights into minimisation of

This research is conducted in collaboration with Cubris - A Thales Company.

C. Paternina-Arboleda and S. Voß (Eds.): ICCL 2019, LNCS 11756, pp. 397–412, 2019.
https://doi.org/10.1007/978-3-030-31140-7_25

energy consumption in railway timetables are a key aspect in reducing CO_2 emissions, while sustaining a business opportunity of the consequently reduced costs.

The problem has attracted attention in recent years. Sicre et al. [10] use Pareto frontiers for optimisation to redistribute slack time of a journey. Gupta et al. [7] use a linear programming optimisation model to optimise energy consumption. The approach takes many constraints into account, including trip time, dwell time, headway, cross-over, connection, and total time. The model is applied to optimise a metro spanning a full-service period of one day. The worst-case optimisation gave a 19.27% improvement, while the best case gave a 21.61% improvement. Cucala et al. [5] suggests a fuzzy linear programming model taking the driver's behavioural response into account. The paper furthermore models uncertainty in delays as fuzzy numbers. The approach was tested on a high-speed line in Spain, which achieved a decrease of 6.7% in energy consumption compared to the timetable in service. Although all aforementioned papers achieved acceptable results in a decrease of energy consumption, the methods were modelled using simulated data. As such, the data does not reflect real world restrictions and conditions of a running railway network. This will inevitably create deviating predictions of energy consumption compared to actual measurements, making it insufficient for a decision support tool.

Scheepmaker et al. [9] distributes the available slack time uniformly. The paper is based on timetables from the Dutch railway and achieved a 7.2% decrease in energy consumption. In practice, a uniform distribution would only be sufficient in situations where enough slack time is available. With less slack time available, the paper would achieve worse results than papers [5,7,10], since it does not prioritise the distribution to minimise energy consumption.

In this paper, we propose a decision support tool to give timetable planners insight into energy consumption for a given timetable to address the problem mentioned above. To our knowledge, we contribute the first end-to-end analysis of real data from rolling stock and show how it can be used to provide an optimised timetable based on the train drivers' actual behaviour. The result is an interactive and intuitive decision support tool letting the timetable planner adjust the optimised timetable according to their domain knowledge and experience.

The remainder of this paper is structured as follows. Section 2 formulates the problem and Sect. 3 proposes an approach to solve it. Section 4 evaluates the approach and its design choices. Last, Sect. 5 concludes and discusses future work.

2   Problem Formulation

Railway timetable planning is a complex process that includes several hard and soft constraints. The hard constraints such as network capacity, rolling stock availability and headway are known by the timetable planner, as opposed to the soft constraints such as punctuality and energy consumption. Though, it is infeasible to include all constraints to the timetable-optimisation. Train timetabling,

in general, is NP-Hard [3] due to the combination of several constraints and requirements. To reduce complexity, one compromise is to consider the soft constraints in a decision support tool and let the timetable planner make valid adjustments according to the hard constraints. The estimation of soft constraints can be approached either theoretically or operationally. The latter is the focus of this paper and requires data from train operation.

To this end, we have collaborated with Cubris [4]. Cubris develops a driver advisory system named GreenSpeed. GreenSpeed improves punctuality and reduces energy consumption. GreenSpeed has been deployed for multiple train operating companies. Regarding optimisation of their customers' operation, Cubris can with permission gain access to massive quantities of data produced by GreenSpeed over the years. Data based on more than 140,000 train runs were accessible for this paper. This data contains information about how train operators have been driving the trains. The data includes precise energy consumption and exact arrival and departure times as shown in Fig. 1(a). For this paper, we assume that the trains already are equipped with GreenSpeed. All data from Cubris is normalised and anonymised.

As mentioned in the introduction, the key idea is to give insights into the energy consumption of a timetable. Based on actual data, an intuitive decision support tool is proposed to give timetable planners insight into the energy efficiency of current timetables. Comparing a timetable to another is one of the approaches available when deciding how a timetable should be adjusted. To give the user an intelligent way of evaluating the energy efficiency of a timetable, a recommendation of the optimal timetable can be visualised, while enabling the user to adjust the timetable to match hard constraints. The user's adjustments will be restricted within the range of actual data to provide the highest accuracy of actual energy consumption. The recommendation is limited to redistribute the slack time given in the timetable only. The decision support tool only modifies one timetable at once with the data that identical timetables can provide.

3   Solution Approach

We propose a five-step approach to the solution: (1) data reduction, (2) outlier detection, (3) regression modelling, (4) optimisation of energy functions, and (5) decision support tool. We describe these steps in the following sections.

3.1   Data Reduction

To be able to predict the energy consumption with high accuracy, the data has to be pre-processed. The pre-processing removes invalid data such as sections with no actual energy consumption, arrival and departure times, and train formation specified. Then the data is reduced to a subset containing only the relevant features which have an impact on energy consumption. A plot showing a section with no reduction is omitted due to the high number of data points.

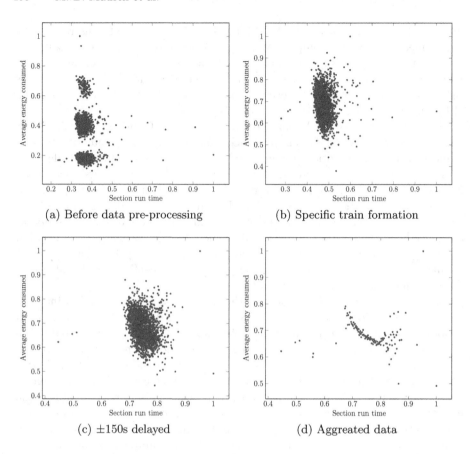

Fig. 1. Data reduction process.

First, the planned section run time affects the slope of the energy function, and the distribution of points being relative to the planned arrival time. The planned run time is the time between departure at the previous station and arrival at the next station. Figure 1(a) shows the result of only looking at one section run time. A detailed comparison and argumentation for this choice can be found in Sect. 4.1. Second, different train formations consume different amounts of energy based on the specifications of the train[1] and how many trains and carriages are linked together. As seen in Fig. 1(a), the data points are clustered in three groups, each representing different train formations. Figure 1(b) shows the same plot after section run time and train formation has been constrained to only include identical values respectively. It should be noted, that due to a high number of data points, the figure shown has been sampled to fewer points. Third, our experiments in Sect. 4.1 show that outliers often are a result

---

[1] Examples are engine type, length, and weight.

of drivers with a high accumulated delay. Based on the experiments, an interval of ±150 s from the planned arrival time was introduced. Figure 1(c) shows the impact of removing highly delayed train runs. Last, to avoid having multiple energy consumptions per run time, the run times are aggregated and the energy consumption is averaged for each run time. Figure 1(d) shows the final result.

A trend is now clearly visible in the plotted data. Not surprisingly, the trend shows that if the train driver has less time to drive the section, the train will consume more energy. To further improve the accuracy and quality of the data, additional methods for outlier detection will be discussed next.

3.2   Outlier Detection

In Sect. 3.1, we discarded points that did not belong to the context of interest by setting constraints on the values. The objective of this section is to remove data points which do not conform to the trend. In other words, the outliers are removed. Outliers in this domain include extraordinary driving conditions. This includes bad weather, malfunctions and signal errors. If not removed, the outliers reduce data quality and lower prediction accuracy. We use unsupervised machine learning techniques for outlier detection. Especially density-based techniques are relevant because of the dense main cluster as shown in Fig. 2. This completely disqualifies widely used centroid-based clustering algorithms. Therefore, the density-based clustering algorithm DBSCAN [6] was chosen for the solution. DBSCAN's robustness to noise makes it a good fit. DBSCAN can use several distance measures; in this work, the Euclidean distance is used.

The data for each section will vary when it comes to the distance between points which roots in the section run properties, i.e. the section to be run and the train formation. Normalisation even out a lot of the density variations, but extreme outliers will also affect the perceived density of the main cluster if not caught by the reduction in Sect. 3.1. When the data has been clustered, noise and all clusters not being the main cluster are removed.

Figure 2 shows an example of detected outliers with DBSCAN. However, DBSCAN only produces a good output if the parameters ε and $MinPts$ are chosen properly. The ε-parameter is the maximum distance between two points to be assigned to the same cluster, and Sect. 4.2 describes how we arrived at $\varepsilon = 0.08$ as a value that provides a satisfactory result. The reader should note that a satisfactory outlier removal does not mean a perfect outlier removal, as the latter cannot be defined properly.

The other parameter to be chosen is $MinPts$. $MinPts$ is the minimum number of data points to form a cluster. This parameter is less sensitive in this use case, though, it cannot be too large. If the parameter value is too large, many points will be considered noise and will be discarded, because there are not enough data points to form the main cluster in low-frequency data sets. If the parameter value is too low, DBSCAN might make a lot smaller local clusters alongside the main cluster. Setting it too low, will not affect the main cluster at all, and the small clusters are filtered out anyway. Therefore, the $MinPts$-value is set to be 10.

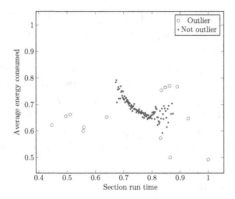

Fig. 2. DBSCAN, $\varepsilon = 0.08$, $MinPts = 10$.

3.3 Regression Modelling

Our goal is to predict the energy consumption for a given amount of time spent running a section. We use regression for this purpose. A regression model is calculated for each section on a journey such that it can be used for optimising the time spent with regards to energy consumption in Sect. 3.4.

By visual inspection of Fig. 1(d), the formation of the points looks like it follows a trend of a second-degree polynomial. This makes good sense if we consider the formula for kinetic energy in classical mechanics:

$$E_k = \frac{1}{2}mv^2, \tag{1}$$

where m is the mass, v is the velocity and E_k is the kinetic energy. If the train driver has less time to drive the section, they will have to drive with an increased velocity to keep up with the timetable. Hence, the negative change in time to drive the section will increase the consumed energy quadratically. Figure 3 shows second-degree polynomials fitted to four different datasets.

Figure 3 shows that the regression model will predict a rise in energy consumption after the vertex. Based on the formula for kinetic energy, a rise in energy consumption is only expected when the velocity is negative. For this reason, the energy function should be discarded after the vertex.

3.4 Optimisation of Energy Functions

From regression, an energy function was obtained for each section in a timetable. Each section is denoted by $s_1, s_2, ...s_n$, where n is the number of sections contained in a timetable. The energy function for each section is denoted by $e_1(t_1), e_2(t_2), ..., e_n(t_n)$, where t_n is the section run time.

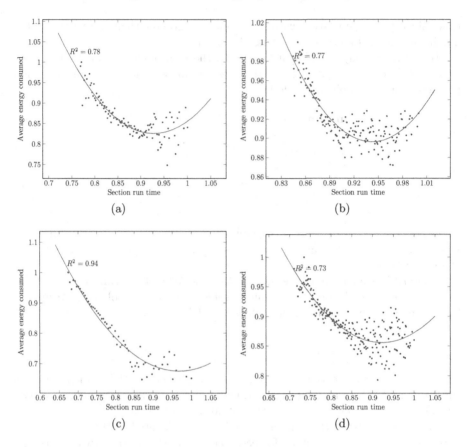

Fig. 3. Average energy consumption and run time for section runs without outliers, fitted with a second-degree polynomial. Example with four different data sets.

Objective Function. Given the energy function for each section, the objective function is the sum of energy functions describing the total energy consumption of the timetable, denoted E:

$$E(t) = \sum_{i=1}^{n} e_i(t_i) \tag{2}$$

Constraints. From the domain described in Sect. 2, the constraints (3b), (3c), and (3d) have been identified. Allowing the section run times to be increased until the minimum has been reached for all $e_i(t_i)$ in E will result in the best-optimised timetable. However, the cost of doing this is a new timetable which total time of the journey is increased. Equation (3b) ensures that the total run time is equal to the current timetable. Equation (3c) constraints each section by a minimum run time, since physical limitations of the train and tracks restricts the speed of the train. Equation (3d) constraints each section by a maximum run

time, since the optimisation variable cannot be greater than the vertex of the quadratic energy function, as discussed in Sect. 3.3.

To output an energy-efficient timetable, the sum of energy functions for the timetable described by Eq. 2, is sought to be minimised. Hence, the optimisation problem can be established as follows:

$$\textbf{minimise } E(t) \tag{3a}$$

$$\textbf{subject to } \sum_{i=1}^{n} t_i = \sum_{i=1}^{n} C_i \tag{3b}$$

$$t_{i_{min}} \leq t_i, \qquad 0 < i \leq n \tag{3c}$$

$$e_i'(t_i) \leq 0, \qquad 0 < i \leq n, \tag{3d}$$

where t is the vector of decision variables $t = (t_1, t_2, ..., t_n)$, n is the number of sections in the timetable, and C is the current timetable containing the run time of each section.

Since $E(t)$ is a sum of convex quadratic functions, the optimisation problem (3a) is a convex quadratic programming problem that can be solved to optimality by a standard commercial solver such as CPLEX.

3.5 Decision Support Tool

The purpose of this section is to present a user interface (UI) which supports timetable planners in making energy-efficient decisions. This is achieved by visualising the cost in energy consumption of adjusting the slack time.

The central part is the timetable configurator, located in the middle of the screen shown in Fig. 4. It shows for each section on the journey two sliders; the lower one with two handles to adjust the departure time and arrival time, and the upper one is fixed and shows the original configuration of the timetable. Initially, the lower slider is configured to show the energy-optimised timetable. When the timetable planner adjusts the energy-optimised timetable with their domain knowledge, the changes are reflected in the estimated difference in energy consumption of the section, which is shown just above the sliders. The estimated energy consumption for the whole journey is shown in the upper right portion of the screen. This enables the timetable planner to use energy consumption as a measure of the overall performance of the timetable, by studying how sections contribute differently to the total energy consumption.

4   Experimental Results

The purpose of our experiments is to evaluate each step conducted in the data reduction and the effect of the parameters for outlier detection. In addition, optimised timetables are compared to the original timetables to evaluate the distribution of slack time and energy reduction. The experiments are based on data from actual train runs conducted in a one year period.

Fig. 4. View for configuring the timetable.

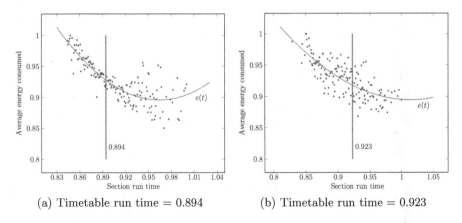

(a) Timetable run time = 0.894   (b) Timetable run time = 0.923

Fig. 5. Different run times affecting the energy consumption of section runs between two stations.

4.1   Data Reduction

Section Run Time. A section can be a part of multiple timetables with differing planned section run time. Figure 5 shows how two different section run times of the same section affects the slope of the energy function. The same figure shows that the slope of the energy functions is different. The separation of planned run times is therefore important since the slope of the energy function determines the cost of changing the run time. Another issue is that the distribution of points is relative to the planned run time. However, this is not visible in Fig. 5 probably due to the difference between the planned run times is too small to have an effect.

Train Formation. The length and weight of the train depends on the number and type of carriages. Different formations of trains thereby consume different amounts of energy. Figure 6(a) and (b) show typical examples of the energy consumption of two different train formations. From the figures it is noticed that the train formation in Fig. 6(a) to the left has a steeper energy consumption compared to the train formation in Fig. 6(b). It impacts energy consumption especially if an early arrival time is chosen for Fig. 6(a). The cost of choosing an arrival time for Fig. 6(b) is not as high. The separation of train formation is therefore important, even though both train formations are in the same range of energy consumption.

Level of Delay. The delay constraint is chosen to be ±150 s to remove outliers. This choice is based on an evaluation of the following delays: 30, 90, 150, 210, and 270 s for five different sections. Figure 7(a) shows a plot for an example with all levels of delay. The figure shows that a large portion of outliers are to be removed, while Fig. 7(b) shows that the energy trend is preserved after the removal.

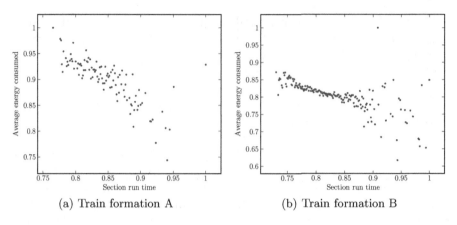

Fig. 6. Different train formations affecting the energy consumption for section runs between two stations.

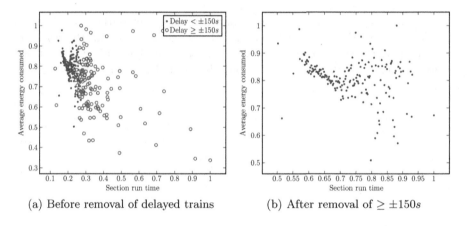

Fig. 7. Level of delay affecting the energy consumption for section runs between two stations.

4.2   Parameter-Tuning of ε in DBSCAN

Four ε-values were chosen: 0.05, 0.08, 0.1 and 0.15. Each value was applied to five data sets extracted from different sections with varying properties, including train formation, data frequency, the visual shape of the data, planned run time, and the density. A total of 20 plots were inspected. If DBSCAN found more than one cluster, the biggest cluster was marked as "not outlier" and everything else as "outlier". The data was outlier detected after data reduction and aggregation.

The value 0.05 marks too many points as outliers in at least two obvious cases as shown in Fig. 8(a) and (b). In Fig. 8(a), an important part of the trend on the upper left side was marked as outliers. In Fig. 8(b) only a small fraction of the original data is remaining.

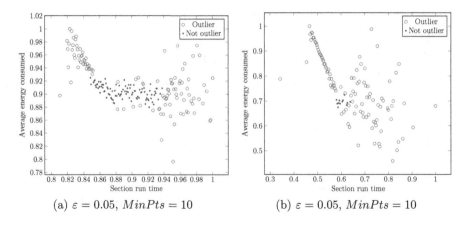

(a) $\varepsilon = 0.05$, $MinPts = 10$ (b) $\varepsilon = 0.05$, $MinPts = 10$

Fig. 8. Comparison of $\varepsilon = 0.05$ on two different data sets.

On the other hand, the ε-value 0.15 barely removes any outliers, as shown in Fig. 9(a) and (b). Here, it is evident that the sparse area around the centre of Fig. 9(a) should have been reduced a bit more to avoid noise. Figure 9(b) shows outliers around $(0.7, 0.2)$ and $(0.75, 0.55)$ that would interfere too much with further data analysis.

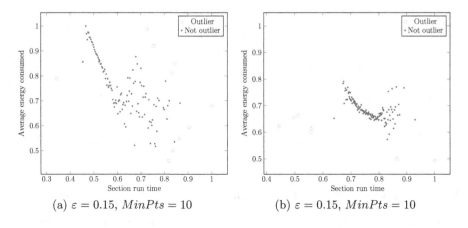

(a) $\varepsilon = 0.15$, $MinPts = 10$ (b) $\varepsilon = 0.15$, $MinPts = 10$

Fig. 9. Comparison of $\varepsilon = 0.15$ on two different data sets.

Based on Fig. 10, the ε-parameter is chosen to be 0.08, as it removed a satisfactory amount of noise compared to $\varepsilon = 0.1$. The noise is mainly generated by the behavioural response of the train driver using too much time on the section, consuming more energy to keep up with the timetable. That also proves the limitation of the solution; it should only be used for smaller changes in the timetables as bigger changes will lead to more significant deviations in the prediction due to the behavioural noise.

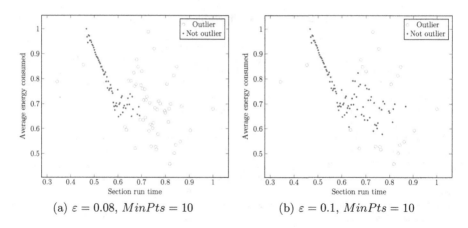

(a) $\varepsilon = 0.08$, $MinPts = 10$ (b) $\varepsilon = 0.1$, $MinPts = 10$

Fig. 10. Comparison of different ε-values on the same data set.

4.3 Optimisation Results

The solution proposed in Sect. 3, will be evaluated based on two examples from a currently working timetable. Both examples were calculated in less than a microsecond using an Intel Core i7-8700K CPU at 3.7 GHz with 32 GB of RAM. The evaluation will compare the actual energy consumption of the current timetable to the energy consumption in the suggested timetable. The comparison will show how the redistribution of the slack time will achieve a decrease in energy consumption by changing the arrival and departure times of each section. Table 1 shows this comparison. It can be seen that the solution suggests taking all the slack time of the last section from S7 to S8, and 13 s from section S6 to S7. This gives a total of 4 min and 28 s to distribute to other sections. One of the sections receiving more slack time is S4 to S5. By adjusting the slack time from 1 min and 58 s to 3 min and 50 s, the energy consumption of that section decreased by 33.07%. Another section which benefitted from an increase in slack time is the section from S2 to S3, which resulted in a decrease in energy consumption by 11.71%. Looking at the average total energy consumption of the

Table 1. Optimised timetable comparison.

	Current timetable			Optimised timetable			
	Run time hh:mm:ss	Slack hh:mm:ss	Avg. energy consumption	Run time hh:mm:ss	Slack hh:mm:ss	Avg. energy consumption	Difference %
S1–S2	00:11:30	00:01:26	0.753	00:11:57	00:01:58	0.739	−1.86
S2–S3	00:06:00	00:01:36	0.333	00:06:38	00:02:13	0.294	−11.71
S3–S4	00:07:30	00:02:46	0.495	00:08:09	00:03:25	0.464	−6.26
S4–S5	00:05:00	00:01:58	0.251	00:06:52	00:03:50	0.168	−33.07
S5–S6	00:08:00	00:00:31	0.493	00:08:50	00:01:21	0.444	−9.94
S6–S7	00:18:30	00:02:00	0.994	00:18:18	00:01:47	1.000	0.60
S7–S8	00:28:30	00:04:15	0.936	00:24:16	00:00:00	0.952	1.71
Total	01:25:00	00:14:32	4.255	01:25:00	00:14:32	4.061	−4.56

Table 2. Optimised timetable comparison.

	Current timetable			Optimised timetable			
	Run time hh:mm:ss	Slack hh:mm:ss	Avg. energy consumption	Run time hh:mm:ss	Slack hh:mm:ss	Avg. energy consumption	Difference %
S8–S9	00:07:00	00:01:11	0.630	00:05:49	00:00:00	0.648	2.83
S9–S10	00:07:30	00:00:52	0.917	00:06:38	00:00:00	0.849	−7.42
S10–S11	00:03:00	00:00:48	0.347	00:03:04	00:00:51	0.342	−1.44
S11–S12	00:05:00	00:00:47	0.829	00:04:17	00:00:04	0.880	6.15
S12–S13	00:05:30	00:00:59	0.708	00:05:26	00:00:54	0.712	0.56
S13–S14	00:04:00	00:00:24	0.467	00:05:04	00:01:28	0.358	−23.34
S14–S15	00:03:30	00:00:35	0.605	00:03:49	00:00:54	0.564	−6.78
S15–S16	00:06:30	00:02:47	1.000	00:05:09	00:01:26	0.955	−4.50
S16–S17	00:03:30	00:00:38	0.544	00:04:42	00:01:50	0.376	−30.88
S17–S18	00:03:30	00:00:36	0.595	00:04:06	00:01:11	0.518	−12.94
S18–S19	00:02:30	00:00:18	0.357	00:02:46	00:00:34	0.323	−9.52
S19–S20	00:05:30	00:00:20	0.786	00:05:49	00:00:38	0.755	−3.94
S20–S21	00:03:00	00:00:24	0.450	00:03:26	00:00:50	0.385	−14.44
S21–S22	00:01:30	00:00:00	0.000	00:01:30	00:00:00	0.000	0.00
S22–S23	00:02:00	00:00:00	0.000	00:02:00	00:00:00	0.000	0.00
S23–S24	00:05:00	00:00:56	0.546	00:04:40	00:00:36	0.569	4.21
Total	01:08:30	00:11:35	8.781	01:08:30	00:11:35	8.234	−6.23

timetable from S1 to S8, the energy consumption of the current timetable can be decreased by 4.5%. Of course, this decrease is only possible if the suggested timetable is implemented without modification. However, the decision support tool allows the timetable planner to adjust the suggested timetable to fulfil external requirements. For example, it might not be possible for all timetables to remove all the slack time to S8 since this is a central station having trains nonstop arriving or departing. These constraints are not taken into account in this paper. The decision support tool gives the timetable planner the insights to which sections are expensive in energy consumption to take into consideration when modifying a timetable. Table 2 shows a timetable containing two sections with no data available. These sections will have an energy consumption of 0 and are constrained to have the same run time as the current timetable. As also seen in Table 1, the solution removes all the slack time from the sections leading to S8. Though, removing all the slack time from a section comes with a high risk. The train cannot be delayed when departing from S8 or on the way towards S9. The train cannot drive any faster, since removing all the slack time sets the run time to be the lowest possible. The suggested timetable shows that for example the section from S16 to S17 and the section from S13 to S14 should be prioritised when distributing slack time to minimise the total energy consumption. The energy consumption of the current timetable can be decreased by 6.23%. The solutions show that the optimised timetables are able to save up to 33.07% energy on a single section and up to 6.23% for a complete timetable.

5   Conclusion and Future Work

This paper has introduced an intuitive decision support tool to support timetable planners in making energy-efficient timetables. The tool gives a recommendation based on actual data. The goal was to give timetable planners insight into how train timetables can be energy-optimised based on actual data. To our knowledge, our approach is the first based on actual data from rolling stock.

With this approach, our results show that optimised timetables can save up to 33.07% energy on a single section and up to 6.23% for a complete timetable. Solutions are computed in less than a microsecond.

In future work, we plan to improve the optimisation to produce a more realistic optimum. One improvement is to utilise data from differently configured run times in the timetables to improve the prediction of energy consumption. Another improvement is the constraints for the optimisation. As seen in the results, a solution could choose to remove all slack time for a section. The optimisation model can be given a minimum slack time constraint to solve this. Known constraints such as availability of tracks and platforms could also be taken into account in the optimisation.

Another point for future work is to improve the outlier detection. Alternatives to DBSCAN exist which do not need ε-parameter to be tuned. OPTICS [1] and hierarchical clustering are possible candidates for further investigation. OPTICS build on the same principles of DBSCAN, but it evaluates the local density relative to the individual data points. However, OPTICS needs a maximum ε to reduce computation time. On the other hand, Hierarchical clustering is not density-based, thus, avoids the ε-parameter altogether.

References

1. Ankerst, M., Breunig, M.M., Kriegel, H.P., Sander, J.: Optics: ordering points to identify the clustering structure. In: Proceedings of the 1999 ACM SIGMOD International Conference on Management of Data, SIGMOD 1999 pp. 49–60. ACM (1999). https://doi.org/10.1145/304182.304187
2. Barrow, K.: Laying the foundations for energy-efficient traction. Int. Rail J. (2019). https://www.railjournal.com/in_depth/energy-efficient-traction/. Accessed 08 June 2019
3. Caprara, A., Fischetti, M., Toth, P.: Modeling and solving the train timetabling problem. Oper. Res. **50**(5), 851–861 (2002). https://doi.org/10.1287/opre.50.5.851.362
4. Cubris - A Thales Company: (2019), http://www.cubris.dk/. Accessed 26 June 2019
5. Cucala, A., Fernández, A., Sicre, C., Domínguez, M.: Fuzzy optimal schedule of high speed train operation to minimize energy consumption with uncertain delays and driver's behavioral response. Eng. Appl. Artif. Intell. **25**(8), 1548–1557 (2012). https://doi.org/10.1016/j.engappai.2012.02.006

6. Ester, M., Kriegel, H.P., Sander, J., Xu, X.: A density-based algorithm for discovering clusters a density-based algorithm for discovering clusters in large spatial databases with noise. In: Proceedings of the Second International Conference on Knowledge Discovery and Data Mining. KDD 1996, pp. 226–231. AAAI Press (1996). http://dl.acm.org/citation.cfm?id=3001460.3001507. Accessed 08 June 2019
7. Gupta, S.D., Tobin, J.K., Pavel, L.: A two-step linear programming model for energy-efficient timetables in metro railway networks. Transp. Res. Part B: Methodol. **93**, 57–74 (2016). https://doi.org/10.1016/j.trb.2016.07.003
8. International Union of Railways: Technologies and potential developments for energy efficiency and co2 reductions in rail systems (2016). https://uic.org/IMG/pdf/_27_technologies_and_potential_developments_for_energy_efficiency_and_co2_reductions_in_rail_systems._uic_in_colaboration.pdf. Accessed 08 June 2019
9. Scheepmaker, G., Goverde, R.M.: Running time supplements: energy-efficient train control versus robust timetables. In: Tomii, N., Hansen, I., Hirai, C. (eds.) Proceedings of the 6th international conference on railway operations modelling and analysis, RailTokyo2015, pp. 1–20. International Association of Railway Operations Research (2015)
10. Sicre, C., Cucala, P., Fernández, A., Jiménez, J.A., Ribera, I., Serrano, A.: A method to optimise train energy consumption combining manual energy efficient driving and scheduling. Comput. Railways **114**, 549–560 (2011). https://doi.org/10.2495/CR100511

Evaluation of Bottom-Up and Top-Down Strategies for Aggregated Forecasts: State Space Models and ARIMA Applications

Milton Soto-Ferrari[1]([✉]), Odette Chams-Anturi[2],
Juan P. Escorcia-Caballero[3], Namra Hussain[1], and Muhammad Khan[1]

[1] Indiana State University, Terre Haute, IN, USA
milton.soto-ferrari@indstate.edu,
{nhussain1,mkhan12}@sycamores.indstate.edu
[2] Universidad de la Costa, Barranquilla, Colombia
ochams@cuc.edu.co
[3] Universidad del Norte, Barranquilla, Colombia
juane@uninorte.edu.co

Abstract. In this research, we consider monthly series from the M4 competition to study the relative performance of top-down and bottom-up strategies by means of implementing forecast automation of state space and ARIMA models. For the bottom-up strategy, the forecast for each series is developed individually and then these are combined to produce a cumulative forecast of the aggregated series. For the top-down strategy, the series or components values are first combined and then a single forecast is determined for the aggregated series. Based on our implementation, state space models showed a higher forecast performance when a top-down strategy is applied. ARIMA models had a higher forecast performance for the bottom-up strategy. For state space models the top-down strategy reduced the overall error significantly. ARIMA models showed to be more accurate when forecasts are first determined individually. As part of the development we also proposed an approach to improve the forecasting procedure of aggregation strategies.

Keywords: Top-down · Bottom-up · Forecast automation ·
Forecast performance · State space models · ARIMA

1 Introduction

Selecting an appropriate forecasting method for a number of time series is a major concern when making decisions. At the organizational level, forecasts are required as critical inputs to many activities in various business areas such as inventory management, marketing, sales, finance, and accounting [1]. There is a frequent need in business for completely automatic forecasting methods (i.e., forecast automation) that takes into account series characteristics and other features of the data without the need for human interference [2]. Literature propose various selection rules in order to enhance forecasting accuracy. The simplest approach for model selection when evaluating multiples series, involves the identification of a single method which is applied over a

© Springer Nature Switzerland AG 2019
C. Paternina-Arboleda and S. Voß (Eds.): ICCL 2019, LNCS 11756, pp. 413–427, 2019.
https://doi.org/10.1007/978-3-030-31140-7_26

combined series without taking into account the specifications of its own components [3]. The idea behind this approach is known as aggregation, where multiple series are combined into a single series without considering their individual specifications such as trend and seasonality [3–5]. When series are aggregated the overall variability of the combined series is reduced, which may result on a superior forecasting accuracy [5, 6]. In addition, the automation for model selection is simpler with a lower complexity but with the cost of losing the specifications from the individual series [3, 7]. On the other hand, individual selection involves the identification of the best method for each series, but this approach is more computationally intensive [4]. The inquiry in this context is to determine which approach would be more effective, in terms of performance, since small improvements in forecast accuracy can lead to large reductions in inventory and increase in service levels [8–10].

The research about the aggregation level of a forecasting process is referred in the literature as Hierarchical Forecasting [5, 11]. In this setting, two forecasting strategies are typically denoted: The bottom-up strategy (BU) and the top-down strategy (TD). In BU, the forecast is developed for each series individually and then these are combined to generate a cumulative forecast of the aggregated series. This is referred as the cumulative forecast, since it is made up by the combination of the individual forecasts of each series. In TD, series are first aggregated to produce a combined forecast, then the forecast is disaggregated and a derived forecast for each series is established usually by means of proportions. Research about the comparisons between TD and BU is available in [12–21], and the principal objective from the developments is to identify which strategy presents a higher forecast performance. However, the findings about whether TD strategies perform better than BU, or vice-versa, remain debatable. Therefore, improvement of forecast performance using these strategies are contemporary, especially if considerations about forecast automation are part of the analysis [4] since this is substantial when working with a large number of series.

Forecast automation is essential when modelling several time series. Automation methodologies for two of the most broadly forecasting methods, autoregressive integrated moving average (ARIMA) and state space models, are recognized to perform very well with several types of time series [2, 22–26]. More advanced or complex methods of forecasting include machine learning procedures such as: Bayesian neural networks, K-nearest neighbor regression, kernel regression, CART regression trees, and support vector regression [27]. The disadvantage with many machine learning algorithms is that often them appear as black boxes or infinite networks with limited and restricted insights into how the forecasts are produced and which data components are important. These attributes of forecasting are often critical for practitioners [28]. State space and ARIMA models are relatively simple but robust approaches to forecasting that are widely used in business with great success in both academic research, educational competitions, and industrial applications [23, 29, 30].

According to Weller and Crone [31] in a survey of forecasting practices, the exponential smoothing family of models is the most frequently used. Actually, it is implemented almost 1/3 of times (32.1%) in detriment of more advanced forecasting techniques that are only applied in 10% of cases. In general, simpler methods are used 3/4 of times, a result that is consistent with the relative accuracy of such methods in forecasting competitions.

Currently the M-Competitions, now in its four version, have attracted great interest in providing objective evidence of the most appropriate way of forecasting various variables of interest. In this research, the objective is to evaluate performance of BU and TD strategies using only forecast automation of state space and ARIMA models. We selected a set of time series from the M4 competition with the purpose of identify the most accurate forecasting method when implementing state space and ARIMA models in combination with the application of BU and TD strategies. Machine learning methods are not considered in this analysis, since the focus of this research is the implementation of forecast automation of the most widely used methods but in the context of aggregation and cumulative forecasts.

2 Forecast Automation

2.1 State Space Models

Since 1950, exponential smoothing methods have been applied with success to several types of time series [2]. The basic variations of exponential smoothing include: simple exponential smoothing, trend-corrected exponential smoothing or Holt's model, additive damped trend, and Holt-Winters additive and multiplicative methods that might include damped trend errors [32–34]. The usual description of these methods is the component form. Component form of exponential smoothing methods comprise a forecast equation and a smoothing equation for the components [24].

Hyndman et al. [2] developed a statistical framework for all exponential smoothing methods. In this statistical structure each model, referred as state space model, consists of a measurement equation that describes the evaluated data, and state or transition equations that describe how the unobserved components or states (level, trend, seasonal) evolve over time. For illustration, let us denote the formulation for the component and state space form of the most common models.

Simple exponential smoothing (1)

Component form

$$\hat{y}_{t+h|t} = l_t$$
$$l_t = \alpha y_t + (1-\alpha)l_{t-1}$$

State space form

$$y_t = l_{t-1} + e_t$$
$$l_t = l_{t-1} + \alpha e_t$$

Where: $e_t = y_t - l_{t-1} = y_t - \hat{y}_{t|t-1}$
for t=1,...., T, the one-step within-sample
forecast error at the time t. l_t is an unobserved state.

Additive damped trend (3)

Component form

$$\hat{y}_{t+h|t} = l_t + (\emptyset + \emptyset^2 + \cdots + \emptyset^h)b_t$$
$$l_t = \alpha y_t + (1-\alpha)(l_{t-1} + \emptyset b_{t-1})$$
$$b_t = \beta^*(l_t - l_{t-1}) + (1-\beta^*)\emptyset b_{t-1}$$

State space form

$$y_t = l_{t-1} + \emptyset b_{t-1} + e_t$$
$$l_t = l_{t-1} + \emptyset b_{t-1} + \alpha e_t$$
$$b_t = \emptyset b_{t-1} + \beta e_t$$

Where: Damping parameter 0<$\emptyset$<1.
If $\emptyset = 1$, indical to Holt's linear trend
As h$\rightarrow\infty$, $\hat{y}_{T+h|T}\rightarrow l_t + \emptyset b_T/(1-\emptyset)$
Short-run forecasts trend, long-run forecasts
constant.

Holt-winters multiplicative (5)

Component form

$$\hat{y}_{t+h|t} = (l_t + hb_t)s_{t-m+h_m^+}$$
$$l_t = \alpha \frac{y_t}{s_{t-m}} + (1-\alpha)(l_{t-1} + b_{t-1})$$
$$b_t = \beta^*(l_t - l_{t-1}) + (1-\beta^*)b_{t-1}$$
$$s_t = \gamma \frac{y_t}{(l_{t-1} - b_{t-1})} + (1-\gamma)s_{t-m}$$

State space form

$$y_t = (l_{t-1} + b_{t-1})s_{t-m} + e_t$$
$$l_t = l_{t-1} + b_{t-1} + \alpha e_t/s_{t-m}$$
$$b_t = b_{t-1} + \beta e_t/s_{t-m}$$
$$s_t = s_{t-m} + \gamma e_t/(l_{t-1} + b_{t-1})$$

Holt's linear trend (2)

Component form

$$\hat{y}_{t+h|t} = l_t + hb_t$$
$$l_t = \alpha y_t + (1-\alpha)(l_{t-1} + b_{t-1})$$
$$b_t = \beta^*(l_t - l_{t-1}) + (1-\beta^*)b_{t-1}$$

State space form

$$y_t = l_{t-1} + b_{t-1} + e_t$$
$$l_t = l_{t-1} + b_{t-1} + \alpha e_t$$
$$b_t = b_{t-1} + \beta e_t$$

Where:

$$\beta = \alpha\beta^*$$
$$e_t = y_t - (l_{t-1} + b_{t-1}) = y_t - \hat{y}_{t|t-1}$$

Holt-winters additive method (4)

Component form

$$\hat{y}_{t+h|t} = l_t + hb_t + s_{t-m+h_m^+}$$
$$l_t = \alpha(y_t - s_{t-m}) + (1-\alpha)(l_{t-1} + b_{t-1})$$
$$b_t = \beta^*(l_t - l_{t-1}) + (1-\beta^*)b_{t-1}$$
$$s_t = \gamma(y_t - l_{t-1} - b_{t-1}) + (1-\gamma)s_{t-m,}$$

State space form

$$y_t = l_{t-1} + b_{t-1} + s_{t-m} + e_t$$
$$l_t = l_{t-1} + b_{t-1} + \alpha e_t$$
$$b_t = b_{t-1} + \beta e_t$$
$$s_t = s_{t-m} + \gamma e_t$$

Holt-winters damped method (6)

Component form

$$\hat{y}_{t+h|t} = [l_t + (\emptyset + \emptyset^2 + \cdots + \emptyset^h)b_t]s_{t-m+h_m^+}$$
$$l_t = \alpha(y_t/s_{t-m}) + (1-\alpha)(l_{t-1} + \emptyset b_{t-1})$$
$$b_t = \beta^*(l_t - l_{t-1}) + (1-\beta^*)\emptyset b_{t-1}$$
$$s_t = \gamma \frac{y_t}{(l_{t-1} + \emptyset b_{t-1})} + (1-\gamma)s_{t-m}$$

State space form

$$y_t = (l_{t-1} + \emptyset b_{t-1})s_{t-m} + e_t$$
$$l_t = l_{t-1} + \emptyset b_{t-1} + \alpha e_t/s_{t-m}$$
$$b_t = \emptyset b_{t-1} + \beta e_t/s_{t-m}$$
$$s_t = s_{t-m} + \gamma e_t/(l_{t-1} + \emptyset b_{t-1})$$

Where for all cases m denotes the period of seasonality, l_t denotes an estimate of the level of the series at time t, b_t denotes an estimate of the trend of the series at time t, s_t denotes an estimate of the seasonality of the series at time t. The initial states and the smoothing parameters α, β, γ are estimated from the observed data. The smoothing parameters α, β, γ are constrained between 0 and 1 with the purpose of that the equations can be interpreted as weighted averages [35].

For space state models, the letters (E, T, S) denote the three forecast components: "Error", "Trend" and "Seasonality". The notation ETS refers to a three-character string form identifying the method used by the framework terminology. For instance, the first

letter denotes the error type ("A", "M" or "Z"); the second letter denotes the trend type ("N","A","M" or "Z"); and the third letter denotes the seasonal type ("N","A","M" or "Z"). In all cases, "N" = none, "A" = additive, "M" = multiplicative, "Z" = automatically selected, and "A_d" denotes additive damped. Then, for example, "ANN" is simple exponential smoothing with additive errors, and "MAM" is multiplicative Holt-Winters with multiplicative errors. The letter "Z" refers to forecast automation, where for the given the data, a state space model is identified automatically to optimize or minimize errors.

2.2   ARIMA

The class of ARIMA models is broad. It can represent many different types of stochastic seasonal and non-seasonal time series such as autoregressive (AR), moving average (MA), and mixed AR or MA processes, where the baseline might need to be differenced and integrated (I). Box-Jenkins et al. [36] developed a systematic and practical model building method. Using this process ARIMA follows three sequential phases; (i) model identification: create and evaluate the correlograms of the series, their patterns enables the identification of the time series that is represented in the baseline, (ii) model estimation: estimation of the parameter values, and (iii) model diagnosis: development of preliminary forecasts, these forecasts are used to diagnose the identification and estimation stage.

The Box-Jenkins methodology has been proved as an effective and practical time series modeling approach. ARIMA models considers three parameters p, d, q that are represented as: ARIMA (p,d,q) Where: p denotes the model and forecast that are based in part or completely on autoregression. p is the number of autoregressive parameters in the model, d is the number of times the series has been differenced to achieve stationarity, and q is the number of moving average parameters in the model that accounts for random jumps in the time series.

For seasonal ARIMA: ARIMA(p,d,q)(P,D,Q), the uppercase letters have the same meaning as the lowercase letters, but these are referred to seasonal parameters. For illustration, let us denote the formulation for the multiplicative seasonal ARIMA model (p,d,q) x $(P,D,Q)_m$

$$\emptyset_p(B)\emptyset_P(B^m)(1-B)^d(1-B^m)^D y_t = c + \theta_q(B)\Theta_Q(B^m)\varepsilon_t$$

Where :

$$\emptyset_p(B) = 1 - \emptyset_1 B - \ldots - \emptyset_p B^p, \emptyset_P(B^m) = 1 - \emptyset_1 B^m - \ldots - \emptyset_P B^{Pm}$$
$$\theta_q(B) = 1 + \theta_1 B + \ldots + \theta_q B^q, \Theta_Q(B^m) = 1 + \Theta_1 B^m + \ldots - \Theta_Q B^{Qm}$$

$$\tag{7}$$

With m as the seasonal frequency, B is the backward shift operator, d is the degree of ordinary differencing, and D is the degree of seasonal differencing, $\emptyset_p(B)$ and $\theta_q(B)$ are the regular autoregressive and moving average polynomials of orders p and q, respectively, $\Theta_Q(B^m)$ and $\Theta_Q B^{Qm}$ are the seasonal autoregressive and moving average polynomials of orders P and Q, respectively, $C = \mu(1 - \emptyset_1 - \ldots - \emptyset_p)(1 - \emptyset_1 - \ldots - \emptyset_p)$ where μ is the mean of $(1-B)^d(1-B^m)^D y_t$ process and ε_t is a zero mean Gaussian white noise process with variance σ^2.

For this research, the state space methods are estimated using the forecast package for R statistical software described in [25]. The automatic ETS function (AUTO.ETS) is used to estimate the state space model form parameters. The ARIMA function (AUTO.ARIMA) implemented in the same package is also used to identify and estimate the ARIMA models. The AUTO.ARIMA function conducts a stepwise selection over possible models and returns the best ARIMA model. The algorithms are applicable to both seasonal and non-seasonal data, these are illustrated using series from the M4 competition, with the forecasting strategies for TD and BU.

3   Forecast Strategy Analysis

3.1   Data Selection

The research literature on TD versus BU strategies is generally characterized into two categories. The first category assumes that the statistical properties of the sub aggregated time series components are known perfectly. In this framework, both TD and BU forecasting would perform equally well, only when the components are uncorrelated and have identical stochastic structures [37, 38]. The second category, assumes that the generating process is not known a priori and data is constantly updated. When data is constantly updated TD could be developed with a higher efficacy since forecasting is done simultaneously for several different components [21, 39–41]. For our case, the series that are part of the study do not have identical stochastic structures, which is in accordance with most business processes. Since the structure for the series are different to each other, it is expected for the implementation a variation on performance when both strategies are applied to the selected series.

3.2   Data Description

We consider five monthly series from the M4 competition to study the relative forecast performance of TD and BU strategies when aggregated series are considered. In the BU strategy, the forecast for each series is determined individually and then a cumulative forecast is obtained by adding the individual components forecasts. We first determine the forecasts for each series individually and then we evaluate the errors for the cumulative forecast when compared with the aggregated series using AUTO.ETS and AUTO.ARIMA. In the TD strategy, the series are first combined to obtain the aggregated series and then a single forecast is determined from the aggregated series. Forecast performance is evaluated using the combined testing sets.

All series from the M4 competition are divided into training and testing sets. The selected series for this study are defined in the competition dataset as: M19, M20, M21, M22, and M23. The total number of observations for each of the series is equal to 192 months, 174 for the training and 18 for the testing set respectively. We assumed an

ending period for all five series equal to December 2018 (12/2018). In order to evaluate forecast performance for the different strategies, RMSE, MAE, and MAPE are calculated for the testing set. Where:

$$\text{RMSE} = \sqrt{\frac{1}{n-m} \sum_{t=m+1}^{n} (y_t - \hat{y}_t)^2} \qquad \text{MAE} = \frac{1}{n-m} \sum_{t=m+1}^{n} |y_t - \hat{y}_t|$$

$$\text{MAPE} = \frac{1}{n-m} \sum_{t=m+1}^{n} \left| \frac{y_t - \hat{y}_t}{y_t} \right| \text{x} 100 \tag{8}$$

We used the training set to identify the best model for the series, using BU and TD strategies, with AUTO.ETS and AUTO.ARIMA. All evaluations of forecast performance are implemented over the testing set. Figure 1 presents the graphical representations of the selected series.

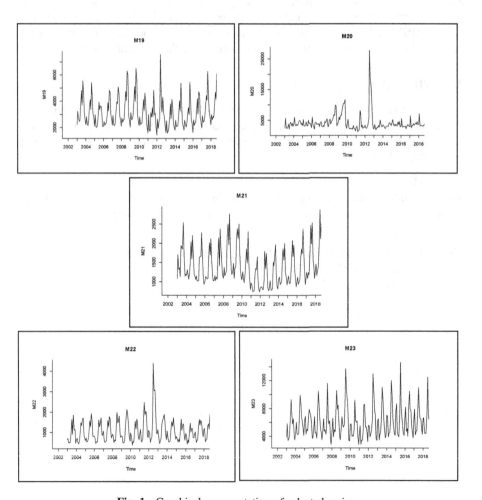

Fig. 1. Graphical representation of selected series.

3.3   BU Strategy

For the BU strategy, we developed the forecast for each of the series individually with the forecast accuracy calculated for each component. Then, all forecasts are combined and the performance is evaluated with the aggregated values from the series. Performance efficiency is calculated using the aggregated testing set. The methodology is developed using AUTO.ETS and AUTO.ARIMA.

ETS - BU

We developed the forecasts for each of the series and then we combined them to evaluate forecast performance of the aggregated series with AUTO.ETS. Figure 2 shows the forecast and performance for each of the components. Figure 4 shows the forecast performance for the aggregated series.

ARIMA - BU

Similar to the ETS-BU approach, we developed the forecasts for each of the series and then we combined them to evaluate forecast performance of the aggregated series with AUTO.ARIMA. Figure 3 shows the forecast and performance for each of the components. Figure 5 shows the forecast performance for the aggregated series.

3.4   TD Strategy

For the TD strategy, the series or components values are first combined and then a single forecast is determined for the aggregated series. Forecast performance is evaluated using the combined testing set. The methodology is developed using AUTO.ETS and AUTO.ARIMA.

ETS - TD

We first combined the series and then a single forecast is determined with AUTO.ETS. Figure 6 shows the forecast and performance for the aggregated series.

ARIMA - TD

Similar to the ETS-TD approach, we first combined the series and then single a forecast is determined with AUTO.ARIMA. Figure 7 shows the forecast and performance for the aggregated series.

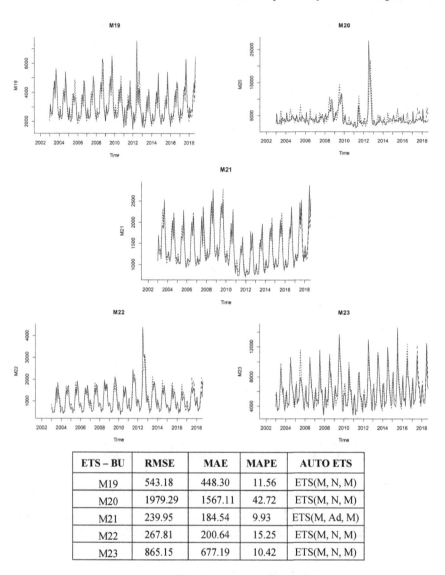

ETS – BU	RMSE	MAE	MAPE	AUTO ETS
M19	543.18	448.30	11.56	ETS(M, N, M)
M20	1979.29	1567.11	42.72	ETS(M, N, M)
M21	239.95	184.54	9.93	ETS(M, Ad, M)
M22	267.81	200.64	15.25	ETS(M, N, M)
M23	865.15	677.19	10.42	ETS(M, N, M)

Fig. 2. ETS-BU individual forecasts

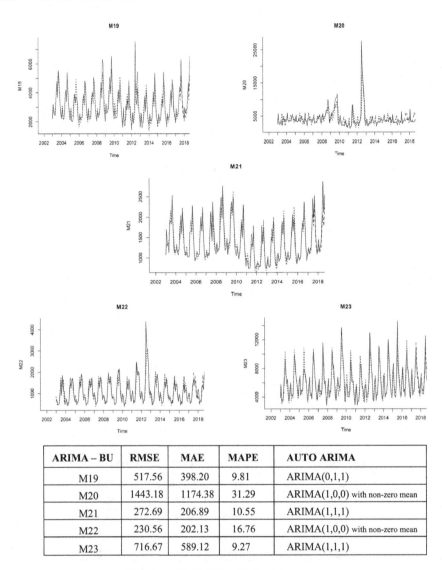

ARIMA – BU	RMSE	MAE	MAPE	AUTO ARIMA
M19	517.56	398.20	9.81	ARIMA(0,1,1)
M20	1443.18	1174.38	31.29	ARIMA(1,0,0) with non-zero mean
M21	272.69	206.89	10.55	ARIMA(1,1,1)
M22	230.56	202.13	16.76	ARIMA(1,0,0) with non-zero mean
M23	716.67	589.12	9.27	ARIMA(1,1,1)

Fig. 3. ARIMA-BU individual forecasts

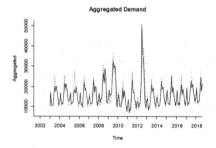

Aggregated ETS – BU	RMSE	MAE	MAPE
Cumulative Forecast	2368.23	2012.99	11.06

Fig. 4. ETS-BU aggregated series forecast

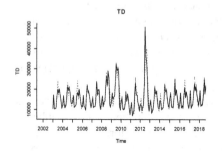

Aggregated ARIMA – BU	RMSE	MAE	MAPE
Cumulative Forecast	1581.34	1344.11	7.66

Fig. 5. ARIMA-BU aggregated series forecast

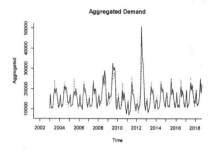

ETS – TD	RMSE	MAE	MAPE
Aggregated Forecast	1548.71	1302.69	7.29

AUTO ETS
ETS(M, Ad, M)

Fig. 6. ETS-TD aggregated series forecast

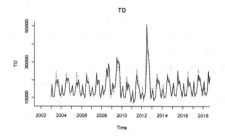

ARIMA – TD	RMSE	MAE	MAPE
Aggregated Forecast	1662.38	1419.60	8.86

AUTO ARIMA
ARIMA(1,0,0)with non-zero mean

Fig. 7. ARIMA-TD aggregated series forecast

4   Discussion and Conclusions

AUTO.ETS showed a higher performance when a TD strategy is applied. On the other hand, AUTO.ARIMA had a greater performance for the BU strategy. For the ETS method, the TD strategy improved significantly the forecast when compared to BU. ARIMA proved to be more accurate when forecasts are determined individually. The best forecast for the study, given MAPE, is ETS with a TD strategy. Table 1 shows the overall results for the strategies.

Table 1. Overall results for the strategies

Strategy	MAPE
(1) ETS-BU	11.06
(2) ARIMA-BU	7.66
(3) ETS-TD	7.29
(4) ARIMA-TD	8.86

ETS-BU (Fig. 2) forecasted significantly higher values for the periods in the testing set of July/2017 to Oct/2017 and July/2018 to Oct/2018 when compared with ARIMA-BU. This inaccuracy of the forecast in these periods increased the error significantly for the ETS-BU strategy. It appears that the forecast strategy ETS-BU does not properly predict the values for this range of months. Given these results it is expected that if the conditions for the series remain constant, this undesirable performance will continue for the following years. For this reason ETS-BU is not recommended as an adequate strategy for the study. On the other hand, the ETS-TD strategy did not show this conduct and it was able to achieve a more accurate forecast when compared with the actual values in the testing set, resulting in a significant lower MAPE. The ETS-TD strategy is the best approach for the study.

Certainly, the most important constrain when applying the proposed forecasting strategies is the arrangement of the data structure in combination with the inherent complexity for the BU strategy of having to calculate each forecast individually. Keeping track of the developed forecasts for each series might prevent researchers of applying the BU strategy. AUTO.ETS and AUTO.ARIMA functions are able to handle as many series as necessary, but the process of aggregating forecasts and evaluating performance is time consuming and in some cases unpractical to apply. If we also include series with different ending and starting times the process becomes more difficult to control. To our knowledge there is not in the literature an automatic procedure to apply the aggregation strategies with several series. The forecast package for R described in [25] details the use of AUTO.ETS and AUTO.ARIMA but before applying these functions with the forecast strategies, the data must be arranged with the appropriate structure in order to be able to perform. In this context, we developed two sets of algorithms presented in Fig. 8, to automatically develop the arrangement of the data and to keep track of the forecasts when applying the strategies.

```
Algorithm 1
start.dates <- matrix(nrow=2, ncol=n)
ending.dates <- matrix(nrow=2, ncol=n)
for (i in 1:n) {
    start.dates[1,i] <- dates.data[1,i]
    start.dates[2,i] <- dates.data[2,i]
    frequen <- frequency
    data <- na.omit(master.data[i])
    observations <- nrow(data)
    cycles <- observations/frequen
    years <- floor(cycles)
    plusperiods <- (cycles - years)* frequen
    ending.dates[1,i] <- start.dates[1,i] + years
    ending.dates [2,i] <- round(start.dates[2,i] + plusperiods - 1, 0)
    if (ending.dates[2,i] > frequen) {
        ending.dates[1,i] <- start.dates[1,i] + years + 1
        ending.dates[2,i] <- ending.dates[2,i] - frequen
    }
}
```

```
Algorithm 2
observations <- 0
for (i in 1:n) {
    data <- as.ts((master.data[i]))
    observations[i] <- length(na.omit(data))
}
minvalue<- min(observations)
Nmiss <- tail(na.omit(master.data),minvalue)
```

Fig. 8. Algorithms for implementation of strategies

The first algorithm deals with the calculations in regards of different ending or starting periods of the series. The second algorithm evaluates the periods available for all series and arrange the structure for forecast calculation. We created both algorithms to be applied on our developments in R. For the algorithms two data frames denominated *master.data* and *dates.data* are generated. The data frame *master.data* contains all the actual values of the series, arranged each one in columns (n). The second data frame *dates.data* has the same number of columns than *master.data* with two rows for each column, the first row has the information of the starting year, for that specific series, and the second row the information about the starting period. The first algorithm will automatically calculate and save the ending date given the frequency, and the second algorithm will arrange the structure to run the forecasts strategies. The second algorithm evaluates if the size of a series is smaller when compared with the others. If that is the case only the tail values that match the size of the smallest series should be considered to calculate the combined forecast. All series must have an equal starting and ending time in order to apply the strategies.

After the application of the proposed algorithms we completed the procedure with the implementation of the different formulations detailed in [25] including the functions "*ts*" and "*as.ts*" to run the forecast automation procedure. The presented algorithms in combination with the automatic forecasting functions enhance significantly the efficiency, coordination, and development of the strategies when several time series are available. We recommend the use of a similar approach to apply the forecasting procedure of aggregation strategies. Finally, future research might include the analysis of machine learning approaches but it is necessary to consider the disadvantages of these methods when analyzing the obtained results.

References

1. Mentzer, J., Bienstock, C.: Sales Forecasting Management: Understanding the Techniques, Systems and Management of the Sales Forecasting Process. Sage Publ, Thousand Oaks (1998)
2. Hyndman, R., Koehler, A., Snyder, R., Grose, S.: A state space framework for automatic forecasting using exponential smoothing methods. Int. J. Forecast. **18**, 439–454 (2002)
3. Fildes, R., Petropoulos, F.: Simple versus complex selection rules for forecasting many time series. J. Bus. Res. **68**, 1692–1701 (2015)
4. Hendry, D., Hubrich, K.: Combining disaggregate forecasts or combining disaggregate information to forecast an aggregate. J. Bus. Econ. Stat. **29**, 216–227 (2011)
5. Widiarta, H., Viswanathan, S.: Forecasting item-level demands: an analytical evaluation of top–down versus bottom–up forecasting in a production-planning framework. IMA. J. Manag. Math. **19**, 207–218 (2008)
6. McLeavey, D., Narasimhan, S.: Production Planning and Inventory Control. Allyn Bacon Inc., Bost (1985)
7. Lutkepohl, H.: Forecasting with VARMA models. In: Handbook of Economic Forecasting, vol. 1, pp. 287–325 (2006)
8. Seifert, M., Siemsen, E., Hadida, A., Eisingerich, A.: Effective judgmental forecasting in the context of fashion products. J. Oper. Manag. **36**, 33–45 (2015)
9. Nenova, Z., May, J.: Determining an optimal hierarchical forecasting model based on the characteristics of the dataset. J. Oper. Manag. **44**, 62–88 (2016)
10. Van der Laan, E., Van Dalen, J., Rohrmoser, M., Simpson, R.: Demand forecasting and order planning for humanitarian logistics: an empirical assessment. J. Oper. Manag. **45**, 114–122 (2016)
11. Zotteri, G., Kalchschmidt, M., Caniato, F.: The impact of aggregation level on forecasting performance. Int. J. Prod. Econ. **93**, 479–491 (2005)
12. Theil, H.: Linear Aggregation of Economic Relations. North-Holl. Publ, Amsterdam (1954)
13. Grunfeld, Y., Griliches, Z.: Is aggregation necessarily bad? Rev. Econ. Stat. **42**, 1–13 (1960)
14. Schwarzkoph, A., Tersine, R., Morris, J.: Top-down versus bottom-up forecasting strategies. Int. J. Prod. Res. **26**, 1833–1843 (1988)
15. Oller, L.: Aggregating problems when forecasting industrial production using business survey data. Ministry of Finance, Economics Department, Helsinky (1989)
16. Ilmakunnas, P.: Aggregation vs. disaggregation in forecasting construction activity. In: Barker, T.S., Pesaran, H. (eds.) Disaggregation Econometric Modelling, pp. 73–86. Routledge, London (1990)
17. Kahn, K.: Revisiting top-down versus bottom-up forecasting. J. Bus. Forecast. **17**, 14–19 (1998)
18. Lapide, L.: New developments in business forecasting. J. Bus. Forecast. **24**, 28–29 (1998)
19. Orcutt, G., Watts, H., Edwards, J.: Data aggregation and information loss. Am. Econ. Rev. **58**, 773–787 (1968)
20. Zellner, A., Tobias, J.: A note on aggregation, disaggregation and forecasting performance. J. Forecast. **19**, 457–469 (2000)
21. Weatherford, L., Kimes, S., Scott, D.: Forecasting for hotel revenue management: Testing aggregation against disaggregation. Cornell Hotel Restaur. Adm. Q. **42**, 53–64 (2001)
22. Makridakis, S., Hibon, M.: The M3-competition: Results, conclusions and implications. Int. J. Forecast. **16**, 451–476 (2000)
23. Chu, C., Zhang, P.: A comparative study of linear and nonlinear models for aggregate retail sales forecasting. Int. J. Prod. Econ. **86**, 217–231 (2003)

24. Ramos, P., Santos, N., Rebelo, R.: Performance of state space and ARIMA models for consumer retail sales forecasting. Robot. Comput. Integr. Manuf. **34**, 151–163 (2015)
25. Hyndman, R., Khandakar, Y.: Automatic time series for forecasting: the forecast package for R. Clayt. VIC, Aust. Monash University. Department Economics and Statistics. 6/7, (2007)
26. Billah, B., King, M., Snyder, R., Koehler, A.: Exponential smoothing model selection for forecasting. Int. J. Forecast. **22**, 239–247 (2006)
27. Admed, N., Atiya, A., El gayar, N., El-shishiny, H.: An empirical comparison of machine learning models for time series forecasting. Econ. Rev. **29**, 594–621 (2010)
28. Sagaert, Y., Aghezzaf, E., Kourentzes, N., Desmet, B.: Tactical sales forecasting using a very large set of macroeconomic indicators. Eur. J. Oper. Res. **264**, 558–569 (2018)
29. Gardner, E.: Exponential smoothing: the state of the art. J. Forecast. **4**, 1–28 (1985)
30. Alvarado-Valencia, J., Barrero, L., Önkal, D., Dennerlein, J.: Expertise, credibility of system forecasts and integration methods in judgmental demand forecasting. Int. J. Forecast. **33**, 298–313 (2017)
31. Weller, M., Crone, S.: Supply Chain Forecasting: Best Practices & Benchmarking Study. Lancaster Cent. Forecast. Technical report (2012)
32. Brown, R.: Statistical forecasting for inventory control. McGraw Hill, New York (1959)
33. Holt, C.: Forecasting trends and seasonal by exponentially weighted averages. Int. J. Forecast. **20**, 5–13 (1957)
34. Winters, P.: Forecasting sales by exponentially weighted moving averages. Manage. Sci. **6**, 324–342 (1960)
35. Makridakis, S., Wheelwright, S., Hyndman, R.: Forecasting: Methods and Applications. Wiley, New York (1998)
36. Box, G., Jenkins, G.: Time Series Analysis: Forecasting and Control. Holden-Day, San Francisco (1976)
37. Rose, D.: Forecasting aggregates of independent ARIMA processes. J. Econ. **5**, 323–345 (1977)
38. Lutkepohl, H.: linear transformations of vector ARMA processes. J. Econ. **26**, 283–293 (1984)
39. Barnea, A., Lakonishok, J.: An analysis of the usefulness of disaggregated accounting data for forecasts of corporate performance. Decis. Sci. **11**, 17–26 (1980)
40. Dangerfield, B., Morris, J.: Top-down or bottom-up: aggregate versus disaggregate extrapolations. Int. J. Forecast. **8**, 233–241 (1992)
41. Fliedner, G.: An investigation of aggregate variable time series forecast strategies with specific sub aggregate time series statistical correlation. Comput. Oper. Res. **26**, 1133–1149 (1999)

Multi-Stage Recovery Resilience: A Case Study of the Dique Canal

Daniel Romero-Rodriguez[1](✉) ⓘ, Alex Savachkin[2], Weimar Ardila-Rueda[2] ⓘ,
Alvaro Sierra-Altamiranda[2] ⓘ, and Julio-Mario Daza-Escorcia[3] ⓘ

[1] Department of Industrial Engineering,
Universidad del Norte, Barranquilla, Colombia
`hromero@uninorte.edu.co`
[2] Department of Industrial Management and Systems Engineering,
University of South Florida, Tampa, Florida, USA
`alexs@usf.edu`, `{weimar,amsierra}@mail.usf.edu`
[3] Program of Industrial Engineering, Konrad Lorenz University Foundation,
Bogotá D.C, Colombia
`juliom.dazae@konradlorenz.edu.co`

Abstract. Disasters may cause severe damage to a community with long
lasting consequences that will impede a short recovery. In high magnitude events the total recovery time is measured in months or years. These
long recoveries are performed in multiple stages and different recovery rates. Traditional resilience metrics have been designed to study
strictly increasing recovery functions, which is not a valid assumption
for the multi-stage recovery case. This paper defines a model to measure
resilience for a multi-stage recovery scenario, where the recovery process is performed in two or more stages with possibly different recovery
rates. A linear approximation metric is proposed to improve resilience
level estimation accuracy. The new metric is tested on a case study of the
Dique Canal breach in Colombia occurred in 2010. The adapted resilience
metric is combined with an optimization model to maximize the average performance on multi-stage scenarios, enabling decision makers to
decide the best strategy per stage for a predefined budget. A comparative analysis confirms that the proposed model offers a better resilience
estimation than previous linear average performance metrics.

Keywords: Resilience metric · Multi-stage recovery ·
Natural disaster · Dique canal

1 Introduction

Throughout history, disasters have caused significant social and economic losses
worldwide every year. Events such as Hurricane Katrina [16], Hurricane Sandy
[27], and The Haiti earthquake [18] are evidence of the negative impact in communities and companies in the affected areas. In order to prepare, anticipate,
respond and recover to these events, resilience is a relevant concept for potentially

© Springer Nature Switzerland AG 2019
C. Paternina-Arboleda and S. Voß (Eds.): ICCL 2019, LNCS 11756, pp. 428–442, 2019.
https://doi.org/10.1007/978-3-030-31140-7_27

affected systems. A wide variety of resilience definitions have been discussed in different areas, where the main common ideas are the system's ability to absorb and recover from a disruptive event at desired performance levels [5,24,26]. A more specific definition in the disaster management area is as follows: "The ability of social units to mitigate hazards, contain the effects of disasters when they occur, and carry out recovery activities in ways that minimize social disruption and mitigate the effects of future earthquakes (events)" [6]. This definition encompasses the resilience dimensions of absorptive and restorative capacities that will be included in our analysis of social systems facing natural disasters.

The importance of resilience in disaster management is that the preparation and recovery phases are planned to minimize the community losses. In the recovery process a common assumption is to consider that the performance is improving continuously from the disrupted state [9]. This framework is valid for small recovery time horizons, but it may not fit long recoveries with different recovery rates and transitions without improvement between stages. Events such as the 2011 Japan earthquake [14] and hurricane Katrina [7], where the recovery time has been measured in years are evidence of the relevance of multi-stage recovery analysis. These models refer to the scenario when a community is disrupted by a natural disaster and the aftermath losses are significant. Thus, the recovery time is measured in months or years. This research seeks to model these types of scenarios and provide tools to perform an accurate resilience estimation and optimal decision making in the recovery stages.

The main research objectives are to develop a framework for the multi-stage scenario and address the lack of resilience measurements and optimization models in the disasters management field. A new metric is developed through a comparative analysis to identify pros and cons of former metrics to fill the current single stage recovery models' gaps. Therefore, after a literature review of the used approaches to measure resilience in disaster management, a linear resilience metric for multi-stage recovery is developed. The new model is evaluated and compared with former metrics in a case study based on the 2010 Dique Canal breach in Colombia.

2   Literature Review

The literature review analyzes common approaches in the study of system's resilience during natural disruptive events, and how flexible are those metrics to capture multi-stage recovery scenarios. Furthermore, a discussion on case studies explores the potential of implementing resilience metrics to assist the decision making process before and after the occurrence of a disruptive event in multi-stage recovery situations. The discussion is divided in two subsections: resilience measurement in disasters management, and case studies.

2.1   Resilience Measurement in Disaster Management

The models to estimate resilience in disaster management include metrics to capture this property under different natural disasters. A classification of resilience

metrics depending of their structure is proposed as average performance, probability, multidimensional and time dependent [25]. The average performance metric or resilience triangle is one of the most recognized tools for resilience estimation. This model quantifies the average performance of a relevant system outcome from the moment a disruptive event occurs until the system has fully recovered. The first model in this category was proposed for communities resilience under seismic events [6].

$Q(t)$: Performance function
T_{OE}: Time of event occurrence
T_{LC}: Control Time

$$R = \int_{T_{OE}}^{T_{OE}+T_{LC}} \frac{Q(t)}{T_{LC}} \, dt \tag{1}$$

The metric from Eq. 1 is a theoretical estimation of a system's average performance as a resilience value. Multiple performance functions have been suggested to increase model flexibility. Features such as non-linear recovery [9], multiple disruptive events [30] and stochastic behavior [2] have been included to improve model accuracy. A linear approximation for Eq. 1 was suggested by Zobel [28] to simplify the estimation of the average performance as a resilience metric.

X: Loss of performance level
T: Recovery Time
T^*: Control Time

$$R = 1 - \frac{XT}{2T^*} \tag{2}$$

The closest reference to the multi-stage scenario is [3]. Performance based resilience metrics have been widely applied in the engineering literature. Miller-Hooks et al. [20] suggest the expected performance level after a disruptive event in transportation networks. Chang [8] models resilience as the probability that the performance does not fall below a minimum service level and the recovery time is lower than a standard value. Time dependent resilience metrics have been suggested to keep track of systems recovery after a disruption. These metrics capture the percentage of recovery at time t [4,13,23].

Another approach for resilience measurement in the disaster management field is the estimation and combination of multiple dimensions. The first step is to identify relevant variables of community resilience. The most commonly evaluated dimensions are ecological, social, economic, institutional, infrastructure, and community competences [11]. The identification of the critical dimension has been extended to the pre-disaster, response and recovery stages to provide a broader social resilience framework [17]. Asadzadeh et al [1] propose a methodology where the significant variables are weighted and combined into a single resilience metric. A more general branch of research is the stochastic multi-stage optimization [12], which can be implemented in multi-stage resilience problems.

2.2   Case Studies

The case studies analysis is performed by categorizing them according to three factors. The first factor is the identification of the system under analysis; whether it is a community or critical infrastructure. The second element is the type of disruptive event; whether the natural disaster was originated by an earthquake, a hurricane or a storm. The third factor refers to the tools that were implemented by the authors; whether the case is about resilience measurement, optimization or simulation models.

From Table 1, it is observed that most of the studies focus on resilience measurement. Authors have developed and implemented their own metrics without a comparative analysis with other models to evaluate the differences and the consistency of these metrics. Two studies considered strategies in two different stages by adding optimization tools to improve resilience levels. In Miller-Hooks et al [20] a two phase optimization model selects the best set of strategies for the preparation and recovery. Similarly, Mackenzie et al [19] maximize resilience by choosing strategies for static (X) and dynamic (T) phases. However, none of the case studies assumed or discussed the possibility of multiple stages in the recovery process.

Table 1. Case studies classification

System	Disruptive event	Type of study	Paper
Water delivery system	Earthquakes	Measurement	[8]
Tans-Oceanic telecommunication	Earthquakes	Measurement, Optimization	[21]
Hospitals network	Earthquakes	Measurement	[10]
Electric power network	Superstorm Sandy	Measurement, Optimization	[19]
Power transmission grid	Hurricane	Measurement, Simulation	[22]
New Orleans economy	Hurricane Katrina	Measurement	[3]
Transportation network	Earthquakes	Measurement, Optimization	[20]
Air transport network	Hurricane	Measurement	[15]
Theran Metropolitan Area	Earthquakes	Measurement	[1]

Numerous models have been proposed and extended for resilience measurement, but none of these metrics focus on long recoveries due to severe natural disasters that may cause a critical damage to communities, private and public systems. The non-linear models are capable of measuring different recovery rates without segmenting the analysis on stages. There is a gap in the identification and measurement of resilience in multi-stage recovery scenarios, which encompasses multiple recovery rates or stages. There are metric classifications in terms of the linearity or non-linearity of the recovery, but there are no discussions about the significance of the recovery time length. The models that fit a short recovery may not work to capture a long recovery with multiple rates. Therefore, a general metric that captures both situations will provide a better and more flexible assessment of system resilience levels.

3   Multi-stage Recovery Model

The system recovery profile under natural disasters or big impact events may not fit the traditional curves of modeling a disruption and its recovery process. In previous models a one-stage continuous recovery with linear and non-linear rates was assumed [6,10,28]. These types of recoveries belong to the category of short recovery, which include those scenarios where a system returns to the initial or desired state in days or weeks. An example of these events are tropical storms that affect a city for one or two days paralyzing some activities, but eventually, systems will recover soon after the natural event has dissipated. Alternatively, there are some disruptive events from which systems will take longer periods to recover. Such event profiles belong to the multi-stage recovery category and the total time will be measured in months or years. An example of these types of events are high impact natural disasters that are capable of devastating entire communities by causing severe damage to the population and infrastructure networks. Figure 1 compares a traditional single stage scenario with a multi-stage recovery.

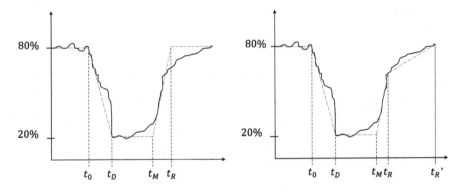

Fig. 1. Single-stage model VS multi-stage recovery

As observed in Fig. 1 an assumption of linear recovery using a single stage and strictly increasing functions will lead to an inaccurate estimation of systems' resilience when the recovery takes a long time and it has two or more recovery stages with different recovery rates. A multi-stage recovery is triggered by at least one of the following factors:

1. **Lack of preparation:** Low static resilience and lack of anticipation to certain events will lead to severe disruptions with long lasting consequences.
2. **Scarce resources:** The unavailability of financial resources, construction materials, food, medicines, water, electricity and other key items, will impede the system's response and recovery in a timely manner.
3. **High magnitude events:** Strong natural disasters may severely damage a prepared or unprepared system. High intensity natural events such as earthquakes, hurricanes and tsunamis are examples of events that may cause mass destruction.

The number of recovery stages are set by decision makers in such a way that a stage ends when the system has recovered up to a predefined performance level. In the last stage (n) the system reaches the initial or desired state. An example is a system with an initial performance level of 99%, after a disruption the total loss is estimated to be 15% which means that the performance right after the event is 84%. It has been defined that the first stage will be completed when the system has reached a performance of 90% and the second stage ends when the system is fully recovered. The activities that are performed during each stage and the defined performance level per stage will depend on the type of disaster and resources availability. In those situations where a natural disaster strikes a community, usually, the first recovery stage includes the following activities: (i) people have been rescued and evacuated from the affected area, (ii) areas have been cleaned up, (iii) water and electrical services are backed up. Consequently, the last stage will be completed when the area is fully available for living and the community is allowed to resume their normal activities.

In order to develop a multi-stage resilience metric, the first step is to define the relevant variables that should be included in the model. The total time (T) is a key dimension in the resilience analysis [2,6,8,28], and it is defined as the time that elapses from the disruption occurrence until the system is fully recovered. This variable includes the performance drop time (T_d), the minimum performance time (T_m) and the recovery time (T_r).

$$T = T_d + T_m + T_r \tag{3}$$

In Eq. 3, T_d has been assumed equal to zero in natural disasters which is interpreted as an instantaneous performance drop [9]. In contrast other models assume a progressive drop after a disruptive event [29]. Under this condition $T_d > 0$, systems do not reach the minimum performance instantaneously. In the case of T_m, a common assumption is that a system begins to recover once it is in the disrupted state at the minimum performance. Thus, this component is set equal to zero in most of the cases [28]. The third element, the recovery time T_r, is generally assumed to be greater than zero $T_r > 0$. Therefore, this element plays a key role in resilience metrics estimation and is considered the most important variable of the multi-stage recovery scenario. The total recovery time is equal to the summation of the recovery time at stage i (tr_i).

$$T_r = \sum_{i=1}^{n} tr_i \tag{4}$$

An initial attempt to measure and improve systems resilience in the multi-stage recovery case is to minimize the total recovery time from Eq. 4. Even though the total time is a significant dimension, it is not enough to characterize resilience by itself. In the multi-stage scenario due to the multiple recovery rates it is possible that when comparing two systems under the same disruptive event the total time is lower for system 1, $T_1 < T_2$, but resilience is higher for system 2, $R_1 < R_2$.

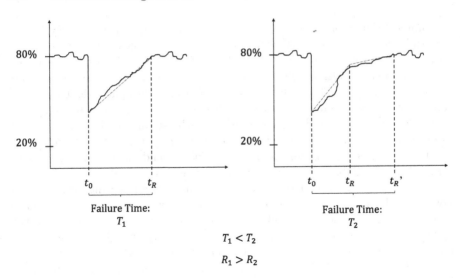

Fig. 2. Comparasion (better fit)

The example in Fig. 2 illustrates the need of adding more dimensions in the resilience assessment process. Thus, the total performance loss (X) is included to guarantee a multi-dimensional analysis that captures different recovery rates. The factor k_i is defined as the percentage of the initial loss recovered at stage i, then the system reaches the initial state at stage n when $\sum k_i = 1$.

$$X = \sum_{i=1}^{n} k_i X \tag{5}$$

A general resilience metric should be able to capture short and long recovery scenarios. We propose a linear approximation model to measure resilience based on the average performance concept. The new model is developed by plugging and adjusting Eqs. (3) and (5) into (2).

$$\begin{aligned} R = 1 - \frac{X}{2T^*}\Big\{ t_d + 2\sum_{i=1}^{n}[tm_i(1 - \sum_{j=1}^{i-1}k_j)] \\ + \sum_{i=1}^{n}k_i tr_i + 2\sum_{i=1}^{n}[tr_i(1 - \sum_{j=1}^{i}k_j)]\Big\} \end{aligned} \tag{6}$$

The general linear approximation can measure short and long recoveries with multiple stages or recovery rates. In the case of a single stage recovery the Eq. 6 is simplified by defining $n = 1$ and $k_1 = 1$. Therefore, the metric is equivalent to the Eq. 2.

The new metric enables decision makers to visualize the average performance or resilience levels when different recovery rates are evaluated. For a better decision making process the metric is extended to a multi-stage recovery optimization model with budget constraint. The objective is to maximize the overall resilience

by calculating the recovery time per stage for a given budget. This model is a tool for planning the recovery phase given an initial loss, then X is a parameter and tr_i are decision variables. The parameter tr'_i is the initial time to recover based on previous experience and without additional recovery strategies. The variable Y_{il} is the time reduction in stage i, and γ_i is the cost of the action l. The total available budget to invest in time reduction is B. The multi-stage recovery linear optimization model is given by the following set of equations:

X: Initial performance loss
t_d: Drop time
tm_i: Minimum recovery at stage i
tr_i: Recovery time at stage i
k_j: Recovery percentage at stage j

$$max \quad R = 1 - \frac{X}{2T^*}\Big\{t_d + 2\sum_{i=1}^{n}[tm_i(1 - \sum_{j=1}^{i-1}k_j)]$$

$$+ \sum_{i=1}^{n}k_i tr_i + 2\sum_{i=1}^{n}[tr_i(1 - \sum_{j=1}^{i}k_j)]\Big\}$$

$s.t$

$$tr_i = tr'_i - Y_i \qquad\qquad \forall_i$$

$$\sum_{i=1}^{n}\gamma_i Y_{il} \leq B$$

$$Y_i \leq \tau_i \qquad\qquad \forall_i$$

For a specific budget, the optimization model calculates the optimal time reduction that maximizes the resilience or the average performance over T^*. Given that the loss improvement is specified as a parameter by policy makers then the model calculates tr_i, which in essence is an estimation of the recovery rate per stage i.

$$r_i = \frac{k_i X}{tr_i} \tag{7}$$

From Eqs (7) and (6) is inferred that for the recovery efforts higher priority should be assigned to early stages in order to maximize system resilience. This result suggests the possibility of developing resilience metrics based on the recovery rate weighted average for the multi-stage recovery scenario.

A numerical example is developed to illustrate the implementation of the new resilience metric and the optimization model. For a natural disaster in a community the parameters after the disruption are given in Table 2.

In Fig. 3 the performance metric during the recovery stage is segmented in three stages due to the changes in the recovery rate and experts opinion. The resilience level in this base scenario without an additional investment is 81.16%. In order to evaluate the investment impact on the resilience levels the linear optimization model is implemented for the scenarios where 150 k and 250 k are the available budgets.

Table 2. Example parameters

Parameter	Value
X	0.4
T^*	80 days
t_d	10 days
t_m	5 days
k_1, k_2, k_3	0.4, 0.35, 0.25
tr'_1, tr'_2, tr'_3	15 days, 30 days, 15 days

Table 3. Scenarios comparison

Scenario	R	T
No investment	81.6%	75 days
150 k investment	85.7%	64.6 days
250 k investment	88%	50.6 days

The scenario comparison from Table 3 and Fig. 4 are evidence that even though the cost to reduce the recovery time in the initial stages is higher the optimal solution will try allocate the money in the initial stages to maximize the system resilience level. Therefore, for an equivalent cost to reduce the stages' recovery time the model will assign the resources ascending from the first stage to the last stage until the budget is spent.

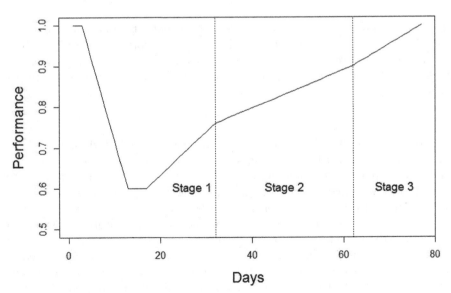

Fig. 3. Disruption profile numerical example

In the optimal solution for a given budget, the recovery rates are increased for one or more stages. In Fig. 4 the recovery rate is increased in stages 1 and 2 for the 150 k and 250 k scenarios. The third stage is improved in the 250 k scenario. By increasing recovery rates the loss is reduced, thus the resilience level is increased.

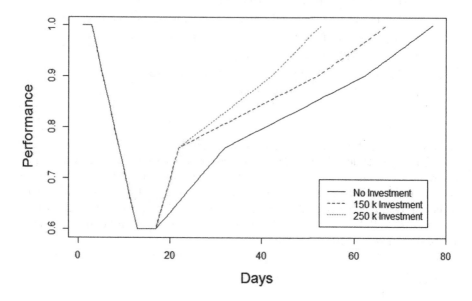

Fig. 4. Disruption profile numerical example

The system's resilience improvement is higher than the values reflected by the measurements. Since the drop time and the minimum performance are assumed to be constants for the scenario comparison, if this area is removed from the analysis then the improvement percentage in the recovery zone increases.

The benefits of the new metric go beyond a better linear approximation of resilience level. It supports decision makers to start planning the recovery phases by adjusting the performance improvement rates when strategies are put in place to reduce the total loss. The linearity of the proposed optimization model assures a low complexity when calculating an optimal solution.

4   Case Study: Dique Canal Breach

In 2010 a breach in the Dique Canal flooded an agricultural area severely affecting five riverside towns in Colombia: Suan, Campo de la Cruz, Santa Lucia, Manati, and Candelaria. The long lasting consequences affected most of the population in these locations due to damage of crops, cattle, and basic infrastructure. Table 4 describes a timeline of the events from the disruption up to the expected recovery.

Table 4. Dique canal timeline

Date	Event
30-Nov-10	Dique Canal breaches
13-Jan-11	80% of the flood water in Santa Lucia has receded
25-Jan-11	The breach is completely fixed
27-Jan-11	95% of the flood water in Santa Lucia has been removed
19-Feb-11	30% of the flood water in Campo de la Cruz has receded
14-Mar-11	82% of the flood water has receded
01-Jun-13	A plan to fully recover the canal is presented
Dec-18	Expected date for a fully recovered canal

From Table 4, the timeline confirms that given the lack of preparation and the event magnitude the total time to recover will be measured in years. This case requires an analysis based on a multi-stage recovery model to fit the varying recovery rates over long periods. The analysis is performed using the habitability as the performance metric. This index is an estimation of the percentage of the area that is available for normal living in each town based on flood water levels and basic infrastructure availability.

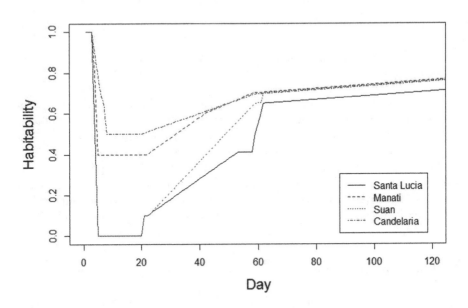

Fig. 5. Habitability after Dique Canal breach

In Fig. 5 it is observed that the initial impact was different for the affected towns. The discrepancy was not due to disaster preparation or static resilience,

it was caused by the geographic location. Those towns that were closer to the Dique canal suffered the biggest drop in their performance levels because the water flooded a bigger area and it resided for a longer period.

A comparative analysis is performed among three equivalent metrics: the theoretical resilience metric from Eq. 1, the linear approximation from Eq. 2 and the proposed metric for multi-stage recovery from Eq. 6. The theoretical metric was obtained with the information after the disruption, so it is an exact value of the average performance. Meanwhile, the other two metrics are approximations based on experts' assessment and press coverage.

Table 5. Case study metrics error

Town	Avg. Perf.	New metric
Santa Lucia	31.97%	3.91%
Manati	10.21%	5.26%
Suan	28.26%	4.61%
Candelaria	6.08%	5.32%
Avg	19.16%	4.77%

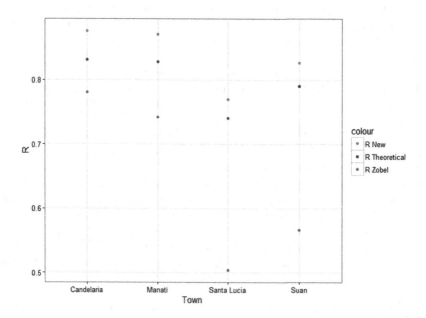

Fig. 6. Resilience metrics comparison

Table 5 contrasts the errors of the average performance metric and the multi-stage metric. The gap is bigger in the one-stage average performance. Furthermore in Fig. 6 is shown that the new metric was closer to the theoretical resilience in comparison with the one-stage linear approximation. it can not be guaranteed

that the new metric will outperform the one-stage linear metric in every single scenario because it depends on the quality of the parameter estimation. In the special case of Santa Lucia where the recovery has three stages, the two-stage metric performed better than the one-stage model. This result suggests that the more the stages number is increased the more flexible and accurate the metric will be to fit the performance function.

The case study was evaluated after the event occurrence, but it is expected from a practitioner's perspective that the measurements and optimization model deployment should be executed before or immediately after the event, during the planning stage. Thus, since the measurement is a forecast of system resilience based on expert's opinion a thorough parameter estimation of ongoing and future events is needed to improve model accuracy.

5 Conclusions

The notion of system recovery is extended to include scenarios with multiple stages and long time recoveries. The importance of such scenarios is to improve the accuracy of resilience estimation by adding multiple stages and rates in the recovery process. A new linear approximation metric was proposed based on the resilience triangle concept. The new metric is evaluated and compared in a case study, where the multi-stage metric outperforms the single-stage measurement, suggesting that as the number of stages increases the bigger the accuracy gap will be.

A linear optimization model was developed to maximize system resilience levels subject to a budget constraint. This model improves average performance by investing resources in the recovery, but the model was simplified by assuming that there are no variables that affect the initial loss. The focus on the recovery phase reduces the model's complexity which otherwise would be non-linear. Our model can be extended to include strategies to minimize the initial loss and the effects of the interaction between X and T.

In future research, there are two major areas to be expanded. The first area is the development of new metrics with multi-stage recovery properties but with a different paradigm than the average performance notion. The second area is to develop and evaluate multi-stage recovery optimization models with stochastic inputs to capture the uncertainty in the parameter estimation. Additionally, dynamic programming models are suggested to evaluate the sequential nature of the decisions that are made in every stage.

Acknowledgements. The authors thank the referees for their helpful comments that significantly contributed to improving the quality of the paper. The work by Daniel Romero-Rodriguez has been partially supported by the Colombian Government through the COLCIENCIAS-scholarship Convocatoria 2015-CC1129579108 and cofinanced by the Universidad del Norte through the XVII Convocatoria Interna de Investigación UNINORTE. The work by Julio-Mario Daza-Escorcia has been partially supported by the Colombian Government through the COLCIENCIAS-scholarships CONV-617-2013-CAP3-CC72357251.

References

1. Asadzadeh, A., Kötter, T., Zebardast, E.: An augmented approach for measurement of disaster resilience using connective factor analysis and analytic network process (f'anp) model. Int. J. Disaster Risk Reduction **14**, 504–518 (2015)
2. Ayyub, B.M.: Systems resilience for multihazard environments: definition, metrics, and valuation for decision making. Risk Anal. **34**(2), 340–355 (2014)
3. Ayyub, B.M.: Practical resilience metrics for planning, design, and decision making. ASCE-ASME J. Risk Uncertainty Eng. Syst. Part A: Civil Eng. **1**(3), 04015008 (2015)
4. Baroud, H., Ramirez-Marquez, J.E., Barker, K., Rocco, C.M.: Stochastic measures of network resilience: applications to waterway commodity flows. Risk Anal. **34**(7), 1317–1335 (2014)
5. Bhamra, R., Dani, S., Burnard, K.: Resilience: the concept, a literature review and future directions. Int. J. Prod. Res. **49**(18), 5375–5393 (2011)
6. Bruneau, M., et al.: A framework to quantitatively assess and enhance the seismic resilience of communities. Earthquake Spectra **19**(4), 733–752 (2003)
7. Burton, C.G.: A validation of metrics for community resilience to natural hazards and disasters using the recovery from hurricane katrina as a case study. Ann. Assoc. Am. Geogr. **105**(1), 67–86 (2015)
8. Chang, S.E., Shinozuka, M.: Measuring improvements in the disaster resilience of communities. Earthq. Spectra **20**(3), 739–755 (2004)
9. Cimellaro, G.P., Reinhorn, A.M., Bruneau, M.: Framework for analytical quantification of disaster resilience. Eng. Struct. **32**(11), 3639–3649 (2010)
10. Cimellaro, G.P., Reinhorn, A.M., Bruneau, M.: Seismic resilience of a hospital system. Struct. Infrastruct. Eng. **6**(1–2), 127–144 (2010)
11. Cutter, S.L., et al.: A place-based model for understanding community resilience to natural disasters. Global Environ. Change **18**(4), 598–606 (2008)
12. Fan, Y., Schwartz, F., Voß, S., Woodruff, D.L.: Stochastic programming for flexible global supply chain planning. Flex. Serv. Manuf. J. **29**(3–4), 601–633 (2017)
13. Henry, D., Ramirez-Marquez, J.E.: Generic metrics and quantitative approaches for system resilience as a function of time. Reliab. Eng. Syst. Saf. **99**, 114–122 (2012)
14. Hosoya, K.: Recovery from natural disaster: a numerical investigation based on the convergence approach. Econ. Model. **55**, 410–420 (2016)
15. Janić, M.: Modelling the resilience, friability and costs of an air transport network affected by a large-scale disruptive event. Transp. Res. Part A: Policy Pract. **71**, 1–16 (2015)
16. Jonkman, S.N., Maaskant, B., Boyd, E., Levitan, M.L.: Loss of life caused by the flooding of new orleans after hurricane katrina: analysis of the relationship between flood characteristics and mortality. Risk Anal. **29**(5), 676–698 (2009)
17. Khalili, S., Harre, M., Morley, P.: A temporal framework of social resilience indicators of communities to flood, case studies: wagga wagga and kempsey, NSW, australia. Int. J. Disaster Risk Reduction **13**, 248–254 (2015)
18. Kolbe, A.R., et al.: Mortality, crime and access to basic needs before and after the haiti earthquake: a random survey of port-au-prince households. Med., Conflict Survival **26**(4), 281–297 (2010)
19. MacKenzie, C.A., Zobel, C.W.: Allocating resources to enhance resilience, with application to superstorm sandy and an electric utility. Risk Anal. **36**(4), 847–862 (2015)

20. Miller-Hooks, E., Zhang, X., Faturechi, R.: Measuring and maximizing resilience of freight transportation networks. Comput. Oper. Res. **39**(7), 1633–1643 (2012)
21. Omer, M., Nilchiani, R., Mostashari, A.: Measuring the resilience of the transoceanic telecommunication cable system. IEEE Syst. J. **3**(3), 295–303 (2009)
22. Ouyang, M., Dueñas-Osorio, L., Min, X.: A three-stage resilience analysis framework for urban infrastructure systems. Struct. Saf. **36**, 23–31 (2012)
23. Pant, R., Barker, K., Ramirez-Marquez, J.E., Rocco, C.M.: Stochastic measures of resilience and their application to container terminals. Comput. Ind. Eng. **70**, 183–194 (2014)
24. Ponomarov, S.Y., Holcomb, M.C.: Understanding the concept of supply chain resilience. Int. J. Logistics Manag. **20**(1), 124–143 (2009)
25. Rodriguez, D.R.: Physical and Social Systems Resilience Assessment and Optimization. Ph.D. thesis, University of South Florida (2018)
26. Rose, A.: Defining and measuring economic resilience to disasters. Disaster Prev. Manag.: Int. J. **13**(4), 307–314 (2004)
27. Rosenzweig, C., Solecki, W.: Hurricane sandy and adaptation pathways in new york: lessons from a first-responder city. Global Environ. Change **28**, 395–408 (2014)
28. Zobel, C.W.: Representing perceived tradeoffs in defining disaster resilience. Decis. Support Syst. **50**(2), 394–403 (2011)
29. Zobel, C.W., Khansa, L.: Quantifying cyberinfrastructure resilience against multi-event attacks. Decis. Sci. **43**(4), 687–710 (2012)
30. Zobel, C.W., Khansa, L.: Characterizing multi-event disaster resilience. Comput. Oper. Res. **42**, 83–94 (2014)

Author Index

T0235675

Printed in the United States
By Bookmasters